Petr Vopěnka

New Infinitary Mathematics

Charles University
Karolinum Press 2022

KAROLINUM PRESS
Karolinum Press is a publishing department of Charles University
Ovocný trh 560/5, 116 36 Prague 1, Czech Republic
www.karolinum.cz

Cover by Jan Šerých

Set by Šárka Voráčová and Alena Vencovská
Printed by Karolinum Press
First English edition

A catalogue record for this book is available from the National Library of the Czech Republic.

ISBN 978-80-246-4663-3
ISBN 978-80-246-4664-0 (pdf)

Contents

Editor's Note

The original reason for this book was the consensus that Vopěnka's mathematical and philosophical contributions made after he left mainstream set theory should be available in English. Bringing the book to publication has taken ten years for the following reasons: first Vopěnka wrote another manuscript in Czech[1] subsequently translated by Hana Moravcová and Roland Andrew Letham, called *The Great Illusion of the Twentieth Century Mathematics*. However, it turned out that the translation of some parts of the text needed more relevant mathematical expertise and Alena Vencovská took on the task of making it correct. The author used the opportunity to extend and modify the book considerably. He worked on it until his sudden death in 2015. The result was twofold: more publications in Czech, namely the four-volumed work *New Infinitary Mathematics*,[2] along with *Prolegomena to the New Infinitary Mathematics*,[3] and a parallel English text with additions to the original book translated by Vencovská. The Czech and English versions differed little from each other, except that the order of the material was different, and Vopěnka left some parts out from the English version. In particular, he did not include what are now the first two chapters, and some sections throughout. This present version does include these initial chapters (on the theological foundations of Cantor's set theory and on its rise and growth, the former translated by Václav Paris) but it does not include all that is in the Czech version.

[1] Petr Vopěnka, *Velká iluze Matematiky XX. stoleti a nové základy* (Plzeň: Západočeská univerzita v Plzni a Nakladatelství Koniáš, 2011).

[2] Petr Vopěnka, *Nová infinitní matematika* (Praha: Karolinum, 2015).

[3] Petr Vopěnka, *Prolegomena k nové infinitní matematice* (Praha: Karolinum, 2013).

Editor's Introduction

About the Author

Petr Vopěnka grew up in the former Czechoslovakia, where he was born in 1935 (to parents who both taught mathematics at a secondary school). He enjoyed scouting in his youth and often remembered times spent at camps. In a way he remained true to the values he formed early on all through his life. Personal integrity, faith in truth prevailing over deceit, loyalty to friends, great love for his troubled country and an unshakeable commitment to his work were some of his most striking characteristics. To this, one needs to add that he loved to laugh.

For much of his life, Czechoslovakia was ruled by the communists: they took over in 1948, and education during Vopěnka's teenage years bore the stamp of Stalinism. Vopěnka reminisced about being asked to take turns in a whole day of reading funereal poems on the school radio upon Stalin's death in 1953, and he arrived in Prague later the same year to study mathematics in a city overlooked from a hill by Stalin's 16-meter-high statue. Fortunately, mathematics is relatively immune to ideological manipulation and Vopěnka remembered his student years and his teachers fondly.

His early research was mainly in topology and he wrote his master's thesis under the supervision of Eduard Čech, an eminent topologist and geometer, whose name lives for example in Čech cohomology and Čech-Stone compactification. Vopěnka used to say that Čech "showed him how to do mathematics". The research that he engaged in at that time concerned compact Hausdorff spaces and their dimensions.

Soon after graduation, Vopěnka started to teach mathematics at Charles University and he remained there for most of his professional life. Quite early on, he developed an interest in mathematical logic, championed in Czechoslovakia by Ladislav Rieger who wrote about the subject for the Czech mathematical community and ran a seminar on set theory. Vopěnka participated and, after Rieger's untimely death in 1962, took over as its organiser to provide strong and inspired leadership for Czechoslovak mathematical logicians. Vopěnka published work on nonstandard interpretations of Gödel-Bernays set theory based on using the ultrapower construction and then in collaboration with the seminar participants he contributed substantially to the exciting discoveries following Gödel and Cohen's groundbreaking work on the consistency and independence of the continuum hypothesis and the axiom of choice. Due to the Iron Curtain, communication with other mathematicians working in the area was limited and some results obtained independently in Prague came later than those in the West, but others remain credited to the Prague group. By all accounts it was as vibrant and fruitful a period as can be – Alfred Tarski wrote about the community in these words:[4]: "I do not know if there is at this point another place in the world, having as numerous and cooperative a group

[4] Quoted in Antonín Sochor, "Petr Vopěnka (born 16. 5. 1935)," *Ann. Pure Appl. Logic* 109 (2001): 1–8.

of young and talented researchers in the foundations of mathematics."

This lasted some years, but then two factors caused it to fall apart. One was that Vopěnka became very sceptical about the role that set theory, as it was, could have in truly explaining the phenomenon of infinity and in serving as a foundation for mathematics. It mattered to him; he did not wish to explore that intricate and bewitching maze any further so he started to look for alternatives. Paradoxically, the one concept which is today perhaps most strongly associated with Vopěnka within this area arose as he was abandoning the subject, when he proposed what became to be known as Vopěnka's principle. This yields a strong large-cardinal axiom that Vopěnka said he believed he could prove to be contradictory, suggesting it merely to make the point that investigating consequences of more and more set-theoretical axioms made little sense. However, Vopěnka's argument that it was contradictory contained an error, and interest in the axiom prospered outside of Czechoslovakia. Tightening controls within the country again limited communication with the West for academics like Vopěnka so it was some years later that he learnt with surprise that this principle was still alive and well established.

The other factor that contributed to the demise of this golden era of mainstream set theory in Prague were the political events – the 1960s brought a gradual thaw of orthodox communism leading to Prague Spring in 1968. This however was followed by the August 1968 invasion whereby the Warsaw Pact armies put an end to it. Some of Vopěnka's collaborators, in particular Tomáš Jech and Karel Hrbáček left the country, and most of the others sought their own independent paths. Vopěnka, who prior to 1968 had joined the efforts led by Alexander Dubček to reform communism and had gained some influence in running the Faculty of Mathematics and Physics at his university, would not support the official line after the invasion and might well have been forced to leave the university along with many other academics in similar positions. He was allowed to stay to do research, although his contact with students was very restricted. Many years later when he learnt that he owed this good fortune to the intervention of the Soviet mathematician P. S. Alexandrov, he used to joke that had he known how powerful a protector he had, he would have been braver (standing up to the suffocating pressure of the Czechoslovak communist "normalisation" of the 1970s and 1980s). In fact, he was one of the few who did stand up to it in any way that seemed possible.

At this turning point, Petr Vopěnka along with Petr Hájek wrote a book on semisets,[5] exploring set theories obtained by modifying the usual von Neumann-Bernays-Gödel axioms for classes and sets so that sets can have subclasses that are not themselves sets (proper semisets). Apart from the importance of semisets for forcing, Vopěnka's new motivation was investigating other ways in which the phenomenon of infinity could be captured mathematically, better reflecting how we encounter infinity when thinking about the world, often as a part of a large finite set. The book did not dwell on this aspect though and

[5] Petr Vopěnka and Petr Hájek, *The Theory of Semisets* (Prague: North Holland and Academia, 1972).

focused on providing a careful formal development of the theory of semisets and on showing its suitability for finding models of set theory via forcing.

Vopěnka then moved on to formulate a different set theory, which he hoped would capture his intuition about infinity in a better way. It was an intuition gained through much reflection on what we understand by infinity and how we see the world, influenced mainly by Bolzano and Cantor's writings, by discussions surrounding the birth of set theory and by the philosophy of Husserl and Heidegger (for many years there was a weekly seminar taking place in Vopěnka's office devoted to the study of their work). Semisets were a step in the right direction, but Vopěnka wished to formulate a new theory from the position of a mathematician free of any commitment to the current view of infinity; to develop mathematics as it might have been developed if satisfactory axioms for infinitesimals had been found before mathematics took its present course.

This led to what he called the alternative set theory. It contains sets and classes; sets alone behave as classical finite sets but they may contain subclasses which are not sets (semisets). Unlike Cantorian set theories, alternative set theory admits only two types of infinity: the countable infinite and the continuum. This is not a necessary requirement of such a set theory, it could be constructed otherwise, but Vopěnka's motivation was to keep only what could be justified by some intuition other than intuition arising purely from the study of set theory itself; for him it meant just the infinities associated with natural numbers or with the real line. A crucial principle in Vopěnka's alternative set theory is the Axiom of Prolongation, related to the phenomenon of the horizon (understood in a very general sense). It reflects the intuition that something seen to behave in a certain well-defined way as far as the horizon will continue to do so beyond the horizon.

Mathematically, the theory is close to the concept of nonstandard models of natural numbers underlying nonstandard analysis. However, from a foundational point of view there is a considerable difference since in nonstandard analysis infinitesimals are complicated infinitary objects whilst in AST some exist just as rational numbers do. Formulating a theory that allows mathematical analysis to be practiced in a way in which it was conceived by Leibniz, that is as a calculus with infinitesimals, was indeed one of Vopěnka's objectives. This had not been done within the alternative set theory at the time, and Vopěnka returned to the task in this book.

Vopěnka succeeded in assembling another group of enthusiastic mathematicians, who wanted to work with him and develop AST. One unfailingly supportive and faithful collaborator from before also joined him in the endeavour, Antonín Sochor. Interesting results were obtained, first within the Prague circle and later on also at other places in the world, but overall its impact was relatively small. In particular, investigations of alternative set-theoretical universes was restricted to what Vopěnka called a limit universe (as opposed to a witnessed universe). In a limit universe no "concrete" set such as the set of natural numbers less than $67^{293^{159}}$ can contain semisets but in a witnessed universe some can. The witnessed universes correspond to Vopěnka's intuition, but their theory is classically inconsistent (Vopěnka envisaged some approach involving the convincingness of proofs).

The first comprehensive account of AST appeared in 1979 in a monograph by Vopěnka.[6] In 1980 there was supposed to be a Logic Colloquium in Prague where AST would surely have been widely discussed and whatever stand logicians would have taken, its ambition to lead to new foundations for mathematics would have attracted more attention. However, shortly before the Colloquium was due to start the communist regime revoked the permission for it to take place, because the logic community was calling for the release of an imprisoned Czech logician and the regime feared the negative publicity. The next Logic Colloquium in Prague had to wait eighteen years, nine years after the Velvet Revolution. Vopěnka was an honorary chairman and his opening words are very telling, both of the man and the bygone times:

> "Ladies and Gentelmen, I am very happy to be able to welcome you to Prague. French historian Ernst Denis once wrote that in Prague every stone tells a story. As you walk across the Charles Bridge, pause to remember Tycho Brahe and Johannes Kepler who used to stroll there over 400 years ago as well as Bernard Bolzano two centuries later. I am sure that you too will fall in love with this old, inspiring, majestic, but also tragic city."
>
> These were the words with which I had planned to welcome participants of Logic Colloquium '80 which was cancelled by the communist government. The totalitarian regime was afraid that the participating mathematicians would call for the release of their colleague, mathematician Vaclav Benda, who was serving a five year prison term. He was imprisoned for publicly drawing attention to politically motivated prosecution of those opposing the regime. For us, Czech mathematicians, the cancellation meant even deeper isolation from our colleagues abroad. But we never doubted that even though mathematics is very beautiful, freedom is even more so. Logic Colloquium '98 will now commence.

By this time Vopěnka had entered yet another stage in his professional life. After the demise of communism in 1989 he had served as the Minister of Education in the new democratic government, throwing all his passion and energy into trying to reform the education system, with mixed success. After completing his term of office, he returned to academia but devoted himself mainly to the history and philosophy of mathematics. He wrote several books, in Czech, most notably *The Corner Stone of European Learning and Power* (Úhelný kámen evropské vzdělanosti a moci, 1998), *Narration about the Beauty of Neo-baroque Mathematics* (Vyprávění o kráse novobarokní matematiky, 2004) and *Meditations on the Foundation of Science* (Meditace o základech vědy, 2001). In 2004 he was awarded the Vize 97 prize by The Dagmar and Vaclav Havel Foundation designated by the charismatic Czech playwright president for "significant

[6] Petr Vopěnka, *Mathematics in the Alternative Set Theory* (Leipzig: Teubner, 1979).

thinkers whose work exceeds the traditional framework of scientific knowledge, contributes to the understanding of science as an integral part of general culture and is concerned with unconventional ways of asking fundamental questions about cognition, being and human existence." Vopěnka continued his work in the same spirit, eventually returning again to mathematics to describe his stand on its foundations.

About the Book

There are some features in Vopěnka's work which it is useful to highlight. Vopěnka wrote extensively (in Czech) about ancient Greek geometry and its development throughout the centuries and about the origins and assumptions of set theory. It was essential for him to understand what mathematicians were doing, and he always wanted to see beyond the formal side of it: proving theorems from axioms did not suffice. He needed to know why anything should be assumed and this led him to formulate his own philosophical standpoint and develop his own terminology.

This is particularly important for his arguments about sets, which he discusses in this book. He explained his positions in detail for example in his book *Meditations on the Foundations of Science.*[7] It was influenced by the philosophy of Edmund Husserl and his followers but Vopěnka adapted the phenomenological program in his own way. The starting point are phenomena we encounter; from those we create objects by conceding them a "personality". It does not matter what is the character of the phenomenon in question, it could be something we perceive or remember or just think. When we single out some objects from those previously created, we can collect them together and when we consider them thus collected and without their various properties and interrelations, we make a collection of objects. Thus collections are determined exclusively by the presence of the objects belonging to them: belonging is not graded, and an object either belongs or not. When we consider a collection as an object, that is concede a personality to it, we make it into a class. The difference is that a collection is a multiplicity of objects but a class is a single object. As an object, it can belong to other collections. A class is uniquely determined by its members and, conversely, it uniquely determines the collection of its members although this can be in various ways and it may not be possible simply to list the members. A set is a class such that the collection of its members is sharply defined. For non-sharply defined collections, Vopěnka refers to examples like the numbers of grains taken from a heap of sand that still leave a heap. A semiset is defined to be a class which is a subclass of some set but not itself a set (where a class X is a subclass of a class Y if all members of X belong to Y).

Apart from collections, Vopěnka uses the notion of domains. He writes

> When talking about people, we often think not only of people who are alive at that moment or have lived in the past, but also of those who are yet to be born or even of those who have never been born nor

[7] Petr Vopěnka, *Meditace o základech vědy* (Prague: Práh, 2001).

> ever will be. The extension of the concept of people is therefore not a collection, but a domain of all people. A domain is not a totality of existing objects (regardless of the modality of their existence); it is the source and simultaneously also a sort of container into which the suitable emerging or created objects fall. Naturally, every collection of objects can be interpreted as a domain, albeit an exhausted one. By actualising a domain we mean exhausting the domain, that is, substituting this domain by a collection of all the objects that fall or can fall into it.[8]

Thus it is some way from a domain to a set, and the questions of whether a domain can be actualised and whether this would yield a set is of fundamental importance.

In *Meditations*, Vopěnka gives an explanation of abstract objects and then he says:

> Abstract objects are the building blocks of the remarkable world of abstract mathematics. The modality of their being is some special, separated (abstract), and yet changeable being. These phenomena arise from nothingness by the strength of our will and their being culminates when they are captured in our minds. If we stop thinking them, they do not perish; just the modality of their being decreases. As if the nothingness slowly absorbed these phenomena but was no longer able to absorb them completely. Hence it at least hides them under the ever-condensing cover of emptiness from which they again surface when we remember them. We will refer to this idiosyncratic being of abstract objects as existence.[9]

It is in this light that we need to understand his arguments about Cantor's set theory and about the existence of the set of natural numbers. (By Cantor's set theory Vopěnka means any considerations based on Cantor's ideas, be it within the most commonly used ZFC – Zermelo Fraenkel set theory with the axiom of choice, or GB – Gödel Bernays set theory, within which he himself worked in the 1960s, or some other system based on the same approach to infinity.) For abstract objects with certain properties to exist, it must be possible to think them so at the very least there cannot be an apparent contradiction in them. But that is not all: we as finite beings should not really be able to think beyond the finite. So what is it that gives us the confidence to do so?

Vopěnka went further back, and started by asking how Euclidean geometry was possible. He argued that the mathematics of the ancient Greek world, that is, the way in which people thought about it, appropriated the capabilities of the Olympian gods to grasp the unchanging truth in the changing world. He refers to Zeus, or to a superhuman, as the performer of ancient (Greek) geometry. Zeus can extend a straight line further than any limit we may come up with

[8] Petr Vopěnka, *Prolegomena k nové infinitní matematice* (Prague: Karolinum, 2015).

[9] Petr Vopěnka, *Meditace o základech vědy* (Prague: Práh, 2001).

and he can see how a straight line approaches to touch a circle. Still, he does not wield absolute power and he could not hold in his mind all that there is. Such power does however belong to the God of medieval scholastic philosophy and using it made Cantor's set theory possible. It was in fact Bolzano, half a century before Cantor developed his theory of infinity, who came up with a proof of the existence of an infinite set (the only proof ever given, as Vopěnka used to say). Bolzano's proof is discussed in Section 1.3. Accordingly, Vopěnka sometimes refers to God, or to a God-man,[10] as the performer of the classical (modern) geometry and mathematics, on the grounds of it being based on Cantor's set theory. Faced with the question of how to perform mathematics now, Vopěnka notes that in the twenty first century, theological support is no longer there and he proposes his New Infinitary Mathematics, in the spirit of the alternative set theory.

The book has the following structure: Part I is a historical, philosophical and mathematical introduction. The author discusses the history of approaches to infinity up to the time when actually infinite sets became an integral part of mathematics. He shows how fundamental a role theological considerations played in enabling Bolzano and Cantor to produce work that established actually infinite sets as a legitimate object of study. Then he outlines the development of the basic ideas of set theory, focusing on the intuition that guided those early pioneers of set theory before the axiomatic frameworks found their final forms. He argues informally, attempting to capture the spirit of what appeared in the early days as the best way to build set theory; this includes the Axiom of Choice. Finally he argues that stripped of the support of medieval rational theology, we lose more than just certainty that actually infinite sets exist. To wit, assuming the actual existence of the set of all natural numbers (identified with their von Neumann's representations) leads, via the ultrapower construction and the ultraextension operator, to another set of all natural numbers containing all the previous ones and more, which is absurd. Although only some of the obvious questions and objections to this argument are answered in Vopěnka's text, one of his aims was to provoke a debate, and there is much that can be said. Part II proposes a new framework for mathematics while carefully motivating why it should be built in this way. The crucial concepts are those of natural real world, natural infinity and horizon. Mathematically, it is similar to the alternative set theory although there are differences, for example nothing corresponds to the axiom of two cardinalities which is adopted therein. Vopěnka saw his theory as an open challenge to be developed further; in particular he felt that predicate calculus may not be the only tool with which to study it. However, he did not investigate this further. In part III the author seeks to provide rigorous foundations for the development of the infinitesimal calculus on the basis of his theory. This is similar to Abraham Robinson's treatment of calculus in non-standard analysis, but Vopěnka's aim is to resurrect the original intuition that guided Leibniz, and to work with infinitesimals that actually exist as finite objects, without the need for them to be representatives of other, infinitary ob-

[10] See page 215.

jects. Part IV is shorter than the others and it is devoted to the real numbers. The four parts mirror the four volumes of the Czech version of *New Infinitary Mathematics*, with the first part including also some of *Prolegomena*. The circumstance that Vopěnka was simultaneously preparing this book and the Czech version[11] should explain some repetitions and variations in the present volume, although effort has been made to minimise them.

Regarding Vopěnka's style, it is useful to note that he frequently specifies the default meaning of symbols or letters at the start of various chapters or sections to be valid within those chapters or sections (or even during a section, to be valid until the end) and he does not necessarily repeat this when the symbols are used in theorems etc.

In the process of arranging for the publication of this work in English, some serious objections were raised, most notably the failure of the author to engage with the more recent scientific and philosophical literature and relate his thoughts to it. This is justified and could be damning, but there is much to redeem the book. It is a serious attempt by a leading mathematician to re-work the foundations of mathematics at a time when many mathematicians prefer to divorce their subject from the obligation to understand its own foundations. Cantorian set theory has in general been taken to provide such a foundation but the fact that there appears to be no one *true* classical set theoretical universe has made this hard to uphold. In a recent article[12] Akihiro Kanamori writes:

> Stepping back to gaze at modern set theory, the thrust of mathematical research should deflate various possible metaphysical appropriations with an onrush of new models, hypotheses, and results. Shedding much of its foundational burden, set theory has become an intriguing field of mathematics where formalized versions of truth and consistency have become matters for manipulation as in algebra. As a study couched in well-foundedness ZFC together with the spectrum of large cardinals serves as a court of adjudication, in terms of relative consistency, for mathematical statements that can be informatively contextualized in set theory by letting their variables range over the set-theoretic universe. Thus, set theory is more of an open-ended framework for mathematics rather than an elucidating foundation.

Still, some mathematicians and certainly philosophers of mathematics worry about the truth. Interesting as it would be, this book does not engage in a discussion of how it relates to such literature. Rather, it tries to find the truth from the position of a mathematician in the early 21st century, who spent a lifetime thinking about foundational issues, who is aware of the big metaphysical/theological assumptions behind the current framework and who searches for what is left when we give them up, relying just on human intuition and ability to make sense of the world.

[11] See page xi.

[12] Akihiro Kanamori, "Set Theory from Cantor to Cohen," in *Handbook of the Philosophy of Science; Philosophy of Mathematics,* ed. Andrew Irvine (Elsevier, 2007).

Part I

Great Illusion of Twentieth Century Mathematics

... the boy began to delight in his daring flight, and abandoning his guide, drawn by desire for the heavens, soared higher. His nearness to the devouring sun softened the fragrant wax that held the wings: and the wax melted: he flailed with bare arms, but losing his oar-like wings, could not ride the air...

Ovid, *Metamorphoses* VIII, 185–235.

Chapter 1

Theological Foundations

1.1 Potential and Actual Infinity

The notion of infinity came to be narrowed and made more precise already around the time of the first clashes between Christian theology and ancient science. During that period, the meanings of this term that refer to indefiniteness, elusiveness and uncertainty were suppressed. Later, modern science – along with the influential parts of philosophy and theology associated with it – rid the notion of infinity of such meanings completely. Therefore, we too (unless otherwise stated) will in this chapter give the name infinity only to infinity with a classical interpretation; that is, the infinity that is still associated with this name in scientific circles.

Our primary encounter with infinity interpreted in this way occurs when dealing with sharply defined infinite events, that is, when drawing on some constant, precisely determined, but inexhaustible possibilities. This happens when we repeatedly add one to a number, when we repeatedly extend a line by a given length in classical geometric space, and so forth. This form of infinity has been given the name **potential infinity**. This is because infinity has an influence on such events, it seizes them, and they submit to it. On the other hand, by **actual infinity** we understand the form of the phenomenon of infinity shown in a work created by the exhaustion of all relevant inexhaustible possibilities. For example, if we say that the sequence of natural numbers 1, 2, 3, ... is potentially infinite, then we mean that only finitely many of these numbers are created at any given point, but it is always possible to create more. But if we say that this sequence is actually infinite, then we mean that all natural numbers have already been created (and so no more can be created). Similarly the aforementioned repeated line extension is potentially infinite; the work that emerges after all these possible extensions is no longer a segment, but a straight line (or a half-line), in which the infinity governing this work shows in its victorious, actual form.

Already in the very definition of actual infinity there is an almost too patent contradiction. It demands that what is inexhaustible be exhausted, that the infinite come to an end. That is, what is infinite should in a sense become also finite. Nevertheless, actual infinity was not universally rejected as a logically contradictory concept unacceptable by pure reason. From time to time it crept into the deliberations of certain thinkers.

Modern European science did not, of course, flood with over-confidence in actual infinity. As far as it could, it circumvented it – more or less successfully – by means of potential infinity, whose classical interpretation was universally accepted.

The problem of actual infinity however did come up almost constantly with insistent urgency in connection with interpretations of the Christian God. The problem of actual infinity was thus – and in fact it still is so today – primarily a theological problem, even though since the middle of the nineteenth century modern mathematics has appropriated, modified, and absorbed into itself the part of theology that developed around it. But before this happened, European thinking struggled very hard with this problem. We will look at some important milestones of this struggle in the chapter that we are currently opening.

1.1.1 Aurelius Augustinus (354–430)

Among the thinkers who elevated the Christian God to dizzying heights above all the grandest pagan gods, St. Augustine occupies the leading place. He was the one who, for the greatness and glory of the resplendent majesty of God, looked into the bottomless depths of absolute infinity and decided the battle between the Christian God and this all-consuming and all-creation-defeating depth in God's favour. He did so above all in the eighteenth chapter of the twelfth book of his famous work *De civitate Dei*, bearing the telling title "Against Those Who Assert that Things that are Infinite Cannot Be Comprehended by the Knowledge of God." For illustration, we offer the following excerpt from Marcus Dods' 1887 translation[1] with the simple note that in Augustine's terms, number usually means the count of natural numbers from one to a given natural number.

> As for their other assertion, that God's knowledge cannot comprehend things infinite, it only remains for them to affirm, in order that they may sound the depths of their impiety, that God does not know all numbers. For it is very certain that they are infinite; since, no matter of what number you suppose an end to be made, this number can be, I will not say, increased by the addition of one more, but however great it be, and however vast be the multitude of which it is the rational and scientific expression, it can still be not only doubled, but even multiplied. Moreover, each number is so defined by its own properties, that no two numbers are equal. They are therefore both unequal and different from one another; and while they are simply finite, collectively they are infinite. Does God, therefore, not know

[1] Augustine, *The city of God,* transl. Marcus Dods (Buffalo NY: Christian Literature Publishing, 1887).

> numbers on account of this infinity; and does His knowledge extend only to a certain height in numbers, while of the rest He is ignorant?

1.1.2 Thomas Aquinas (1225–1274)

The great Church teacher St. Thomas Aquinas dealt with the relationship between God and infinity predominantly in connection with God's power. This theologian, perhaps the most influential theologian, subjected the power of God to the necessities of reason and hence, without being willing to admit it, subordinated God to reason. This organic involvement of God in the humanly-sought and reason-ruled order of the real world significantly stimulated and nurtured the development of modern European natural science and science in general, yet at the same time it also surrendered God to it.

St. Thomas Aquinas justified the restriction of God's power on account of God having only active possibilities, and therefore active power (being omnipotent in this sense), and not the possibilities of passiveness, and therefore the power of passivity, for God is pure being. The passive possibility of something is – somewhat loosely speaking – its being something that it could be, but by its very nature it can no longer be. For example, if a bird takes off, then it had the option not to take off, but if it has already taken off, then it has lost that opportunity; it is no longer an active or feasible possibility, but only a passive one.

In the article entitled "Whether God Can Do Not to Be Past," found in the first volume of the *Summa Theologica*, St. Thomas Aquinas says quite openly:

> ...there does not fall under the scope of God's omnipotence anything that implies a contradiction. Now that the past should not have been implies a contradiction. For as it implies a contradiction to say that Socrates is sitting, and is not sitting, so does it to say that he sat, and did not sit. [...] Whence, that the past should not have been, does not come under the scope of divine power. [...] Thus, it is more impossible than the raising of the dead; in which there is nothing contradictory, because this is reckoned impossible in reference to some power, that is to say, some natural power; for such impossible things do come beneath the scope of divine power.[2]

In short, God's power over the real world, that is, over created beings and ideas imprinted onto this world, is limited by the law of logical contradiction and only by it. Thus, the limits of God's power are only the necessities of pure reason.

At a time when the whole world was inside a crystal spherical surface, it was not difficult to gain the conviction that in such a world, actual infinity does not occur. Similarly, it is certain that no right-minded person would object to the fact that no created being is able to see actual infinity. The expected objection is that actual infinity can be discovered at least among the ideal phenomena, and

[2] Thomas Aquinas, *Summa Theologica* 1.1–26.

namely in the mathematical sciences, since geometers are in the habit of saying "let's take this infinite line." St. Thomas Aquinas dismisses this objection in the first part of the *Summa Theologica* as follows:

> A geometrician does not need to assume a line actually infinite, but takes some actually finite line, from which he subtracts whatever he finds necessary; which line he calls infinite.

In other words, geometry does not need to get mixed up in God's preserve, which is actual infinity; it has no reason to do so. After all, no geometrician can see that far.

On the other hand, especially in the cultivation of mathematics, we often encounter infinity, even if only in its potential form. Thomas Aquinas briefly mentions this in picking apart the seventh question of the third volume of the *Summa Theologica*, when he writes:

> If we speak of mathematical quantity, addition can be made to any finite quantity, since there is nothing on the part of finite quantity which is repugnant to addition. But if we speak of natural quantity, there may be repugnance on the part of the form... And hence to the quantity of the whole there can be no addition. [He means, apparently, the sky – that spherical crystal surface].

With these words St. Thomas Aquinas indirectly suggests that it is in the cultivation of the mathematical sciences that we can, through reason, touch – but only timidly – the greatness of God. And also the other way round, as Boethius (480–524) claims, knowledge of things Divine cannot be acquired by anyone completely devoid of mathematical training.

By depriving the real world of actual infinity, granting the right to handle it only to God and denying it to created beings, St. Thomas Aquinas greatly simplified the problem of actual infinity. He transferred it to the exclusive competence of God and so opened up the search for its solutions in a realm beyond human intellectual cognition. Consequently the only question was whether God could really know actual infinity, that is, to show how he knows it in a way that eschews the contradiction contained in the very concept of actual infinity, which appears in it when we approach it from potential infinity. In other words, it is necessary to show how the supreme Christian God overcomes actual infinity. To this end, it helps to be aware of the differences between the intellectual cognitive capacities of God and man. These are pointed out by St. Thomas Aquinas in the first book of the *Summa Contra Gentiles,*[3] where he writes that:

> our intellect does not know the infinite, as does the divine intellect. For our intellect is distinguished from the divine intellect on four points which bring about this difference. The first point is that our intellect is absolutely finite whereas the divine intellect is infinite.

[3] Thomas Aquinas, *Summa Contra Gentiles* 1.69.14.

> The second point is that our intellect knows diverse things through diverse species. This means that it does not extend to infinite things through one act of knowledge as does the divine intellect. The third point follows from the second. Since our intellect knows diverse things through diverse species it cannot know many things at one and the same time. Hence, it can know infinite things only successively by numbering them. This is not the case with the divine intellect which sees many things together as grasped through one species. The fourth point is that the divine intellect knows both the things that are and the things that are not, as has been shown [in one of the earlier chapters of the *Summa contra Gentiles*].

1.1.3 Giordano Bruno (1548–1600)

Three dialogues about actual infinity in the real world are contained in the slim book *De l'infinito universo et Mondi* (*The Infinite Universe and the World*), published in Venice in 1584 by Giordano Bruno. In them, this educated Dominican gradually reveals weaknesses in the teachings of his famous predecessor and confrer, St. Thomas Aquinas (already a saint at that time and declared a church teacher) about the fact that there can be no actual infinity in created things. Aware that Thomas's explanation limits the expressions of God's power and thereby also God's power itself, Giordano Bruno's book opens the way for actual infinity to enter even into the material component of the real world.

Thomas's evidence of the impossibility of actual infinity in the real world is taken almost literally from Aristotle and that is why Giordano Bruno attacks Aristotle. He prefers not to mention Thomas at all out of caution. Yet, this does him no good.

Unlike St. Thomas Aquinas, Giordano Bruno has already, with full consciousness, taken the fateful step of modern science, which consists in inserting the real world into classical geometric space. This step was so captivating that until the time of Riemannian geometry, that is, until the middle of the nineteenth century, it was irreversible.

According to Aristotle, real space extends all the way to that spherical crystal surface dotted with fixed stars. There is no place behind it. This spherical surface is the boundary of the real world and is its place. If it did not exist, there would be no place left after it.

Such a ridiculously small piece of work as would be the real world of Aristotle and St. Thomas Aquinas, however, not only would not correspond to God's infinite goodness, but it would also not reflect God's infinite majesty, and most importantly – it would not be worthy of God's infinite power, which in its unattainable magnitude can in fact be manifested only by an actually infinite deed. These reasons – which point in favour of an actually infinite number of different suns and planets in the universe, many of which appear as stars in the night sky – are noted by Giordano Bruno in the following words:[4]

[4] Rather than offering a literal translation of Bruno's Italian dialogues, the English translations presented here seek to preserve the interpretation of J. B. Kozák's Czech edition, used

> Infinite majesty is incomparably better represented in innumerable individuals than in those which can be counted and are finite. God's inaccessible countenance must be reflected in an infinite image in which, as innumerable members, the worlds [meaning the sun and the planets] are as innumerable as the other worlds [meaning different from our Sun and our Earth]. For the same reason that there are innumerable degrees of perfection that must develop the disembodied majesty of God in a bodily manner, there must be innumerable individuals such as these great bodies (of which the Earth is one, our divine mother, which gave birth to us and nurtures us, and which will not try again). Infinite space is needed to hold their quantity without number. Therefore, as it is good that this world exists [the Earth is understood and life on it], that it could be and can be, so too is it good that there really are, could be and can be, innumerable similar worlds to this one [...]
>
> Why should we and can we claim that God's goodness, which can be communicated to an infinite number of things and poured out into infinity, would want to be greedy and withdraw to nothingness, because everything finite is nothing in relation to infinity? Why do you want this center of divinity, which can expand endlessly in the form of an infinite sphere (if it can be so expressed), remain as if unwilling, somewhat sterile, rather than communicating as a creative father, a sublime one and beautiful? Why should it communicate itself to a lesser extent, or to put it better, not communicate at all, rather than being, according to the nature of His magnificent potency, everything? Why should the infinite creative ability be useless, be deprived of the possibility of the existence of innumerable possible worlds? Why should the seriousness of God's image be dimmed, which would have to shine in an unreduced mirror and in an infinite or cosmic way? Why should we claim something that causes such inconsistencies and destroys so many principles of philosophy without in any way benefiting laws, faith, or morals? How do you want God to be limited in potency as well as in activity and effects [...]

Giordano Bruno not only pointed out that a finitely large real world in infinite space would not correspond to God's goodness, majesty and power, but in the following words he also made it clear that it was doubtful about any God who would not manifest himself in an actually infinite work, whether he could manifest himself in this way at all; in other words, whether his creative power was not limited by actual infinity.

> For all those reasons, therefore, that this world of ours, if it is conceived as finite, can be said to be appropriate, good, necessary, all

by Vopěnka. [Translator's note.]

> other innumerable worlds can be said to be appropriate and good; for the same reasons, omnipotence does not deny their existence. And if we don't recognize them, this omnipotence could be accused of not wanting or not being able to make them such, and so leaving a void (or, if you don't want to use the word void, leaving infinite space); this would not only diminish the infinite perfection of being, but also the immeasurable majesty of the effective cause in things created, if created, or in dependent things, if they are eternal. Why should we believe that an agent who can do an infinite amount of good has made it limited? And if he has made it limited, why should we believe that he has the opportunity to make it infinite, when for him the possibility and the realizing activity merge into one?

Giordano Bruno remained true to this conviction until his death. He openly claimed it during his third interrogation before the Inquisition tribunal in 1592, where he said:

> There is an infinite universe that is the result of God's infinite power, for I consider it unworthy of God's goodness and power for the Deity to give birth to one finite world when, in addition to this world, he could give rise to another and infinitely many others. So be it; I have declared that there is an infinite number of individual worlds, similar to this world of our Earth, which I believe with Pythagoras is a moon-like star, other planets, and other stars, of which there are infinitely many, and that all these celestial bodies are innumerable worlds, which then bring together infinite universality in infinite space; and this is called the infinite universe, in which there are innumerable worlds. Thus there are two infinities [he means in actuality, in the real world]: the infinity of the magnitude of the universe and an infinite number of worlds; from which indirectly follows a rejection of truth according to faith.

"Because he remained stubborn not only in this delusion but also in other delusions, on February 16, 1600, servants of justice released him from the dungeon of the Holy Inquisition and took him to the Campo de' Fiori in Rome, where stripping him and tying him to a stake, they burned him alive with his tongue in a vice. Two Dominicans and two Jesuits persuaded him until the last moment to renounce the obstinacy in which he finally completed his miserable unhappy life" – was written in a report by the Brotherhood of the Severed Head of John the Baptist and in other reports from that time. On the same square, three years later, all his books and writings were proclaimed as banned.

1.1.4 Galileo Galilei (1564–1654)

In 1638, one of the last works of the then already famous pioneer of modern European science, Galileo Galilei, was published in the Netherlands. This work is *Discorsi e dimonstrazioni matematiche, intorno a due nuove scienze*, containing four dialogues and a short appendix. The conversations in these dialogues are between Salviati, Sagredo and Simplicio. The first takes Galileo's view and instructs the other two, with Simplicius being less understanding than Sagredo, so Salviati has to explain everything to him a little more broadly.

Although the whole of Galileo's book is remarkably stimulating, what interests us at the moment is only a short excerpt from the first dialogue, included therein really only to illustrate the views expressed there; namely in relation to Simplicio's claim that there are more points on a longer line than on a shorter one. Since what Galileo demonstrates here is now very often explained in a distorted way, let us present this excerpt in a more-or-less literal translation.[5]

> SALV. These are the some of the difficulties that derive from the conversation we have with our finite intellect around infinities, giving them those attributes that we give to things that are finished and ended; which I think is inconvenient, because I believe that these attributes of largeness, smallness and equality do not agree with infinities, of which one cannot be said to be greater or less or equal to the other. As an example I offer a case that has already occurred to me, which for a clearer explanation I put to Mr. Simplicio, who has raised the difficulty. I suppose you know very well which numbers are square, and which are non-squares.
>
> SIMP. I know very well that the square number is the one that arises from the multiplication of another number in itself: and so four, nine, etc., are square numbers, the one being born from the two, and this from the three, multiplied by themselves.
>
> SALV. Very well: and you know further that as the products are called squares, the producers, that is, those that multiply, are called sides or roots; the others, which do not arise from numbers multiplied in themselves, are otherwise not squares. So if I say that all numbers, including squares and non-squares, be more than just squares, I will certainly say a true proposition: isn't it so?
>
> SIMP. We cannot say otherwise.
>
> SALV. If I go on to ask then, how many square numbers there are, I can answer truthfully that there are as many as there are their roots, since it happens that every square has its root, every root its square, nor any square has more than one root, nor any root more than a single square.
>
> SIMP. So it is.
>
> SALV. But if I ask how many roots there are, it cannot be denied that there are not as many as all numbers, since there is no number that is

[5] Galileo Galilei, *Discorsi e dimonstrazioni matematiche, intorno a due nuove scienze* (Leiden: Louis Elsevier, 1638). The English translation presented here is a literal version of the Italian. [Translator's note.]

> not the root of some square; and given this, it will be appropriate to say that square numbers are as many as all numbers, since there are as many as their roots, and roots are all numbers: and even at the beginning we said, all numbers are much more than all squares, being the most not square. And yet the multitude of squares is always decreasing with greater proportion, as more numbers are passed; because up to a hundred there are ten squares, which is the same as saying the tenth part to be squares; in ten thousand only the hundredth part are squares, in one million only the thousandth: and even in the infinite number, if we could conceive it, it should be said, there are as many squares as all the numbers together.
>
> SAGR. What then does this show?
>
> SALV. I do not see that we can come to any other decision than to say that all the numbers are infinite, the squares infinite, their roots infinite, nor the multitude of squares be less than that of all numbers, nor this greater than that, and ultimately, that the attributes of equal, greater, or lesser, make no sense in the infinite, but only in the finite quantities.

Today we would say that the set of all squares of natural numbers can be mapped by a one-to-one mapping onto the set of all natural numbers, and that, consequently, both of these sets have the same **cardinality**. However, Galileo does not mention any such concept. His reasoning is merely meant to be a deterrent to the difficulties and mistakes we might get tangled up in if we talked about infinity using the concepts we have created for finite quantities. He recommends we not talk about the actual infinity at all.

As can be seen, Galileo's advice is unequivocal: Infinity yes, but potential, because in this way we only ever have to actually deal with the finite, and therefore we stand on solid ground. Reflections on actual infinity are dangerous because we do not have the ability to come up with suitable concepts for this infinity, while the concepts created in the study of finite phenomena are not applicable to it.

1.1.5 The Rejection of Actual Infinity

The problem of actual infinity was, from the perspective of the cold intentions of modern European science, at best, a marginal problem. The gaping chasm between man and God, created by the demolition of the Baroque superstructure of the real world, sharply separated science from theology. In this process the problem of actual infinity, thanks to its incomprehensibility by man, fell almost exclusively into the scope of theology. Science saw it as a question impossible to decide, and therefore uninteresting for an active Western European. **Benedict Spinoza** (1632–1677) mentions actual infinity's inaccessibility to either human reason or even human will only as if in passing. He does so in discussing the implications of the forty-ninth proposition in the second volume of his *Ethics* (that is, in examining questions that are more urgent for science):

> If it be said that there is an infinite number of things which we cannot perceive, I answer, that we cannot attain to such things by any thinking, nor, consequently, by any faculty of volition.[6]

Thus, for most of the seventeenth century, influential figures in the nascent modern sciences did not even need to know Galileo's warning words to see clearly not only that potential infinity would suffice for their purposes, but also that considerations of actual infinity unnecessarily, unjustifiably, and dangerously, therefore inadmissibly, exceeded the defined scope of scientific study. For this reason, even Descartes did not pose the question of actual infinity, and **Thomas Hobbes** (1588–1679) wrote in the third chapter of Leviathan:

> Whatsoever we imagine is finite. Therefore there is no idea or conception of anything we call infinite. No man can have in his mind an image of infinite magnitude; nor conceive infinite swiftness, infinite time, or infinite force, or infinite power. When we say anything is infinite, we signify only that we are not able to conceive the ends and bounds of the thing named, having no conception of the thing, but of our own inability. [...] No man therefore can conceive anything, but he must conceive it in some place; and endued with some determinate magnitude; and which may be divided into parts...[7]

These words are, moreover, an eloquent testimony to the origin of the abyss that has opened up in the minds of Western European scholars between man and God. More than by anyone else, its terrifying presence was felt by one of the most famous thinkers of the time, a Catholic but an opponent of the Jesuits, **Blaise Pascal** (1623–1662). In a remarkable document with the apt title *Pensées*, we find the following confession:

> On beholding the blindness and misery of man, on seeing all the universe dumb, and man without light, left to himself, and as it were astray in this corner of the universe, knowing not who has set him here, what he is here for, or what will become of him when he dies, incapable of all knowledge, I begin to be afraid, as a man who has been carried while asleep to a fearful desert island, and who will wake not knowing where he is and without any means of quitting the island.[8]

According to Pascal, man is midway between nothingness represented by zero and infinity, which belongs only to God. There is an insurmountable infinite abyss on both sides, the infinity of which – the same in the direction of greatness as in the direction of insignificance – shows man only its potential form. In his treatise *Of the Geometrical Spirit*, Pascal wrote:

[6] Benedictus de Spinoza, *Ethics*, trans. R. H. M. Elwes (London; G. Bell and Sons, 1887).

[7] Thomas Hobbes, *Hobbes's Leviathan.* Reprinted from the Edition of 1651 (Oxford: Clarendon, 1909), 23.

[8] Blaise Pascal, *Pascal's Pensées; or, Thoughts on Religion*, trans. Gertrude Burford Rawlings (Mount Vernon, N.Y.: Peter Pauper Press, 1900), 7.

> For however quick a movement may be, we can conceive of one still more so; and so on ad infinitum, without ever reaching one that would be swift to such a degree that nothing more could be added to it. And, on the contrary, however slow a movement may be, it can be retarded still more; and thus ad infinitum, without ever reaching such a degree of slowness that we could not thence descend into an infinite number of others, without falling into rest.
>
> In the same manner, however great a number may be, we can conceive of a greater; and thus ad infinitum, without ever reaching one that can no longer be increased. And on the contrary, however small a number may be, as the hundredth or ten thousandth part, we can still conceive of a less; and so on ad infinitum, without ever arriving at zero or nothingness.
>
> However great a space may be, we can conceive of a greater; and thus ad infinitum, without ever arriving at one which can no longer be increased. And, on the contrary, however small a space may be, we can still imagine a smaller; and so on ad infinitum, without ever arriving at one indivisible, which has no longer any extent.
>
> It is the same with time. We can always conceive of a greater without an ultimate, and of a less without arriving at a point and a pure nothingness of duration.
>
> That is, in a word, whatever movement, whatever number, whatever space, whatever time there may be, there is always a greater and a less than these: so that they all stand betwixt nothingness and the infinite, being always infinitely distant from these extremes.[9]

A sharp and open rejection of actual infinity occurred also in the island empire. **John Locke** (1632–1704) went so far in this matter that he did not even grant man the opportunity to create a positive idea of infinity. According to him, the phenomenon of infinity is a mere disposition, that is, a tendency to some action. He writes about how one acquires the idea of infinity in the seventeenth chapter of his *Essay Concerning Human Understanding*:

> Every one that has any idea of any stated lengths of space [...] how often soever he doubles, or any otherwise multiplies it, he finds that after he has continued his doubling in his thoughts, and enlarged his idea as much as he pleases, he has no more reason to stop, nor is one jot nearer the end of such addition, than he was at first setting out. The power of enlarging his idea of space by farther additions remaining still the same, he hence takes the idea of infinite space.[10]

[9] Blaise Pascal, "Of the Geometrical Spirit," trans. Orlando Williams Wight. The Harvard Classics: *Blaise Pascal* (New York: Colier and Son, 1910), 436–37.

[10] John Locke, "An Essay Concerning Human Understanding," in *The Works of John Locke in Nine Volumes,* 12th ed. (London: Rivington, 1824), 1:195.

John Locke effectively abolished infinite space as something inaccessible to man, replacing it with mere spatial endlessness, and thus expelled actual infinity from space (even if filled only with an emptiness that many think is mere nothing) and replaced it with potential infinity. He did the same in other cases of infinity, with time and numbers for example.

1.1.6 Infinitesimal Calculus

The rejection of actual infinity, however, was not as simple a matter for modern European science as it seemed to the first pioneers of this tendency of thought. A conflict with this phenomenon occurred in the last decades of the seventeenth century in connection with the discovery of infinitesimal calculus. The immense effectiveness of this new (higher) mathematics was evident from the first works of its discoverers, **Isaac Newton** (1643–1727) and **Gottfried Wilhelm Leibniz** (1646–1716). The range over which infinitesimal calculus was then developed by the Bernoulli brothers at the turn of the seventeenth and eighteenth centuries and by many other mathematicians during the eighteenth century, including in the foremost place **Leonhard Euler** (1707–1783), could no longer leave anyone in doubt that science had acquired a tool that it could not renounce even for the most compelling philosophical reasons.

In infinitesimal calculus, several separate streams of mathematics and their various conceptions intertwined. We will not worry about unraveling these influences here. We will also not reveal in detail those phenomena which became the basis for the formation of the geometric intuition of this higher mathematics, on which calculations could then be based and which provided meaning to these calculations. We will merely point out that the basis of the geometric attitude on which infinitesimal calculus rests, is founded on the relationship between the ancient and classical geometric worlds (these concepts will be discussed in more detail later). As long as mathematicians, even if only unconsciously, based their reasoning on this relationship, they used and developed this calculus correctly. However, once they suppressed this relationship, they got into difficulties and made mistakes.

In itself, in fact, the basic concept of infinitesimal calculus, which is the concept of an infinitely small quantity, would not need to bring classically interpreted actual infinity into mathematics. Not even if the true nature of this higher mathematics – which we have briefly mentioned – remained unconscious, as, after all, it has done to this day. To wit, infinitely small quantities could be interpreted, for example, as objects of a purely speculative nature. This means that they could be interpreted as something that does not occur in the geometric or real world, but which, if added to these worlds, provides a foundation for calculations, the usefulness of which is undeniable. In other words, infinitesimal quantities could be understood similarly to imaginary numbers, for instance, as already Leibniz mentions. In short, differential calculus did not call for the necessity of actual infinity in mathematics, and mathematicians subconsciously felt this.

In contrast, in integral calculus, based on additions of infinitely many infinitesimal quantities, the clash of mathematics with actual infinity seemed inevitable. This was not so much because of what quantities were added, but because of the fact that the number of terms to be added was infinitely large. This opened up a new field of activity for a special current of mathematics, known as the number magic, which is a peculiar part of a much broader stream, namely the mathematics of singular phenomena.

1.1.7 Number Magic

Until this point, number magic had focused on finding various peculiarities related to only a few specific (mostly natural) numbers or often even related to a single number. Now, however, this research passion rather than sober science spread to the search for arithmetic oddities related to infinity. And it was this infinitary number magic that brought actual infinity into mathematics; infinitesimal calculus only made this task easier.

Infinitary number magic had for some time been a recognized part of mathematics and even such prominent mathematical personalities as Euler have, from time to time, succumbed to its allure. However, it was the part of mathematics with which, at a time when mathematics was trying to stick closely to science, mathematics transcended science. Its findings, if they concerned science at all, were judged very cautiously by scientists.

In infinitary number magic, which focuses primarily on the arithmetic of infinite number series, both the extraordinary investigative power of number magic and the misery of the unfettered passions operating in it appear in their naked form.

For our purposes, a well-known geometric series will provide a sufficient picture of the variety of things that can be done with infinite sums of number series.

If we let

$$S = a + aq + aq^2 + aq^3 + \dots$$

then

$$qS = aq + aq^2 + aq^3 + \dots$$

and hence $S = a + qS$. Consequently,

$$S = \frac{a}{1-q}.$$

As long as the absolute value of q is less than 1, this is a generally recognised result. However, its validity was frequently accepted even when the absolute value of q was greater than 1. Then for example for $a = 1$, $q = 2$ we obtain

$$-1 = 1 + 2 + 4 + 8 + 16 + \dots$$

In spite of the obvious lack of sense of such assertions, Euler himself declared that

$$\dots + \frac{1}{a^3} + \frac{1}{a^2} + \frac{1}{a} + 1 + a^2 + a^3 + \dots = 0,$$

since by the above equality we have

$$1 + \frac{1}{a} + \frac{1}{a^2} + \ldots = \frac{a}{a-1},$$

$$a + a^2 + a^3 + \ldots = \frac{a}{1-a}$$

and

$$\frac{a}{1-a} + \frac{a}{a-1} = 0.$$

This statement by Euler is a clear demonstration of the kind of irresistibly seductive and clear-reason-defeating passion that infinite number magic awakened.

Undoubtedly more interesting and instructive, however, is the infinitary number magic performed in connection with Euler's number e, which we define today by the rule:

$$e = \lim_{n\to\infty} \left(1 + \frac{1}{n}\right)^n.$$

If we set $e = \left(1 + \frac{1}{\infty}\right)^\infty$ and expand this according to the binomial theorem we obtain

$$e = \sum_{k=0}^{\infty} \left(\frac{1}{\infty}\right)^k \binom{\infty}{k}.$$

Since

$$\binom{\infty}{k} = \frac{1}{k!}\infty(\infty-1)(\infty-2)\ldots(\infty-k+1)$$

and since for every finite n we have $\infty - n = \infty$, the above yields

$$\binom{\infty}{k} = \frac{\infty^k}{k!} \quad \text{and hence} \quad e = \sum_{k=0}^{\infty} \frac{1}{k!},$$

$$e^x = \left(1 + \frac{1}{\infty}\right)^{x\infty} = \sum_{k=0}^{\infty} \left(\frac{1}{\infty}\right)^k \binom{x\infty}{k} = \sum_{k=0}^{\infty} \frac{1}{k!} x^k \frac{\infty^k}{\infty^k} = \sum_{k=0}^{\infty} \frac{x^k}{k!}.$$

A student following these calculations might sigh, what an easy job the mathematicians of the olden days had compared to us, and they will surely not be able to resist the temptation to calculate that for an infinitely small x, that is, $x = \frac{1}{\infty}$, we have

$$e^x = \left(1 + \frac{1}{\infty}\right) \quad \text{and hence} \quad \frac{e^x - 1}{x} = \infty \frac{1}{\infty} = 1.$$

For us however, nothing remains but to reflect what a pity it is that such calculations are no longer allowed today and that we must no longer teach them to students. Not because they should be correct, but because they are an important source of inspiration. After all, only now, after getting acquainted with them, do we realize how it was possible that Euler could have discovered such laws, how it could even have occurred to him that something like this was true.

Such, however, is the fate not only of the infinitary number magic, but of number magic in general. It has a great power for discovery, but it is uncritical of what it brings to light. It is on account of this uncritical dimension that the age of plain reason has no understanding for it.

1.1.8 Jean le Rond d'Alembert (1717–1783)

> Not only does philosophy abandon to the ignorant subtlety of barbaric centuries these imaginary objects of speculation and disputes, with which our schools still resound; it even refrains from dealing with questions the object of which may be more real, but the solution of which is no more useful to the advancement of our knowledge.[11]

In the enlightened eighteenth century, for which Jean le Rond d'Alembert thus established the subject and boundaries of thought in the *Essay on the Foundations of Science,*[12] magical juggling with actual infinity was, of course, an upsetting and scandalous phenomenon. Aware, however, that deep mathematical truths were hidden in not a few of the results presented in this way — especially those of his contemporary Euler — d'Alembert became convinced that the cause of all the improprieties associated with infinitary number magic was inadmissible calculations with actual infinity. Therefore, he set as a criterion for the correctness of these results the possibility of arriving at them using only potential infinity. In the section entitled "On the Metaphysical Assumptions of Infinitesimal Calculus" contained in the appendix to the *Essay on the Foundations of Philosophy*, he then suggests in the following way how the whole issue can be set right.

> In order to form exact notions of what geometers call infinitesimal calculus, we must first fix in a very clear way the idea we have of infinity. [...] We can also see from this notion that the infinite, such as analysis considers it, is properly the limit of the finite, that is to say the term to which the finite always tends without ever arriving there, but which we can suppose that it always approaches more and more, although it never reaches it. Now it is from this point of view that geometry and analysis consider the infinitesimal quantity, well understood. An example will serve to make this clear.
>
> Suppose this sequence of fractions continues to infinity, $\frac{1}{2}, \frac{1}{4}, \frac{1}{8}, \frac{1}{16}, \ldots$ etc., and so on, always decreasing by half. Mathematicians say and prove that the sum of this sequence of numbers, if we suppose it pushed to infinity, is equal to 1. This means, if we want to speak

[11] Translated from the French of Jean le Rond d'Alembert, "Essai sur l'élémens de philosophie," in *Oeuvres completes de D'Alembert* 1 (Paris: A. Belin, 1821), 131.

[12] This quotation appears in the "Essai sur l'élémens de philosophie," not, as Vopěnka suggests, in D'Alembert's encyclopedia article "Élémens des sciences." [Translator's note]

> only according to clear ideas, that the number 1 is the limit of the sum of this series of numbers; that is, the more numbers we take in this sequence, the closer the sum of these numbers will approach to equalling 1, and that we can approach as close as we want. This last condition is necessary to complete the idea attached to the word limit. [...]
>
> Likewise when we say that the sum of this sequence 2, 4, 8, 16, etc. or any other that goes on increasing, is infinite, we want to say that the more terms we take from this sequence, the greater the sum will be, and that it can be equal to as large a number as we want.
>
> Such is the notion that must be formed of the infinite, at least in relation to the point of view from which mathematics considers it; a clean idea, simple, and resistant to attack.
>
> I am not examining here whether there are indeed infinite quantities actually existing; if space is really infinite; if duration is infinite; if there is a really infinite number of particles in a finite portion of matter. All these questions are foreign to mathematicians' infinity, which is absolutely, as I have just said, only the limit of finite quantities; of which limit it is not necessary in mathematics to suppose the real existence; it suffices only that the finite never attains it.
>
> Geometry, without denying the existence of actual infinity, therefore does not suppose, at least not necessarily, the infinite as really existing; and this consideration alone suffices to resolve a large number of objections which have been raised against mathematical infinity.
>
> We ask, for example, if some infinities are not greater infinities than others, if the square of an infinite number is not infinitely greater than this number? The answer is easy for the geometrician: an infinite number does not exist for him, at least necessarily; the idea of the infinite number is for him only an abstract idea, which expresses only an intellectual limit which no finite number ever reaches.[13]

In a straightforward and still recognized way, d'Alembert then explains not only the limit of a function at a point, the derivative or quantities of different orders, but also the definite integral of a function and the integral of a differential equation. The concept of limit introduced by him thus became a basic concept henceforth of what became simply mathematical analysis, no longer mathematical analysis of infinitesimal quantities. These departed from mathematics – albeit reluctantly and disruptively – for a much longer time even than actual infinity.

Through the notion of limits, d'Alembert thus purified mathematics of phenomena contrary to the spirit of modern European science, namely of actual infinity and infinitesimal quantities. Over time, however, it turned out that he

[13] D'Alembert, "Essai sur l'élémens de philosophie" in *Oeuvres completes de D'Alembert* 1, 288–289,

had cleaned it up too much in this way. This, however, will be treated in more detail in our discussion of infinitesimal calculus.

Finally, it is useful to recall that the notion of limits, which retained a privileged position in mathematical analysis even after the rapid entry of actual infinity into mathematics in the late nineteenth century, was originally introduced by d'Alembert in order that actual infinity might be excluded from mathematics.

1.2 The Disputation about Infinity in Baroque Prague

In those places where baroque Catholicism was celebrating a clear victory during and after the Thirty Years' War, the hitherto promising buds of the new science did not find a fertile ground for flourishing. From Copernicus' Toruń, Tycho's and Kepler's Prague, and to a large extent also from Galileo's and Bruno's Italy, the spirit of nascent modern science had relocated to the countries along the banks of the English Channel. For Prague Catholic scholars, especially the Jesuits in the Clementine, the very subject of study was God himself, while the real world was only a necessary complement, rather a means, a suitable way than a subject of study. By contrast, the leading figures of modern European science now quite openly took the opposite direction. Their gaze clung to the real world, which was the genuine object of their study, while God, as the glorious and powerful creator of this world, the guarantor of the truth about it and the guarantee of the meaningfulness of scientific endeavor, was a helper in need, but not a subject of study.

It is worth emphasizing that with all the doubts and possible objections to Giordano's interpretation of the real world, the irrefutable fact remains that the act of penetrating the vault of heaven and expanding the real world into the entirety of geometric space was one of the highlights of the Baroque. At the same time, however, let us keep in mind that it was not science, but theology, that broke through the crystal spherical surface surrounding the then real world.

1.2.1 Rodrigo de Arriaga (1592–1667)

Our gaze will now focus on the very brain of the Czech Baroque, who dwelled within the walls of the Prague Clementinum and who was brought to our attention in Stanislav Sousedík's revelatory study, Rodrigo de Arriaga: Contemporary of J.A. Comenius. After studying the philosophical and theological writings produced at the time in Prague, Sousedík found that the leading spirit of the Clementine College was the Spanish Jesuit Rodrigo de Arriaga. In order to bring this Jesuit closer (along with the environment among the professors of the Clementine College more generally), we will now draw out some of the data and suggestions of Sousedík's study.

The aim of the Jesuit effort was not so much to bring only the Czechs back into the lap of the Roman Church, but also the whole of Europe. Nevertheless, the conversion of the Czechs played a key role in achieving this aim. This was not only because victory in this country would be sweeter than in others

— for though initially devoted to Rome, it was the first country to renounce obedience, and from it all kinds of heresy had penetrated throughout Europe in various visible and invisible ways. Much more importantly, it was because the country had became a hotbed of diverse views thanks to its pre-White-Mountain religious freedom. Studying and struggling with these views was an invaluable preparation for the expected later clashes.

The Jesuits were convinced that only their militant, internally disciplined, and purposefully working order was able to bring this struggle to a successful end; after all, it was precisely for this purpose that this order was founded and shaped. So as not to be disturbed by other religious orders, they succeeded – despite the protests of the Dominicans and Franciscans – in having the Charles University entrusted to them in 1623.

The professors of the Clementine College differed significantly from those ordinary Jesuits who converted the rural population to Catholicism. The task of these professors was not to suppress unwanted opinions by burning the books in which they were found, but on the contrary to know, study and consider which of these ideas were acceptable and which were not, and then to defeat the unacceptable ones in the spiritual field by the power of thought. The Jesuits knew full well that after the final military victory, their main rivals would not be rural peasants and city merchants, but scholars of the calibre of Leibniz. To this end, it was necessary for there to be almost complete freedom of research within their group, which was manifested externally by moderate restraint, relatively ample tolerance, and even support for the things that might usefully be taken over, even though they were born on Protestant soil.

A group of professors was formed in Prague, which although small was very effective and powerful in its research. The head of this group was the excellent Spanish scholar Rodrigo de Arriaga, who set out for Bohemia in 1625, was solemnly declared a doctor of theology in Prague in January 1626, and shortly thereafter began teaching.

Both of Arriaga's major works, the *Cursus philosophicus*, published in 1632, and *Disputationes theologicae* — the eight volumes of which began to be published after 1663 — are marked by intellectual independence, novelty, and often audacity. In them Arriaga certainly shows himself worthy of the task assigned to him.

Arriaga's philosophy and theology are discussed in detail in literature. We will now demonstrate only two excerpts from the *Philosophical Course*. The first is from the preface to this work and is intended to prepare the reader for the fact that the book was created on the basis of free thinking, and not by copying from older books.

> In this book, I am not talking about theological problems [...], but about the questions that arise when exploring this natural world around us. On these questions, the ancient philosophers expressed various views. In posing them, of course, they were starting out not from God's revelation [...], but with sensory experience, that is, with what they saw, heard, touched, and tasted. However, their

> views were elevated to inviolable dogmas by their interested followers. I will not have to convince you, dear readers, that we today have the same five senses with which we can perceive as well as they did. Nor does it seem that reason disappeared from the world with the death of Plato or Aristotle. And in terms of our quantity of empirical knowledge, we are much better off today compared to them. [...] After all, new discoveries are taking place day by day, of which no one had any idea at the time [...]. If you read and study my Philosophy Course without an intolerant passion for older philosophy, you will find that some of the conclusions, which it has sometimes been tantamount to sacrilege to doubt, are based on such poor reasons that you will be surprised that they could have been believed for so many centuries.[14]

Arriaga, therefore, did not have to adhere to Aristotle or other church-recognized thinkers. Not only that, however, at the time of Galileo's trial he took a seemingly impartial, but in fact favourable opinion of Galileo, as a reader willing to read between the lines, so to speak, will immediately find out from the following passage.

> Some years ago, certain scholars began to completely overturn the system of heavens, based on the results of careful observations by certain mathematicians and astronomers, who achieved them with wonderful new instruments, especially telescopes. Some state that the heavens are gaseous, others that they are corruptible and vigorously show that otherwise it is not possible to explain the new phenomena discovered in these years [...]. On the other hand, others, fierce proponents of inherited opinions, try instead to brand this new presumption with audacity or error. It is not my intention here to deal directly with this whole issue or to compare the weight of the arguments that both parties put forward in their favour.

The center of Arriaga's interest was not only the real world itself, or even God's relationship to this world viewed from the outside by an uninvolved human gaze. Rather it was God himself and his intercourse with what man could reach by amplifying the phenomena of our outside world – seen so to speak with God's eyes (indeed, this was the only way possible for a researcher bound up in the spiritual world of the Catholic Baroque). Infinity belongs to the first rank of such phenomena, infinity that is, classically understood.

Arriaga devoted mainly the twenty-third and twenty-sixth disputations in his pivotal work, *Cursus philosophicus*, to this infinity examined by a gaze which sought to be God's gaze, and which was in fact the gaze of the Baroque Catholic God. In this section, we will focus exclusively on the ideas contained in Disputatio XXIII, bearing the title "De Infinito."

[14] Rodrigo de Arriaga, *Cursus philosophicus* (Antwerp, 1632). Quoted here from the Czech translation of Stanislav Sousedík.

At the very beginning of the first section of this treatise, Arriaga distinguishes the following three forms of the phenomenon of infinity.

(a) Infinity as **multitude**, that is, a multitude of units that cannot be counted in a way that the counting ends. In other words, a multitude whose number cannot be expressed by any classical natural number and which therefore will never be counted, neither by any human being, nor by any super-human being, such as an angel or the devil.

(b) Infinity in **size**, that is, one which consists of infinitely many parts, by which is probably meant parts spatially arranged, or temporally arranged, and so on.

(c) Infinity in **intensity**, such as infinite temperature, infinite effort, and the like.

Like some other theologians, Arriaga insures himself – and let us add, prudently – by attributing none of these forms of the phenomenon of infinity to God himself, but only to phenomena emanating from God, whether material, materially mediated, or purely spiritual. We are dealing therefore, with a, so to speak, worldly infinity; God is infinite in a different way. For Arriaga is now not concerned with how God is infinite, but with how God treats and rules over infinity as it is manifestated in phenomena emanating from God.

The reason why the learned Jesuits operating in the Prague Clementinum were forced to deal with the relationship between God and actual infinity, was that they were faced with the following dilemma:

(A) God must necessarily have created an actually infinite number of celestial bodies (Giordano Bruno).

(B) Actual infinity is logically contradictory (Galileo Galilei).

(C) God can only create what is logically without contradiction (St.Thomas Aquinas).

The whole treatise, "De infinito," is devoted to Arriaga's attempt to solve this dilemma. For illustration, we select very briefly from his views.

ad A) In the treatise "De infinito" Arriaga debates with Giordano Bruno, although, for obvious reasons, the name of this heretical Dominican never appears in it. Above all, Arriaga rejects a God in whom potency and action merge. In other words, whatever is non-contradictory is possible for God, meaning it is feasible, even if he only makes happen those parts that he wants to make happen and not all of it. That is, for the sake of clarity, figuratively and very loosely: Anyone can get blind drunk, but a wise person will not do it, and anyone can unjustly harm someone else without reason, but a good person does not do it. God, too, has the opportunity to do things that he will not do, for he is most wise and good.

Arriaga, however, does not reject Giordano's interpretation so completely as to subordinate God to infinity, as he feels that the supreme sovereign Christian God must rule infinity and not be powerless against it. He does not, therefore, indiscriminately reject everything Giordano proclaims, but reflects on his observations and draws from them the impetus for his own interpretations of a God who is active and sovereign over all being arising from him – material and ideal.

The fourth, concluding section of Arriaga's treatise on infinity bears an eloquent title: "God can create the infinite in quantity, size, and intensity."

This determines the degree of Arriaga's agreement and disagreement with Giordano Bruno. According to Arriaga, God can actualise infinity; according to Giordano Bruno he has to.

ad B) Already in the first section of the treatise "De infinito" Rodrigo de Arriaga demonstrates that actually infinite multitude is subject to a different logic than finite multitude. And that in terms of size, there are different infinities. For example (in loose translation):[15]

> One infinity can be larger than another. The proof is obvious, because any infinity (not including God) is composed of two parts, at least one of which is infinite and smaller than the whole. If both were finite, their union would also be finite.

The following example of infinites of different sizes is also instructive in a way.

> Because there is an infinity of possible people, the number of possible hands, eyes, or hairs is greater than the number of these people. [...] Furthermore, if God exterminates three or four people, there will still be an infinite number of them.

The rejection of other objections to actual infinity is based by Arriaga partly on the fact that in terms of size there are different infinities (as was pointed out already in the first section of his treatise), and partly on the possibility of posing these same objections also to potential infinity. Some of these objections, together with Arriaga's rejection of them, will now be presented in a much abbreviated and slightly modified form.

> *Objection:* If infinitely many people were created and Adam was subtracted from this number, the remaining number would be infinite again, which is a contradiction, because the infinite is what nothing else can be added to.
>
> *Answer:* We know that there are different infinities in terms of size. This objection could also be raised against potential infinity. Moreover, the definition of infinity as something to which nothing can be added is wrong.
>
> *Objection:* If God created an infinite number of things, then he would create everything, by which He would be exhausted.

[15] Rodericus de Arriaga, *Cursus philosophicus.*

Answer: There are different sizes of infinity; thus he would not have to be exhausted by creating infinite things. However, even if he created everything, he would not be exhausted, but on the contrary he would be fully outwardly expressed.

Objection: If God created an infinitely long rod and removed two inches from it in the middle and joined the rest, then the rod would remain as long as before.

Answer: This objection can also be turned against potential infinity.

Objection: Imagine two infinitely long sticks, one longer than the other. Then the second is not infinite.

Answer: They are of different sizes of infinity.

ad C) After the Thirty Years' War, when Europe was divided into Catholic and Protestant parts, however, the foremost task of Catholics was to maintain the unity agreed long before at the Council of Trent (1545–1563). For this reason, even the Jesuits began to avoid public disputes returning, in good order, to the teachings of Thomas Aquinas.

As we have already stated, in solving the problem of actual infinity, Arriaga not only did not adhere to Aristotle and Thomas Aquinas, but even refuted their solution to this problem. Therefore, obedient to the call of the times, the Jesuits of Prague began to reevaluate what Arriaga wrote about infinity. **Johann Senftleben** (1648–1693), a professor at what was already the Charles-Ferdinand University in Prague, took on this task most responsibly.[16] He realized that the only way to finish with actual infinity would be to show that the concept contains a logical contradiction. If actual infinity is not internally logically contradictory, then it is feasible for God, and it cannot be ruled out that God has already made it or intends to do it. However, if the very notion of actual infinity is logically contradictory, then God cannot create such a thing, just as he cannot create a triangle without sides and vertices, a round square, and so on.

Senftleben tried to revive some objections to actual infinity by referring in appropriate places to some of the following Euclidean axioms:

Axiom 1. Things which are equal to the same thing are equal to one another.

Axiom 4. If unequals are added to equals, the wholes are unequal.

Axiom 7. Things which coincide with one another are equal to one another.

Axiom 8. The whole is bigger than the part.

Strictly speaking, he did not refer directly to the fourth axiom, but to its following simple logical consequence:

[16] Johann Senfleben, *Philosophia Aristotelica Commentarijs Doctorum Societ.* (Typis Universitatis Carolo-Ferdinandeae in Coll. Soc. Jesu ad S. Clementem, 1685).

> If unequals are subtracted from equals, the remaining parts are unequal.

By using Euclidean axioms in objections to actual infinity, these objections gained weight, as these axioms were generally accepted as utterly unquestionable truths of pure reason.

1.2.2 The Franciscan School

The Jesuit-controlled Charles-Ferdinand University was not, however, the only locus of Prague's Baroque Catholic scholarship. No less important, though less influential, was the Franciscan monastic school of Our Lady of the Snows, in which Scottish philosophy and theology were cultivated. It was raised to a high level by Irish Franciscans, who soon educated a number of successful domestic pupils. Among them, **Vilém Antonín Brouček** in particular stood out, who in 1663 published the first textbook of Scotism in Bohemia, *Domus sapientiae Scoti.*[17]

Although the Franciscan monk **John Duns Scotus** (ca. 1264–1308) was a theological opponent of Thomas Aquinas, the Catholic Church did not forbid his teachings and tolerated them even after the Council of Trent. The Franciscans of Prague thus represented a counterweight to the Jesuits, which was certainly also one of the reasons why they were supported by the then Archbishop of Prague, Cardinal Arnošt Vojtěch of Harrach (1598–1667). The Franciscans of Our Lady of the Snows were able to make extensive use of this position. Here, however, we are only interested in the fact that they rejected Aristotle's and Thomas's objections to actual infinity and took essentially the same position as Rodrigo de Arriaga on this issue. This is expressed in clear and comprehensible words by Vilém Brouček in the second part of the above-mentioned book, from which we excerpt:

> Infinity according to the denotation of the word means that which has no end. Infinity is twofold: absolute and other. Absolute infinity is that which has an infinite essence, and that is only God. And so only God is infinite in all kinds of beings. The second infinity is one that has infinity only in a certain genus. It is both categorical and syncategorical. Categorical infinity is also called actual and it contains an infinite number of real parts; as if infinitely many people or infinitely many angels and the like were created. A syncategorical infinity, also called potential, is one which, although it has only finitely many real parts, can still have more and more of them indefinitely. Such infinity, for example, is in the thoughts of the saints, for the multitude of their thoughts is finite, of possible ones infinite, but it will never be infinite in reality. In what follows, we will focus only on actual infinity, because no one doubts potential infinity.

[17] Vilém Antonín Brouček, *Domus Sapientiae Doctoris subtilis Joannis Duns Scoti* (1663). Details on the activities of the Franciscans in Bohemia are taken from the book Stanislav Sousedík, *Jan Duns Scotus: Doctor Subtilis and his Czech Pupils* (1989).

Note that Brouček attributes the ability to potentially infinite knowledge already to the saints (meaning glorified after death in heaven), that is, to superhumans, as we would say, while Senftleben was later forced to lower God's ability to this level.

According to Brouček, however, it is the sovereign and unlimited rule over actual infinity that is one of those distinctive features that elevates God to dizzying heights above all creatures. This is evident from his following words:

> Although the creation of actual infinity for natural reasons is impossible, the absolute power of God is a guarantee that its actual infinity being is not contradictory. That is what Scotus, Occam and others say. The scope of nature is limited, and therefore something infinite cannot be created by the forces of nature. Actual infinity, however, is not contradictory either in terms of power or in terms of this concept. For God sees an infinite number of people or angels; he imagines them, and therefore can also create them.

In other words, what is without contradiction can be created by God, but also what God sees as possible is without contradiction. Equally important, however, is that only God offers the guarantee of being without contradiction, and therefore of the possibility of actual infinity.

1.3 Bernard Bolzano (1781–1848)

When the Czech writer Karolina Světlá was a little girl, her father pointed out Bolzano to her with the words: "He is a priest who has seen God."[18] We would not be able to capture this man – who penetrated into the depths of infinity – more aptly. He spent his whole life in Prague or in the Czech countryside and thus became a remarkable flower of the Czech Baroque.

1.3.1 Truth in Itself

In order to follow Bolzano's ideas, we must first come to an understanding of the notion of "truth in itself" introduced by him. In the first volume of his most comprehensive work, the *Wissenschaftslehre* (*Theory of Science*), Bolzano seeks to open our understanding with the following explanation:

> Thus, to state it again, I shall mean by a truth in itself any proposition which states something as it is, where I leave it undetermined whether or not this proposition has in fact been thought or spoken by anybody. In either case I shall give the name of a truth in itself to the proposition, whenever that which it states is as it states it to be. In other words I shall give it the name of a truth in itself whenever the object with which it deals actually has the properties

[18] From the celebration of bringing light into the Bernard Bolzano Lodge in the Prague Orient in the ninth year of Czechoslovak independence.

> that it ascribes to it. Thus, for example, the number of blossoms that were on a certain tree last spring is a statable if unknown, figure. Thus, the proposition which states this figure I call an objective truth, even if nobody knows it. Since the present concept is very important, I want to make sure that I am not misunderstood. Hence I shall state the following fairly obvious theorems about truths in themselves. [...]
>
> [Truths in themselves] do not have actual existence, i.e., they are not something that exists in some location, or at some time, or as some other kind of actual thing. Recognised or thought truths have indeed actual existence at a definite time in the mind of the being that recognises or thinks them. They exist as thoughts of some kind that begin at one moment and end at another. But to truths themselves, i.e., the matter of these thoughts, no existence can be ascribed. [...]
>
> Thus, to give another example, the logician must have the same right to speak of truths in themselves as the geometrician who speaks of spaces in themselves (i.e., of mere possibilities of certain locations) without thinking of them as filled with matter, [...].
>
> I said above that a truth in itself is "a proposition which states something as it actually is." These words are not to be taken in their original nor in their ordinary sense, but in a certain higher, more abstract sense. [...] The following is also easily overlooked: if we say that a truth states something as it actually is this phrase must also be taken in a figurative sense. This is so, because not all truths state something that actually is (i.e., has existence). This holds especially of those truths that deal with objects that do not themselves have actuality, e.g., other truths or their constituents, i.e., ideas in themselves. Thus, the proposition "A truth is not something that exists" certainly does not state something that exists, and yet it is a truth.[19]

Bolzano's opening of the domain of truths in themselves recalls Pythagoras' opening of the domain of numbers in the world of ideas.[20]

Even truths in themselves belong to the world of ideas; they are each unique and their existence is permanent, timeless and unchanging. Truths in themselves do not belong to the real world, but like ideal numbers, they extend into this

[19] Bernard Bolzano, *Theory of Science,* trans. Paul Rusnock and Rolf George (Oxford: Oxford University Press, 2014), 1:84–86 (§25, 112–115).

[20] Numbers as such, that is in their solitude, are not real-world phenomena since their being is ideal. Although the numbers lie in the world of ideas, they still appear in the real world. They do not appear in it in absolute purity, but in forms more or less obscured by the phenomena of the real world, through which, so to speak, they shine through. Therefore, discoveries about numbers spontaneously carry over to their occurrences in the real world, and conversely, discoveries about their occurrences in the real world tend to carry across to the world of numbers.

world. Very loosely speaking, a truth in itself is a truth that we interpret as an object, that is, as an independent individual. Interpreting individual truths as objects might be considered an arbitrary act, but such an interpretation is possible and, as it turns out, useful.

In his *Theory of Science*, Bolzano proved by contradiction that there is at least one truth in itself. If there were no truth in itself, then the sentence "There is no truth in itself" would be true, and its idealization would be a truth in itself. Immediately after proving the existence of at least one eternal truth in itself, Bolzano demonstrated that there are infinitely many such truths. In the *Theory of Science*, he proved it in a way that we will present in a somewhat modified, more detailed form.

Let us call P_1 the proposition "There is at least one truth in itself," which we have earlier shown to capture an eternal truth in itself. Then we can gradually create propositions $P_1, P_2, P_3, \ldots$ as capturing various eternal truths in themselves, by saying that P_{n+1} is the proposition: "There is at least one truth in itself, which is different from all the truths captured by the propositions $P_1, P_2, \ldots, P_n$."

Suppose first that P_{n+1} captures no eternal truth in itself; that is, P_{n+1} is sometimes untrue. In such a case, however, the true proposition is: "There is no truth in itself other than that which is captured in one of the propositions $P_1, P_2, \ldots, P_n$." Let us denote this truth by the letter R. We shall prove that R captures a different truth than any of the propositions $P_1, P_2, \ldots, P_n$, which will be a contradiction. Let it not be so then. Evidently R captures a different truth than P_1. Let k be the smallest natural number such that R captures the same truth as P_{k+1} (k is evidently smaller than n). However, the proposition P_{k+1} asserts that there is another truth in itself than that which is captured by one of the propositions $P_1, P_2, ..., P_k$. On the other hand, the proposition R claims that there is no truth in itself other than that which is captured by one of the propositions $P_1, P_2, \ldots, P_n$. As a result, the propositions R and P_{k+1} capture different truths. It remains to prove that P_{n+1} captures a different truth in itself than any of the propositions $P_1, P_2, \ldots, P_n$. Clearly P_{n+1} captures a different truth than P_1. Let k again be the smallest natural number such that P_{n+1} captures the same truth as P_{k+1}. However, the proposition P_{n+1} asserts that there is also a truth other than that captured by the proposition P_{k+1}, as a result of which both of these propositions capture other truths in themselves, which is a contradiction.

In the study *Paradoxes of the Infinite*, published in 1851, that is, after Bolzano's death, Bolzano provides yet another, simpler, proof that there are infinitely many eternal truths in themselves. We will now present this proof again in a slightly modified form. P_1 again indicates the same proposition as in the previous case. However, P_{n+1} denotes the proposition: "Propositions $P_1, P_2, \ldots, P_n$ capture eternal truths in themselves." It is immediately apparent that if $P_1, P_2, \ldots, P_n$ capture eternal truth in itself, then even P_{n+1} captures an eternal truth in itself. At the same time, P_{n+1} captures a different truth in itself than any of the propositions $P_1, P_2, \ldots, P_n$, because the subject of the assertion P_{n+1} is the truths captured by the propositions $P_1, P_2, \ldots, P_n$, while the subject of the assertion P_k, where k is less than or equal to n, is fewer of these truths. Bolzano explicitly points out that if we take some truth P, then the proposition "P is true" is different from P.

The multitude of propositions that capture these truths can, of course, be seen as potentially infinite; after all, Bolzano's proof is in fact a guide to how more and more propositions of this kind can be created.

Actually Infinite Multitudes

Bolzano would not be a true Catholic priest, educated in an environment only very gradually detaching itself from the Baroque spiritual world, if he did not know where the existence of all eternal truths in themselves were elevated to actualised being; that is, if he did not know the one who sees all these truths. In *Paradoxes of the Infinite* he expresses this in the following words:

> I say then: we call God infinite because we are compelled to admit in Him more than one kind of force possessing infinite magnitude. Thus, we must attribute to Him a power of knowledge which is true omniscience, and which therefore comprises an infinite set of truths, to wit, all truths — and so forth.[21]

While Rodrigo de Arriaga and his followers acknowledged God's ability to create some actually infinite multitude, they were unsure whether creating such a thing would be wise, and therefore whether God had exercised that ability. Giordano Bruno thought it would be good if God created an infinite number of suns and planets, and for that reason God did create them. Yet until the end of the eighteenth century, few people shared this view with him. Furthermore, over all such considerations there hovered the doubt whether actual infinity was not logically contradictory and therefore not even in God's power to create. However, Bernard Bolzano demonstrated that God can not only create actual infinity, but that this infinity is an integral part of his spiritual vision, for God immediately sees an actually infinite number of eternal truths in themselves.

Bolzano's choice of the example of an actually infinite multitude was extraordinarily shrewd. If he wanted to, God could create the infinity of natural numbers in his imagination, or infinitely many suns in the real world, and so on. However, we can hardly find a reasonable reason why he should have done so. On the other hand, God must know infinitely many truths, each one individually. He sees the eternal ones immediately, for otherwise he would not be the most wise and omniscient Christian God. It is not for nothing that Bolzano is not looking for the classical actual infinity in the real world, because nature does not have the ability to create actual infinity, as the Prague Scotists said (cited by Vilém A. Brouček, p. 48) and as modern science confirms by doing without this infinity.

[21] Bernard Bolzano, *Paradoxes of the Infinite,* trans. Donald A. Steele (London: Routledge, 1950), 81 (§11).

On the other hand, although the true existence of truths in themselves is constantly fulfilled only in the mind of God, this does not mean that man does not have access to these truths. In *Theory of Science*, Bolzano writes about this:

> It follows indeed from God's omniscience that each truth is known to him and is continually represented in his understanding, even if no other being is acquainted with it or thinks it. Consequently, there actually is no truth which is recognised by nobody at all. This, however, should not keep us from speaking of truths in themselves, since their concept does not presuppose that they must be thought by someone. The fact that they are thought is not contained in the concept of such truths, but it can nevertheless follow from some other circumstance (in this case the omniscience of God) that they must be recognised by God himself, if by no one else.[22]

Strictly speaking, at this point, Bolzano needs God only to guarantee that truths in themselves can be treated as if their existence were a real being, which in itself justifies the possibility of actual infinity. The actually infinite multitude of truths in themselves is however found only in the mind of God.

Bolzano's proof of the existence of actually infinite multitude is utterly unique, and no one has yet provided any other proof of this claim. Bolzano shields himself with God in his proof, which in the eyes of many people could invalidate this proof today. Without this shield, however, one cannot reach classically interpreted actual infinity. It remains to be noted that along with Bolzano's order of truths in themselves, natural numbers also entered the mind of God. In other words:

> **Bolzano proved that the multitude of natural numbers is actually infinite.**

1.3.2 The Paradox of the Infinite

The divinely perceived actually infinite multitude of all natural numbers changed the dilemma of Prague's Baroque theologians (which we mentioned in the above section on "The Disputation about Infinity in Baroque Prague"). According to Giordano Bruno, God created an actually infinite multitude of celestial bodies; according to Galileo actual infinity is contradictory! Bolzano turned this antinomy into a paradox by making light of Galileo's objections.[23] At the same

[22] Bernard Bolzano, *Theory of Science,* 1:85 (§ 25, 113).

[23] If any phenomena come into mutual conflict and if this dispute is unsolvable, if it is so hopeless that there is no way out of it, then we call it an **aporia**. But if it's just an apparent contradiction, then we call it a **paradox**. If it is a conflict that can be turned into a paradox by devaluing some of the phenomena entering into it, then we call it an **antinomy**. Antinomies tend to be a sources of knowledge on account of their urgent challenge to be solved. Some antinomies may even have various useful solutions and thus initiate different paths of knowledge.

time, he did not mean to understand the emergent paradox as an exceptional peculiarity, but rather as something that is characteristic of infinite sets and as something that has a wholly fundamental importance for them. It is for this reason that the study, published by František Přihonský from his estate in 1851, is called *Paradoxes of the Infinite.* In this book, Bolzano gave Galileo's objection the following positive form:

> We now pass on to consider a very remarkable peculiarity which can occur in the relations between two sets when both are infinite. Properly speaking, it always does occur – but, to the disadvantage of our insight into many a truth of metaphysics and physics and mathematics, it has hitherto been overlooked. Even now, when I come to state it, it will sound so paradoxical that we shall do well to linger somewhat over its investigation. My assertion runs as follows: When two sets are both infinite, they can stand in such a relation to one another that: (1) it is possible to couple each member of the first set with some member of the second in such wise that, on the one hand, no member of either set fails to occur in one of the couples; and on the other hand, not one of them occurs in two or more of the couples; while at the same time,[24] (2) one of the two sets can comprise the other as a mere part of itself, in such wise that the multiplicities to which they are reduced, when we regard all their members as interchangeable individuals, can stand in the most varied relationships to one another.[25]

What was originally considered a contradiction thus turned into a paradox, in fact, only by a change of attitude towards this conflict of phenomena. The opposition to actual infinity brought about by the clash of these phenomena, interpreted as a contradiction, gradually turned into a favourable expectation of both a closer explanation of this paradox and of its consequences. This very change of mindset broadened the understanding of the phenomenon of size in the case of infinite sets, which Bolzano pointed out in the following words:

> As I am far from denying, an air of paradox clings to these assertions; but its sole origin is to be sought in the circumstance that the above and oft-mentioned relation between two sets, as specified in terms of couples, really does suffice, in the case of finite sets, to establish their perfect equimultiplicity in members. Whenever, in fact, two finite sets are constituted so that every object a in the one corresponds

[24] In this context, one cannot fail to draw attention to the important role played by the choice of names in gaining knowledge of various kinds. No one can overlook how labouriously Bolzano expresses a relationship that later came to be called a one-to-one mapping of one set onto another. If he also chose this apt name (or a similar one), he would not have needed to speak of "assembling the elements of two sets into pairs, as we have mentioned several times." The name not only serves to shorten and simplify expression, but a well-chosen name speaks, shows and often shines on contexts that would otherwise remain hidden.

[25] Bernard Bolzano, *Paradoxes of the Infinite,* 96 (§20).

> to another object b in the other which can be paired off with it, no object in either set being without a partner in the other, and no object occurring in more than one pair: then indeed are the two finite sets always equal in respect of multiplicity. The illusion is therefore created that this ought to hold when the sets are no longer finite, but infinite instead. The illusion, I say – for a closer study reveals the fact that no such necessity exists, because the grounds upon which this holds for finite sets are bound up precisely with the finitude, and become inoperative for infinite sets.[26]

Bolzano's solution of this antinomy is (briefly expressed) as follows.

In infinite sets, the phenomenon of size appears in two different forms, which are related in some way (which needs to be explored in more detail, as indeed it later was). However, we cannot thoughtlessly interchange these forms, even though we may be tempted to by the fact that in finite sets the two forms of phenomena of size merge. This should not be surprising. After all, material bodies also show different, and little related, forms of the phenomenon of size; for example, one for volumes, another for weights, and still another for hardness and the like.

The form of the phenomenon of size in sets, which is based on Euclid's axiom stating that "Things which coincide with one another are equal to one another" was later given the very apt name of the **cardinality** of the set. Sets A, B have the same cardinality precisely when there is a one-to-one mapping of one onto the other.

In contrast, the form of the phenomenon of the size of a multitude, which is based on Euclid's axiom stating that "the whole is greater than a part," led to the introduction of the concept of **subset**. A set A is a subset of (or also part of) a set B if each element of the set A is also an element of set B. If A is a proper subset[27] of B, then the set B is of course larger than the set A.

No contradiction should be seen in the fact that a proper subset A of an infinite set B can have the same cardinality as the set B, just as there is no contradiction in a body with a smaller volume being as heavy as a body with a larger volume.

1.3.3 Relational Structures on Infinite Multitudes

Entire books could be written about what Bolzano did in mathematics. In our modest selection, however, his emphasis on relational structures on infinite sets as the main bearers of various forms of infinity cannot be neglected. Bolzano wrote about this in *Paradoxes of the Infinite*:

> It only remains to ask, therefore, whether a definition merely of what is to be called an infinite multitude can put us in a position to

[26] Bernard Bolzano, *Paradoxes of the Infinite,* 98 (§22).

[27] A set A is a **proper** subset of a set B is A is a subset of B and there is an element of B that is not an element of A (in other words, B is not also a subset of A).

> determine what the infinite be in itself. That would be the case if it came to light that multitudes were the only things to which the idea of the infinite could be applied in its strict sense – in other words, if it came to light that infinitude were, strictly speaking, a property only of multitudes; or again in other words, if everything we judge infinite is so judged solely because, and solely in so far as, we find it possessing a property which can be considered as infinite multitude. Now to my thinking, this really is the case. Mathematicians use the word plainly in no other meaning [...].[28]

Bolzano therefore urges us to look for some infinite multitude under each form of infinity, by which this form of infinity is evoked. The actualising of this multitude then results in actualising the corresponding form of infinity. According to him, the individual forms of this phenomenon are caused by the different compositions of the respective underlying infinite multitudes, or – as we would say today – by the relational structures of the respective communities of objects.

We abstract a relational structure from a given community of objects by emptying the content of the properties of its members and the connections prevailing between them, leaving only the mere presence of the respective members. This means that we replace a property with sets of those objects that have such a property, and we replace connections with relations, that is, with sets of ordered pairs, triplets, etc. of those objects that enter into the corresponding connections. Still, in order to obtain such relational structures, we do not have to abstract them from some organic communities. They can also be created artificially.

This brings us, all of a sudden, to the mathematics of the twentieth century. Let us note that this mathematics also initially focused mainly on the study of infinite relational structures, that is, those whose underlying sets were infinite. It was not until the second half of the twentieth century that it became clear that finite relational structures were at least as interesting as infinite ones.

Bolzano's method thus became the program of the set-theoretical mathematics of the twentieth century. It is a program whose position among various other well recognized programs is absolutely exceptional, since it lacks precisely what is characteristic of such programs. No one followed it in public, no one referred to it. Most mathematicians did not even know Bolzano's words. And yet they obediently carried out what Bolzano had tasked them with. So it is not a program in the true sense of the word, but rather a prophecy that has been fulfilled, as it was based on a clear and far-reaching insight into things to come. The fact that Bolzano's program for the Neo-Baroque set-theoretical mathematics was not fulfilled consciously does not diminish its value. On the contrary. We can't but wonder that mathematicians followed it nevertheless.

[28] Bernard Bolzano, *Paradoxes of the Infinite,* 79 (§10).

1.4 Georg Cantor (1845–1918)

1.4.1 Transfinite Ordinal Numbers

Almost at the very beginning of his scientific career, Cantor felt the need to count objects falling into some domain of geometric points. In carrying out his study of trigonometric series, published in 1872,[29] he confronted the task of removing all **isolated points** from some subdomain B of the domain of all points lying on the numerical axis. By an isolated point a domain B we mean such a point (number) x belonging to the domain B, for which a positive rational number r can be found such that no point from the domain B falls into the open interval $(x - r, x + r)$ except the point x. Cantor called the domain, which is obtained by removing all isolated points from the domain B, the **derivative** of the domain B. Let $B'(1)$ be the derivative of the domain B and $B(1)$ the domain of all isolated points of the domain B. It is obvious that some points not isolated in the domain B, may become isolated points of its derivative $B'(1)$.

Let us denote $B'(2)$ the derivative of the domain $B'(1)$ and $B(2)$ the domain of isolated points of the domain $B'(1)$ and then $B'(3)$ the derivative of the domain $B'(2)$ and $B(3)$ the domain of isolated points of the domain $B'(2)$ etc.

Cantor found that for each natural number n it is possible to create a subdomain B of the domain of all points lying on the numerical axis, for which the domains $B'(n)$, $B(n)$ are nonempty. If we remove from such a domain B all points falling into one of the domains $B(1), B(2), \dots$, we obtain the domain $B'(\omega)$, which can also be nonempty; the domain of its isolated points, just as its derivatives, etc. may also be non-empty domains.

During the constant continuation of this process, the **transfinite ordinal numbers** began to reveal themselves to Cantor. And they did so very clearly, for the continuation of their transfinite sequence was always obvious; in other words, the regularity of their creation also became apparent.

The first transfinite ordinal number (that is, the first that lies behind all natural numbers) Cantor marked with the Greek letter ω, and to this day we denote this number as such. The initial segment of the ordinal numbers thus has the form below, in which it is clear what numbers should be written in the dotted spaces.[30]

$$
\begin{aligned}
&0, 1, 2, 3, \dots, \omega, \omega + 1, \omega + 2, \dots, \omega + \omega = 2\omega, 2\omega + 1, \dots, 2\omega + \omega = 3\omega,\\
&3\omega + 1, \dots, 4\omega, \dots, 5\omega, \dots, \omega \cdot \omega = \omega^2, \omega^2 + 1, \dots, \omega^2 + \omega^2 = 2\omega^2, 2\omega^2 + 1, \dots,\\
&\omega^2\omega = \omega^3, \omega^3 + 1, \dots, \omega^4, \dots, \omega^5, \dots, \omega^\omega, \omega^\omega + 1, \dots, \omega^\omega + \omega^\omega = 2\omega^\omega, \dots,\\
&3\omega^\omega, \dots, \omega \cdot \omega^\omega = \omega^{\omega+1}, \omega^{\omega+\omega} = \omega^{2\omega}, \dots, \omega^{3\omega}, \dots, \omega^{\omega\cdot\omega} = \omega^{\omega^2}, \dots \omega^{\omega^3}, \dots,\\
&\omega^{\omega^\omega}, \dots, \omega^{\omega^{\omega^\omega}}, \dots
\end{aligned}
$$

After all these numbers $\omega, \omega^\omega, \omega^{\omega^\omega}, \dots$ there is naturally again some first

[29] Georg Cantor, "Ueber die Ausdehnung eines Satzes aus der Theorie der trigonometrischen Reihen." *Mathematische Annalen* 5 (1872): 123–132.

[30] Now it is more common to write the ordinal multiplication the other way round, so $\omega \cdot 2$ for $\omega + \omega$ etc. [Ed]

ordinal number, which we are no longer able to denote with the symbol ω and conventional arithmetical operations. Cantor denoted this number ϵ. Once we have a notation for it, we can continue to create ordinal numbers: $\epsilon+1,\ldots,\epsilon+\omega,\ldots,\epsilon+\epsilon=2\epsilon,\ldots,\epsilon^2,\ldots,\epsilon^\epsilon,\ldots$ etc, etc.

Let us note only in passing that before and shortly after 1880, Cantor obtained a number of then striking results (now well-known) concerning numbering of various domains of points or numbers, and one-to-one correspondences between such domains. He published the results in *Mathematische Annalen* in four articles that came out in 1879, 1880, 1882 and 1883 under the common title "Über unendliche lineare Punktmannigfaltigkeiten."

The most important of these, along with a few others, also appear in letters sent by Cantor to his friend **Richard Dedekind** (1831–1916). They are:

(a) The numbering of all rational and subsequently algebraic numbers by natural numbers (November 1873).

(b) The non-existence of any numbering for all real numbers by natural numbers (December 7, 1873).

(c) A one-to-one correspondence between the domain of points in a line segment and the domain of points in a square (June 20, 1877).

1.4.2 Actual Infinity

It is quite probable that in his contemplations concerning various domains of objects (in particular, of real numbers and points lying in space), Cantor sometimes drew on actual infinity; if not consciously, then at least subconsciously. In his published works, however, and even in the letters sent to Dedekind, he expressed himself in such a way that it could all be interpreted as reflections on potentially infinite domains, and not as collections of infinitely many existing points or numbers.

Under these circumstances, that is, as long as he avoided actual infinity out of caution, Cantor could not consider transfinite ordinal numbers to be numbers in the true sense of the word – for they did not correspond to the order of any existing objects. For him, they were merely signs used to denote various ways of going through objects belonging to a given domain, moreover a countable one.

In a letter to Dedekind of November 1882, Cantor admitted that until recently he had interpreted transfinite ordinal numbers merely as such signs. It is reasonable to assume that the radical change in Cantor's view of ordinal numbers was caused by the following event.

Sometime in late September 1882, Cantor discovered Bolzano's more than 30-year-old book *Paradoxien des Unendlichen* in a bookstore.

Already on a cursory leafing through it, he noticed that Bolzano declared the contradictions into which reflections on actual infinity fall to be only apparent, that is, to be paradoxes. To this end he considered which deliberations made with finite collections were applicable to infinite collections and which were not.

Consequences of the latter are characteristic for actual infinity, and therefore interpretations and reflections on infinite collections must be based on them.

Cantor soon discovered that Bolzano's path to actual infinity, namely his creation of infinitely large quantities, differed markedly from that presented to him by transfinite ordinal numbers; that is, not just by the signs, but (after accepting actually infinite collections) by true infinite numbers. For this reason he returns to the bookstore where he found Bolzano's book and buys another copy for Dedekind. He then sends it to him along with a letter on October 7, 1882. He writes that it is a very remarkable book, and although it omits much of what ought to be in it, even the main point according to Cantor, it still fascinates him.

Cantor acquaints Dedekind with transfinite ordinal numbers and some operations performed with them in a letter dated November 5, 1882. In this letter he describes two procedures for creating these numbers. In particular, the creation of a successor $\alpha + 1$ of the number α, if α is the last number created so far. If there is no largest ordinal number amongst those created so far, then it is necessary to create a number that is behind all those created so far; such numbers are, for example, $\omega, 2\omega, \omega^2$ and the like (these numbers have no immediate predecessor). At the same time, he draws special attention to the case when the collection of ordinal numbers created in this way gains a new cardinality greater than all the previous ones. In other words, he does not forget that it is possible to actualise even ordinal numbers such that the collection of all smaller numbers is uncountable.

In this letter, Cantor also expresses what came to be known as the **continuum hypothesis**, according to which all points lying on a straight line can be numbered with ordinal numbers smaller than the first uncountable ordinal number. In such a case, every infinite subdomain of the domain of all points lying on a line (or in a plane ...) would be either countable or have the cardinality of continuum.

1.4.3 Rejection of Cantor's Theory

Convinced not only of the necessity of actual infinity in mathematics, but also of the fact that he was the one to have gained the insight into the composition of its various sizes, Cantor completed a more extensive work on this issue within the last two months of 1882.

In addition to the results concerning infinite ordinal numbers, the work is also devoted to a kind of germinal defense of these numbers and of actually infinite multitudes. Or, rather, it is devoted to trying to open up this new world to other mathematicians and philosophers. In this work, Cantor wrote, among other things:

> The presentation so far of my investigations in the theory of manifolds has reached a point where its continuation becomes dependent upon an extension of the concept of a real whole number [natural numbers are understood] beyond the present boundaries; in partic-

> ular, this extension goes in a direction in which, to my knowledge, no one has so far looked for it. I find myself dependent to such an extent upon this extension of the concept of number that without it I would hardly be able to make without constraint, the smallest further step forward in the theory of aggregates (sets). May this circumstance serve as a justification, or if necessary an excuse, for the fact that I am introducing seemingly foreign ideas into my reflections. For what is at issue is an extension, or actually a continuation, of the sequence of real whole numbers beyond the infinite; as daring as this may seem, I can nonetheless express not only the hope, but also the firm conviction that in due time this extension will come to be regarded as a thoroughly simple, appropriate, and natural one.
>
> At the same time I am not at all unaware that with this undertaking I am placing myself in some contradiction with widespread notions on the mathematical infinite and views held all too frequently on the nature of numerical magnitude.[31]

Cantor then goes on to express his pleasure over his discovery of ordinal and cardinal numbers in the following words:

> When I conceive of the infinite as I have done here and in my earlier attempts, I derive true pleasure (to which I gratefully yield) from seeing how the whole concept of number, which in the finite only has the background of number-of-elements, in a certain sense splits up into two concepts when we ascend to the infinite: that of power, which is independent of the order given to an aggregate, and that of number-of-elements, which is necessarily tied to a lawful ordering of the aggregate by means of which the latter becomes a well-ordered aggregate. And when I decend again from the infinite to the finite, I see just as clearly and beautifully how the two concepts again become one and flow together to form the concept of the finite whole number.[32]

In a cursory examination of the prominent thinkers of antiquity and the modern age, Cantor did not find anyone on whom to rely firmly in defending actual infinity. The only clue was given to him by the Leibniz's following rare statement, which Bolzano used as a motto for *Paradoxes of the Infinite*:

> I am so in favour of actual infinity that instead of admitting that Nature abhors it, as is commonly said, I hold that Nature makes

[31] Georg Cantor, *Grundlagen einer allgemeinen Mannigfaltigkeitslehre* (1883). The translation cited here is from Georg Cantor, "Foundation of a General Theory of Manifolds" trans. Uwe Parpart, *The Campaigner: Journal of the National Caucus of Labor Committees* 9, no. 1–2 (Jan–Feb, 1976), 70.

[32] Georg Cantor, *Foundation of a General Theory of Manifolds,* 78.

> frequent use of it everywhere, in order to show more effectively the perfections of her Author.[33]

Cantor had no choice but to safeguard his defence of actual infinity with Bolzano. He did so in the following words:

> However, the genuine-infinite [he means actual infinity] as we encounter it, for example in the case of well-defined point aggregates, [...] has found its most decisive defender in a philosopher and mathematician of our century with a most acute mind, Bernhard Bolzano, who has developed his views relevant to the subject especially in the beautiful and substantial essay, *Paradoxes of the Infinite* (Leipzig, 1851). It is the purpose of this essay to demonstrate that the contradictions which skeptics and peripatetics of all times have tried to find in the infinite do not exist at all, if only one takes the trouble (which, of course, is not always altogether inconsiderable) to internalize the concepts of infinity in all seriousness and in accordance with their true content. [...] Bolzano is perhaps the only one who, to a certain extent, assigns the genuine infinite numbers their rightful place; they are frequently spoken on at any rate.[34]

At the same time, however, Cantor has reservations about Bolzano's interpretation of infinitely large numbers.

> However, the actual way in which he [Bolzano] deals with them, without being able to advance any kind of real definition of them, is something about which I am not at all in agreement with him [...] The author lacks both the general concept of power and the precise concept of number-of-elements for a real conceptual grasp of determinate-infinite numbers. Both occur with him in germinal fashion in a number of places, in the form of specialities; it seems to me, however, that he does not work through towards full clarity and distinctness, and this explains many non sequiturs and even several errors contained in this valuable essay. I am convinced that without the two concepts mentioned, there will be no progress in the theory of manifolds...[35]

Cantor was aware that his theory of actually infinite multitudes and numbers might not be positively received by most mathematicians, at least until the moment when they saw its beauty. Therefore, he defended his infinite numbers with the following reference to the freedom of mathematical research.

---

[33] "Je suis tellement pour l'infini actuel, qu'au lieu d'admettre que la nature l'abhorre, comme l'on dit vulgairement, je tiens qu'elle l'affecte partout, pour mieux marquer les perfections de son auteur." (Gottfried Wilhelm Leibniz, *Opera omnia*, ed. L. Dutens (Geneva, 1768), 2/1:243.)

[34] Georg Cantor, *Foundation of a General Theory of Manifolds*, 78.

[35] Ibid.

> Mathematics is entirely free in its development, bound only by the self-evident concern that its concepts be both internally without contradiction and stand in definite relations, organized by means of definitions, to previously formed, already existing and proven concepts. In particular, in introducing new numbers mathematics is obliged only to give such definitions of them as will lend them the kind of determiniteness and, under certain circumstances, the kind of relationship to the older numbers, which in a given case will definitely permit them to be distinguished from one another. As soon as a number satisfies all these conditions, mathematics can and must regard it as existent and real. Here I see the reason [...] why the rational, irrational and complex numbers should be regarded just as much as existent as the finite positive whole numbers. I believe that it is not necessary to fear, as many do, that these principles contain any danger to science. On one hand the designated conditions under which the freedom of the formation of numbers can alone be exercised, are such that they leave extremely little room for arbitrariness. And then every mathematical concept also carries within itself the necessary corrective; if it is unfruitful and inapt this is soon demonstrated by its uselessness, and it will then be dropped because of its lack of success. Any superfluous confinement of mathematical research work, on the other hand, seems to me to carry with it a much greater danger, a danger that is so much the greater as there is really no justification for it that could be deduced from the essence of the science, for the essence of mathematics lies precisely in its freedom.[36]

Cantor's theory of ordinal and cardinal numbers gave rise to much stronger resistance than Cantor had expected. Its main opponent was one of the leaders of German mathematics at the time, Professor **Leopold Kronecker** (1823–1891) of Berlin, who called it humbug. This influential mathematician then waged a relentless and not only ideological struggle against Cantor, by means of which, given his position, he achieved the goal that no mathematician dared publicly to defend Cantor's theory.

Cantor probably felt the bitterness of the damage Kronecker caused him most when, at the turn of 1884–1885, he sent an article to the Swedish mathematical journal *Acta Mathematica*, in which he discussed his theory. Although the editor of this journal was Cantor's friend, the well known mathematician **Gosta Mittag-Leffler** (1846–1927), and although Cantor arranged for him to publish such an article before the end of 1882, Cantor still failed. Mittag-Leffler refused to publish the work, arguing in a letter to Cantor dated March 9, 1885:

> I am convinced that the publication of your new work, before you have been able to explain new positive results, will greatly damage your reputation among mathematicians. I know very well that

[36] Georg Cantor, *Foundation of a General Theory of Manifolds,* 78.

> basically this is all the same to you. But if once your theory is discredited in this way it will be a long time before it will again command the attention of the mathematical world. It may well be that you and your theory will never be given the justice you deserve in our lifetime. Then the theory will be rediscovered in a hundred years or so by someone else, and then it will subsequently be found that you already had it all. Then, at least, you will be given justice. But in this way you will exercise no significant influence, which you naturally desire as does everyone who carries out scientific research. Therefore, I believe that until you are able to present the use of your theory, it will be better for it and for you if you postpone its publication.[37]

Cantor failed to demonstrate the usefulness of infinite ordinal and cardinal numbers for the mathematics of the time, which is not surprising. It did not begin to surface until many years later, mostly after Cantor's death. In other words, Mittag-Leffler's request was unfulfillable for Cantor.

Nevertheless, Cantor was adamant that his infinite numbers made sense, very profound sense. He felt that through the beauty of their magnificent structure, which was gradually emerging before his mental vision, a serious message was coming to him, even though he did not know what it was at first. In a letter to Mittag-Leffler in early 1884, he even wrote that he did not consider himself the creator of these infinitely large numbers, but only the herald of what God had revealed to him. Mittag-Leffler of course, did not take this – and as a serious scientist could not take it – into account when, a year later, he refused to publish Cantor's work on these numbers.

Cantor found little recognition among mathematicians. It seems that not even Dedekind supported him significantly at this time. Out of a seemingly inevitable depressing loneliness, however, he was delivered by an unexpected force. His theory began to attract some theologians and philosophers, whose interest in science, namely the contribution of theology to science, was stimulated by the encyclical *Aeterni Patris*, proclaimed by Pope Leo XIII on August 4, 1879. Among them, the leading position was held by Cardinal J. Franzelin, with whom Cantor corresponded as late as January 1886, not long before Franzelin's death.

The eminent Catholic philosopher **Constantin Gutberlet** (1837–1928) studied the relationship between the mathematical infinity and the absolute infinity of God's very existence. In a work of 1878,[38] he briefly explained Cantor's theory of infinite numbers and clairvoyantly declared that with this theory the study of infinity had entered a new epoch. Gutberlet's subsequent statement,

---

[37] Mittag-Leffler to Cantor, March 9, 1885 in I. Grattan-Guiness, "An Unpublished Paper by Georg Cantor: 'Principien einer Theorie der Ordnungstypen, Erste Mittheilung'." *Acta Mathematica* 124, 102. English translation partially quoted from Joseph Warren Dauben, *Georg Cantor: His Mathematics and Philosophy of the Infinite* (Princeton: Princeton University Press, 1879), 138.

[38] Constantin Gutberlet, "Das Unendliche, mathematisch und metaphysisch betrachtet," in H. Meschkowski, *Das Problem des Unendlichen* (München, 1974).

according to which God himself guarantees the existence of Cantor's infinite numbers, was of key importance.

> But in the absolute mind the entire sequence is always in actual consciousness, without any possibility of increase in the knowledge or contemplation of a new member of the sequence.

These words confirmed Cantor in his gradually arising conviction that it was through God's infinite kindness that he had been allowed to look into the actual infinite multitude present in the sensorium Dei. Chosen in this way to be God's prophet, Cantor considered it his main task to correct the mistake in Thomas Aquinas's teachings about infinity. For Aquinas had held Aristotle's position, rejecting the presence of the actual infinity in created things.

To avoid unnecessary misunderstandings, Cantor parted actual infinity into **absolute** infinity, that is, that which depends on God's very existence and is incomprehensible and therefore inaccessible to man, and **transfinite** infinity, which God made knowable to people at a certain stage of their development. The efforts he made to correct the teachings of Thomas Aquinas then found a positive response from some influential learned theologians.

The conviction that he had set out on the right path, propped up even on the *Summa contra Gentiles* of Thomas Aquinas, and the unwavering faith in the truth of the knowledge he had gained, radiated from the letters that Cantor sent to the Catholic priest Ignatius Jeiler in 1888.

> In any case, the question of the veracity of the Transfinitum must be seriously examined, because if I am right in my assertion about the veracity and possibility of Transfinitum, then (no doubt) there would be a great danger of religious error in taking the opposite view, because "error circa creaturas redundant in falsam de Deo scientiam."[39]
>
> I entertain no doubts as to the truth of the transfinites, which I have recognized with God's help and which, in their diversity, I have studied for more than twenty years; every year, and almost every day brings me further in this science.[40]

Cantor then expressed his satisfaction with his fate in a letter to Charles Hermite (1822–1901) in 1894:

> But now I thank God, the all-wise and all-good, that He always denied me the fulfillment of this wish [to gain a position at the University of Göttingen or Berlin], for He thereby constrained me, through a deeper penetration into theology, to serve Him and His

[39] Thomas Aquinas, *Summa contra Gentiles* 2.3.

[40] Cantor to Jeiler, in J. Bendiek, "Ein Brief Georg Cantors an P. Ignatius Jeiler O.F.M.," *Franziskanische Studien* 47 (1965): 68. English translation partially quoted from Dauben, 147.

> Holy Roman Catholic Church better than I have been able with my exclusive preoccupation with mathematics.[41]

In this period, however, Cantor penetrated not only into theology, but above all into the architectural art needed to build the tower of infinite cardinal and ordinal numbers. This art then required him, among other things, to replace domains of objects he encountered with the collections of all the objects falling to them. Furthermore, he needed to view these collections, which he subjected to various daring operations, as separate individuals, that is, to replace them by the corresponding sets. (Cantor realized the usefulness of interpreting some collections of objects as independent individuals as early as 1883, albeit only in a footnote to a paper from that time.) Of the results he obtained, let us mention at least what came to be known as the **diagonal method**, which was published in the annual report of the German Mathematical Association in 1890–91.[42] After Kronecker's death, *Mathematische Annalen* became accessible to Cantor again. There, in 1895 and 1897 he published articles under a shared title "Beiträge zur Begründung der transfiniten Mengenlehre."[43] It is Cantor's final work, in which his results are published in the final form. Thanks to where it was published, and also because Cantor suppressed the theological influences in it, this study opened Cantor's world to other mathematicians.

Just as up to the nineteenth century, mathematicians appropriated the abilities of Zeus, so now those who practiced mathematics based on Cantor's set theory began to appropriate the abilities of God. What is at stake is not just an ability to observe infinite sets, but also to manipulate them boldly and, last but not least, to actualise various unbranched domains of objects, i.e. to bring all objects belonging to them into existence and thus replace them with collections or even with sets.

What is concerned here, as has been said several times, is God as interpreted by the medieval and, above all, by the Baroque rational theology. At the same time, however, this interpretation of God, who can handle infinite multitudes just as well as the finite ones, has become so familiar in Europe that all educated people, with at least some degree of good will to do so, ultimately understand him well. All educated Europeans could easily acquire the abilities of God interpreted in this way, that is, to understand them and to act as if they had them, although in fact, they have neither these, nor even the abilities of Zeus.

[41] Cantor to Hermite, Jan 22, 1894 in H. Meschkowski "Aus den Briefbüchern Georg Cantors," *Archive for History of Exact Sciences* 2 (1965): 513. English translation qt. from Dauben, 147.

[42] Georg Cantor, "Über eine elementare Frage der Mannigfaltigkeitslehre," *Jahresbericht der Deutsch. Math. Vereinq.* 1 (1890–1891): 75–76.

[43] Georg Cantor, "Beiträge zur Begründung der transfiniten Mengenlehre," *Mathematische Annalen* 46 (1895): 481–512.
Georg Cantor, "Beiträge zur Begründung der transfiniten Mengenlehre (Zweiter Artikel)", *Mathematische Annalen* 49 (1897): 207–246.

For mathematicians who could do this, the possibility to use these abilities was the most important message that Cantor's set theory brought to them; they saw this as its main meaning.

Chapter 2

Rise and Growth of Cantor's Set Theory

2.1 Basic Notions

Cantor's set theory has found recognition from some mathematicians already at the first International Mathematical Congress in Zurich in 1897. Four years later, **David Hilbert** (1862–1943) helped to initiate its soaring development by lectures at the Second International Mathematical Congress in Paris. Shortly afterwards many of the concepts of this theory began to crystalise into the form presented below.

If we single out some previously formed objects (no matter whether consecutively or all at once) then we collect these objects together. If we give these objects the same standing, meaning that we take them out of their contexts then we say that we single out a **collection** of these objects.

The singled-out objects thus participate in the collection by their presence only. Hence collections can differ only through the presence of some object (it belongs to one and not the other). Thus for example if we interpret trees as objects then the trees from a certain wood form a collection of trees provided of course that we do not care that one is taller than another, that some two of them grow closer together, that we ignore their position within the wood, the order in which we focus on them etc. If we add or remove a single tree though then we deal with a different collection.

When talking about people, we often think not only of people who are alive at that moment or have lived in the past, but also of those who are yet to be born or even of those who have never been born nor ever will be. The extension of the concept "people" extends the mere collection of living people (or also dead people). By this word we denote no collection, but a domain of all people. Similarly the word mathematics denotes a certain domain of human activity or knowledge. It is not a totality of all that has been done or discovered in it but it contains much of what will be done or discovered one day or may never be discovered.

A **domain** is not a totality of existing objects (regardless of the modality of their existence); it is the source and simultaneously also a sort of container into which the suitable emerging or created objects fall. Naturally, every collection of objects can be interpreted as a domain, albeit an exhausted one. By actualising a domain we mean exhausting the domain, that is, substituting this domain by a collection of all the objects that fall or can fall into it. If the domain is sharply defined, the corresponding collection can be interpreted as a set.

We understand a **set** to mean a collection which is sharply defined and interpreted as an independent individual, that is, as an object. The objects belonging to this collection are called **elements** of the thus-created set.

We create sets in order to be able to view the many as one. To wit, a collection usually are many objects, the set created out of it is a single object. In consequence this set can be an element of another set.

Only one set can be created from a given sharply defined collection of elements. In other words, although a set is not the collection of its elements, it is uniquely determined by this collection. Also conversely, a set uniquely determines the collection of its elements. The property of being uniquely determined by the collection of its members is called **extensionality**.

The expression $X \in Y$ means that X is a member of the set Y (X belongs to Y) and the expression $X \notin Y$ denotes that X is not a member of the set Y (X does not belong to Y).

The expression $X \subseteq Y$ means that X, Y are sets and every element of the set X is also an element of the set Y. If $X \subseteq Y$ then we say that the set X is a subset of the set Y.

It is easy to see that any sets X, Y, Z satisfy:

(i) $X \subseteq X$
(ii) If $X \subseteq Y$ and $Y \subseteq Z$ then $X \subseteq Z$.

Furthermore, extensionality implies:

(iii) If $X \subseteq Y$ and $Y \subseteq X$ then $X = Y$ (letters X and Y denote the same set).

If A is an object then $\{A\}$ denotes the set such that its only element is the object A.

If A, B are objects then $\{A, B\}$ denotes the set such that its only elements are the objects A, B. Since a set is uniquely determined by its elements, we have $\{A, B\} = \{B, A\}$. If we form a list of some things, we do not increase the number of these things by writing some of them in several times. Hence if $A = B$ then $\{A, B\}$ is a set with a single element and $\{A, B\} = \{A\} = \{B\}$.

If A, B, C are objects then $\{A, B, C\}$ denotes the set such that its only elements are the objects A, B, C. Similarly if A, B, C, D are objects then $\{A, B, C, D\}$ denotes the set such that its only elements are the objects A, B, C, D etc.

To simplify considerations about sets it is useful to introduce the **empty set**, that is a set which has no elements. Since a set is uniquely determined by its elements, there exists only one empty set.

We will fix $\emptyset$ to be a constant denoting the empty set (whereby it is not forbidden to denote the empty set by some other term if convenient).

Note that the empty set is an object and thus it is something and not nothing even though no object is its element. The set $\{\emptyset\}$ whose only element is the empty set contains one element and hence it is not empty: it is non-empty. Hence $\emptyset \neq \{\emptyset\}$, $\emptyset \neq \{\{\emptyset\}\}$.

By the **intersection** of sets A, B we understand the set whose elements are just the objects that belong both to the set A and the set B. We shall denote it $A \cap B$.

We say that sets A, B are **disjoint** if their intersection is the empty set (that means $A \cap B = \emptyset$).

By the **union** of sets A, B we understand the set whose elements are just the objects that belong to the set A or to the set B (possibly to both). We shall denote it $A \cup B$.

By the **difference** of sets A, B in this order we understand the set whose elements are just the objects that belong to the set A and do not belong to the set B. We shall denote it $A \setminus B$.

The intersection, union and difference of two (not necessarily distinct) sets are binary operations on the domain of all sets. Clearly we can agree that these **Boolean operations** are on the domain of all sets sharply defined.

Basic relationships of the intersection, union and difference of sets are captured in rules named after Augustus De Morgan (1806-1871).

De Morgan's rules. Let A, B, C be sets. Then

$$A \setminus (B \cap C) = (A \setminus B) \cup (A \setminus C), \qquad A \setminus (B \cup C) = (A \setminus B) \cap (A \setminus C).$$

We shall refer to sets such that all their elements are sets as **sets of sets** and denote them by calligraphic letters $\mathcal{A}, \mathcal{B}, \ldots$

As before, we will surely agree that the unary operations below (called generalised Boolean operations) yield a set of any set of sets.

By the **union of a set of sets** $\mathcal{A}$ we understand the set whose elements are just the objects that are elements of some set belonging to $\mathcal{A}$. We shall denote it $\bigcup \mathcal{A}$.

By the **intersection of a non-empty set of sets** $\mathcal{A}$ we understand the set whose elements are just the objects that are elements of every set belonging to $\mathcal{A}$. We shall denote it $\bigcap \mathcal{A}$.

It is easily verified that for any two sets A, B we have

$$A \cup B = \bigcup \{A, B\}, \quad A \cap B = \bigcap \{A, B\}.$$

Thus if X, Y, Z are some objects then $\{X, Y, Z\} = \bigcup \{\{X, Y\}, \{Z\}\}$.

The empty set is also a set of sets since its every element is a set. For if somebody points out to us an object that is an element of the empty set then we can easily prove anything we might wish about this object, including it being a set.

Clearly $\bigcup \emptyset = \emptyset$ since if we had $X \in \bigcup \emptyset$ then a set $Y \in \emptyset$ would have to exist such that $X \in Y$ so $X \in Y \in \emptyset$ which is a contradiction.

On the other hand, the assumption $\mathcal{A} \neq \emptyset$ is necessary in the definition of the intersection of $\mathcal{A}$. According to the definition, $\bigcap \emptyset$ would need to be a set containing all objects. For if X is some object then if somebody presents any set Y belonging to $\emptyset$ then we can easily prove to them that $X \in Y$ holds, and hence $Y \in \bigcap \emptyset$.

Still, if we concentrate in some course of study exclusively on elements and subsets of some set V then it may be useful to agree a convention according to which $\bigcap \emptyset$ is also a set and $\bigcap \emptyset = V$.

It is easy to verify the generalised De Morgan rules valid for the generalised Boolean operations.

Generalised De Morgan rules. Let $\mathcal{A}$ be a non-empty set of sets and let A be some given set. Let $\bar{\mathcal{A}}$ be the set of all sets $A \setminus X$ where $X \in \mathcal{A}$. Then

$$A \setminus \bigcup \mathcal{A} = \bigcap \bar{\mathcal{A}}, \quad A \setminus \bigcap \mathcal{A} = \bigcup \bar{\mathcal{A}}.$$

If $A, \bigcup \mathcal{A} \subseteq V$, where V is a set, and provided that we have accepted the convention $\bigcap \emptyset = V$ then these generalised De Morgan rules do not need the assumption $\mathcal{A} \neq \emptyset$.

2.1.1 Relations and Functions

An **ordered pair** of objects x, y, denoted $\langle x, y \rangle$, is a special object that satisfies:

(a) The object $\langle x, y \rangle$ is uniquely determined by the objects x, y and their order.

(b) The object $\langle x, y \rangle$ uniquely determines the objects x, y and their order.

(c) The binary operation that from any two objects x, y in this order creates the object $\langle x, y \rangle$ is sharply defined.

The ordered pair $\langle x, y \rangle$ is not a set. The objects x, y are not its members but entries with the object x as the first entry and y as the second. The set $\{x, y\}$ is an unordered pair of objects since $\{x, y\} = \{y, x\}$. Although this set is uniquely determined by the objects x, y and in turn it uniquely determines these objects, it is not possible to establish which out of the two objects x, y is the first.

If x, y, z, u are some objects then we could similarly consider an ordered triple $\langle x, y, z \rangle$, quadruple $\langle x, y, z, u \rangle$ and so on. For our purposes however it suffices to define

$$\langle x, y, z \rangle = \langle x, \langle y, z \rangle \rangle, \quad \langle x, y, z, u \rangle = \langle x \langle y, z, u \rangle \rangle \quad \text{etc.}$$

If $\phi(x)$ expresses a sharply defined property of object x then $\{x; \phi(x)\}$ denotes the set of all x with property ϕ.

Similarly if $\phi(x_1, x_2, \ldots, x_n)$ expresses a sharply defined relation of objects $x_1, x_2, \ldots, x_n$ then $\{\langle x_1, x_2, \ldots, x_n \rangle; \phi(x_1, x_2, \ldots, x_n)\}$ denotes the set of all $\langle x_1, x_2, \ldots, x_n \rangle$ with $x_1, x_2, \ldots, x_n$ related by ϕ.

By the **Cartesian product of sets** A, B (denoted $A \times B$) we understand the set

$$A \times B = \{\langle x, y \rangle;\ x \in A, y \in B\}.$$

If A, B, C are sets then we let $A \times B \times C = A \times (B \times C)$. Similarly we define also the Cartesian product of there, four, five ... sets. The sets $A \times A, A \times A \times A, \ldots$ will be also denoted $A^2, A^3, \ldots$

We shall say that a set R is a **relation** if every element of it is an ordered pair.

The **inverse relation to a relation** R is defined as follows: $R^{-1} = \{\langle y, x \rangle\}; \langle x, y \rangle \in R\}$.

If R is a relation then $\mathrm{dom}(R)$ (or $\mathrm{rng}(R)$ respectively) denotes the set of all x for which there exists y such that $\langle y, x \rangle \in R$ (or $\langle x, y \rangle \in R$ respectively).

Clearly $R \subseteq \mathrm{rng}(R) \times \mathrm{dom}(R)$.

We shall say that a set R is a **relation on a set** A if $R \subseteq A^2$.

We shall say that F is a **function** if F is a relation and to each $x \in \mathrm{dom}(F)$ there exists a unique y such that $\langle y, x \rangle \in F$; we denote this y by $F(x)$.

If F is a function then $\mathrm{dom}(F)$ is called the domain of F and $\mathrm{rng}(F)$ the range of F.

We say that a function F is **one-to-one** if F^{-1} is also a function (in other words, if $x, y \in \mathrm{dom}(F)$ and $x \neq y$ then $F(x) \neq F(y)$).

If F is a function then by the **restriction** of the function F to a set $D \subseteq \mathrm{dom}(F)$ we understand a function G such that $\mathrm{dom}(G) = D$ and for each $x \in D$ we have $G(x) = F(x)$. We write $G = F|D$.

We say that a function F maps a set A **(in)to** a set B (or that F is a mapping of a set A into a set B) if $\mathrm{dom}(F) = A$ and $\mathrm{rng}(F) \subseteq B$. If $\mathrm{rng}(F) = B$ then we say that the function maps A **onto** B. If F is one-to-one and it maps A onto B then we also say that F is **bijective**.

An ordered (n+1)-tuple $\langle A, R_1, \ldots, R_n \rangle$ where A is a set and $R_1, \ldots, R_n$ are relations (for simplicity, binary) on A is called a **relational structure** (on the set A).

Let $\langle A, R \rangle$, $\langle B, S \rangle$ be relational structures. We say that F is an **isomorphism** of $\langle A, R \rangle$ onto $\langle B, S \rangle$ if the following holds:

(a) F is a one-to-one mapping of the set A onto the set B.
(b) For each $x, y \in A$ we have $\langle x, y \rangle \in R$ just when $\langle F(x), F(y) \rangle \in S$.

If F is an isomorphism of $\langle A, R \rangle$ onto $\langle B, S \rangle$ then F^{-1} is an isomorphism of $\langle B, S \rangle$ onto $\langle A, R \rangle$.

We say that **relational structures** $\langle A, R \rangle$, $\langle B, S \rangle$ are **isomorphic** if there exists an isomorphism F of the structure $\langle A, R \rangle$ onto $\langle B, S \rangle$.

If relational structures $\langle A, R \rangle$, $\langle B, S \rangle$ are isomorphic and also $\langle B, S \rangle$, $\langle C, T \rangle$ are isomorphic then also $\langle A, R \rangle$, $\langle C, T \rangle$ are isomorphic.

We say that F is an isomorphism of a relational structure $\langle A, R_1, \ldots, R_n \rangle$ onto a relational structure $\langle B, S_1, \ldots, S_n \rangle$ if for each $i = 1, 2, \ldots, n$, the function F is an isomorphism of the relational structure $\langle A, R_i \rangle$ onto the relational structure $\langle B, S_i \rangle$.

An isomorphism of a relational structure onto itself is called an **automorphism**.

2.1.2 Orderings

In this section $\langle A, R\rangle$ denotes some relational structure on the set A, where R is a binary relation. We shall continue using letters x, y, z, u, v to denote elements of the set A. In place of $\langle x, y\rangle \in R$ we shall also write xRy.

We shall say that the relation R is on the set

(a) **reflexive** if for each x we have xRx;

(b) **symmetric** if for each x, y we have xRy just when yRx;

(c) **antisymmetric** if for each x, y, whenever both xRy and yRx hold then $x = y$;

(d) **transitive** if for each x, y, z, whenever both xRy and yRz hold then xRz;

(e) **dichotomous** if for each x, y, one of xRy, yRx holds.

If a relation R is on a set A reflexive, antisymmetric and transitive then we say that R is a (partial) **ordering of the set** A, or also that the set A is (partially) ordered by the relation R. If the relation R is moreover on the set A dichotomous then we say that R is a **linear ordering of the set** A, or also that the set A is linearly ordered by the relation R.

In what follows R denotes some ordering of a set A. As usual, we shall write $x \leq y$ in place of xRy and $x < y$ in place of $x \leq y$ and $x \neq y$. (We read this as "x is less than y" or also "y is greater than x".)

Let X be some subset of the set A and let v be an element of the set X. Then we say:

(a) v is the **least** (or also **first**) **element** of the set X if for each $y \in X$ we have $v \leq y$.

(b) v is the **greatest** (or also **last**) **element** of the set X if for each $y \in X$ we have $y \leq v$.

(c) v is the **minimal element** of the set X if for no $y \in X$ we have $y < v$.

(d) v is the **maximal element** of the set X if for no $y \in X$ we have $v < y$.

Clearly, if v is the least element (or greatest element respectively) of a set X then it is also its minimal element (or maximal respectively).

It is easy to see that when the given ordering is linear, then also conversely if v is the minimal element (or maximal element respectively) of a set X then it is also its least element (or greatest respectively).

We assume, up to the end of this section, that $\leq$ is a linear ordering on a set A.

We say that a set X is a **cut** on the set A in the ordering $\leq$ (abbreviated to a cut in what follows) if $x \subseteq A$ and for each $x, y \in A$, if $x \in X$ and $y \leq x$ then $y \in X$.

In other words, a cut X on A contains along with any element x also all elements of the set A that are smaller than x.

By a **proper cut** we understand a cut X such that $X \neq A$.

We can easily verify that if u is some element of the set A then the following sets are cuts:

(i) The set of all x such that $x \leq u$.

(ii) The set of all x such that $x < u$.

These cuts are called **main cuts** and the element u is their **head**. In the former case it is **inner head** and in the latter case, **outer head**.

The following assertions are also trivial.

(i) $\emptyset, A$ are cuts.

(ii) If X, Y are cuts then $X \subseteq Y$ or $Y \subseteq X$.

(iii) If $\mathcal{A}$ is a set of cuts then $\bigcup \mathcal{A}$ is a cut.

(iv) If $\mathcal{A}$ is a set of cuts then $\bigcap \mathcal{A}$ is a cut.

(v) Let $Y \subseteq A$. Then the set of all x for which there exists $y \in Y$ such that $x \leq y$ (or just $x < y$ respectively) is a cut. If $Y = \emptyset$ then this cut is $\emptyset$.

(vi) Let $Y \subseteq A$. Then the set of all x such that for each $y \in Y$ we have $x \leq y$ (or just $x < y$ respectively) is a cut. If $Y = \emptyset$ then this cut is A.

2.1.3 Well-Orderings

We say that a set A is **well ordered** by a relation R (or also that R is a well-ordering of the set A) if the following two conditions hold:

(a) The set A is linearly ordered by the relation R.

(b) Every proper cut on the set A has an outer head in the ordering R.

Proposition 2.1. The condition (b) from the above definition is equivalent to:

(b') Every non-empty subset of the set A has a first (least) element in the ordering R.

Proof. (b) $\Rightarrow$ (b'). Let $\emptyset \neq X \subseteq A$. Let $\bar{X}$ denote the set of all $x \in A \setminus X$ such that for each $y \in X$ we have xRy. $\bar{X}$ is clearly a cut, $\bar{X} \neq A$ and if v is the outer head of the cut $\bar{X}$ then it is easily seen that v is the first element of the set X in the ordering R.

(b') $\Rightarrow$ (b). Let X be a proper cut. Then the set $A \setminus X$ is non-empty and its first element is clearly the outer head of the cut X.

□

Note. If A is well ordered by a relation R and $C \subseteq A$ then the set C is also well ordered by the relation[1] R since if $\emptyset \neq X \subseteq C$ then the first element of the set X in the ordering R of the set A is clearly also the first element of the set X in the ordering R of the set C.

Up to the end of this section, we use A (or B respectively) to denote some given set and R (or S respectively) to denote some well ordering of the set A (or B respectively). Letters x, y, z denote elements of the set A and letters u, v, w elements of the set B.

Proposition 2.2. Let C be a subset of the set A and let F be an isomorphism of the relational structure $\langle A, R\rangle$ onto the relational structure $\langle C, R\rangle$. Then the following hold:

(i) $xRF(x)$ for each x.

(ii) There is no $z \in A \setminus C$ satisfying xRz for each $x \in C$.

(iii) If C is a cut on A in the ordering R then $C = A$.

(iv) If $C = A$ then $F(x) = x$ for each x. (In other words, the only automorphism of A is the identity.)

Proof. For the first three of these simple assertions we use proofs by contradiction.

(i) Assume that the set Y of all x such that $x \neq F(x)$ and $F(x)Rx$ is nonempty and let z be its first element. Then $z \neq F(z)$, $F(z)Rz$ and since F is an isomorphism, also

$$F(z) \neq F(F(z)) \quad \text{and} \quad F(F(z))RF(z).$$

Hence $F(z) \in Y$ but $F(z)$ is less than z, contradiction.

(ii) Assume that $z \in A \setminus C$ and xRz for each $x \in C$. Then (since $F(z) \in C$)

$$z \neq F(z) \quad \text{and} \quad F(z)Rz$$

which contradicts (i).

(iii) Assume that $z \in A \setminus C$. If $x \in C$ then clearly $x \neq z$ and xRz. Hence $F(z) \neq z$ and $F(z)Rz$ which contradicts (i).

(iv) By (i), $xRF(x)$ for each x. Since F^{-1} is also an isomorphism of $\langle A, R\rangle$ onto $\langle A, R\rangle$, (i) gives also $xRF^{-1}(x)$ for each x and hence $F(x)Rx$. It follows that $F(x) = x$.

□

Proposition 2.3. Let F, G be isomorphisms of a relational structure $\langle A, R\rangle$ onto $\langle B, S\rangle$. Then $F = G$, in other words, for each $x \in A$ we have $F(x) = G(x)$.

[1] To be precise, here and in the following proposition, the relation on C which is meant is $R \cap C^2$. Similarly below. [Ed]

Proof. Clearly, G^{-1} is an isomorphism of the relational structure $\langle B, S\rangle$ onto $\langle A, R\rangle$. Let $H(x) = G^{-1}(F(x))$ for each x. It is easily seen that H is an automorphism of the structure $\langle A, R\rangle$. By Proposition 2.2, $H(x) = x$ for each x, that is, $x = G^{-1}(F(x))$. It follows that $G(x) = F(x)$. □

Proposition 2.4. There exists at most one cut C on the set A in the ordering R such that the relational structures $\langle C, R\rangle$, $\langle B, S\rangle$ are isomorphic.

Proof. Assume that C, D are two such cuts and let for example $C \subseteq D$. Clearly C is a cut on the set D in the ordering R. Since the structures $\langle D, R\rangle$, $\langle C, R\rangle$ are isomorphic, we have $C = D$ by Proposition 2.2(iii).

□

Proposition 2.5. Let C be a cut on the set A in the ordering R and let D be a cut on the set B in the ordering S. Let structures $\langle A, R\rangle$, $\langle D, S\rangle$ be isomorphic and also structures $\langle C, R\rangle$, $\langle B, S\rangle$ be isomorphic. Then $A = C$ and $B = D$.

Proof. Let G be an isomorphism of $\langle B, S\rangle$ onto $\langle C, R\rangle$. Let $E = \{G(u); u \in D\}$. Clearly, E is a cut on the set C (and hence also on the set A) in the ordering R; $E \subseteq C \subseteq A$. Since the structures $\langle A, R\rangle$, $\langle E, R\rangle$ are isomorphic, we have $E = A$ and hence also $C = A$ by Proposition 2.2.

□

Proposition 2.6. There exists either a cut C on the set A in the ordering R such that the structures $\langle C, R\rangle$, $\langle B, S\rangle$ are isomorphic or a cut D on the set B in the ordering S such that the structures $\langle A, R\rangle$, $\langle D, S\rangle$ are isomorphic.

Proof. Let $[x]$ (or $[u]$ respectively) denote the set of all y (or v respectively) for which

$$yRx \quad \text{(or } vSu \text{ respectively).}$$

Let C be the set of all $x \in A$ for which there exists a cut Y on the set B in the ordering S such that the structures $\langle [x], R\rangle$, $\langle Y, S\rangle$ are isomorphic.

By Proposition 2.4, for a given $x \in C$ there exists a unique such cut Y. Since x is the inner head of the cut $[x]$, Y also has an inner head; we will denote it $F(x)$. Let D be the set of all $F(x)$ where $x \in C$. It is easily checked that the set C is a cut on the set A in the ordering R and that the set D is a cut on the set B in the ordering S, and also that F is an isomorphism of the structures $\langle C, R\rangle$, $\langle D, S\rangle$.

Assume that $C \neq A$ and $D \neq B$. Let z (or w respectively) denote the outer head of the cut C (or D respectively) in the ordering R (or S respectively). If we set

$$H(x) = F(x) \text{ for } x \in C \quad \text{and} \quad H(z) = w$$

then clearly H is an isomorphism of the structure $\langle [z], R\rangle$ onto $\langle [w], S\rangle$ and hence $z \in C$, contradiction. This means that either $C = A$ or $D = B$. Since the structures $\langle C, R\rangle$, $\langle D, S\rangle$ are isomorphic, the proposition is proved.

□

2.2 Ordinal Numbers

We say that a set A ordered by a relation R has the same **order type** as a set B ordered by a relation S if the relational structures $\langle A, R\rangle$, $\langle B, S\rangle$ are isomorphic.

The order type of the ordering R of the set A (or also of the set A ordered by R) is thus carried solely by this ordering and it is not dependent on the nature of the elements of the set A nor any relations in which they may stand except of the relation R.

If we abstract from some well ordering of some set its order type as such and interpret it as an object, we obtain an **ordinal number**.

To denote ordinal numbers, we shall use lower case Greek letters $\alpha, \beta, \gamma, \ldots$.

The empty set is well ordered. The ordinal number that is the type of this ordering is called zero. We denote it by a permanent constant 0.

Let a set A well ordered by a relation R have type α and a set B well ordered by a relation S type β. We define $\alpha \preceq \beta$ (or $\alpha \prec \beta$ respectively) just when it is possible to form a cut C (or a proper cut C respectively) on the set B in the ordering S such that the relational structures $\langle A, R\rangle$, $\langle C, S\rangle$ are isomorphic; in other words if the cut C ordered by the relation S has the order type α.

The relations $\alpha \preceq \beta$, $\alpha \prec \beta$ defined on ordinal numbers as above clearly do not depend on the choice of the sets A, B and the relations R, S such that the set A ordered by the relation R has type α and the set B ordered by the relation S has type β.

The following proposition follows easily from propositions proved in the preceding section.

Proposition 2.7. (i) $\alpha \preceq \alpha$ (reflexivity).

(ii) If $\alpha \preceq \beta$, $\beta \preceq \gamma$ then $\alpha \preceq \gamma$ (transitivity).

(iii) If $\alpha \preceq \beta$, $\beta \preceq \alpha$ then $\alpha = \beta$ (antisymmetry).

(iv) Either $\alpha \preceq \beta$ or $\beta \preceq \alpha$ (dichotomy).

(v) $\alpha \prec \beta$ just if $\alpha \preceq \beta$ and $\alpha \neq \beta$.

(vi) $0 \preceq \alpha$.

Proof. The assertion (iii) follows from Proposition 2.5, (iv) is a corollary of Proposition 2.6 and (v) follows from Proposition 2.2.

□

The term $[\alpha]$ denotes the set of all ordinal numbers β such that $\beta \prec \alpha$. It is easily verified that the set $[\alpha]$ is well ordered by the relation $\preceq$ and that the order type of this ordering is α.

Proposition 2.8. Let M be some set of ordinal numbers. Then the following hold:

(a) It is possible to create an ordinal number γ such that $M \subseteq [\gamma]$.

(b) The set M is well ordered by the relation $\preceq$.

Proof. (a) Let L denote the set of all ordinal numbers α for which there exists $\beta \in M$ such that $\alpha \preceq \beta$. Clearly $M \subseteq L$. Let γ be the order type of the set L in the ordering $\preceq$. Then $\gamma \notin L$ since otherwise $[\gamma]$ would be a proper cut on the set L in the ordering $\preceq$ which contradicts Proposition 2.2(iii). It means that for each $\alpha \in L$, $\alpha \prec \gamma$ and hence $M \subseteq L \subseteq [\gamma]$. (b) is an immediate consequence of (a).

□

Proposition 2.9. Let γ be the type of the set $X \subseteq [\alpha]$ in the ordering $\preceq$. Then $\gamma \preceq \alpha$.

Proof. Assume that $\alpha \prec \gamma$. Then there exists a proper cut Y on the set X in the ordering $\preceq$ such that its type is the number α. Let $\beta \in X \setminus Y$. Then $Y \subseteq [\beta]$ which contradicts Propositin 2.2(ii).

□

The number **one** is that ordinal number which is the type of the well ordering of a one-element set. To denote it we use a permanent constant 1.

Clearly $[1] = \{0\}$ and if $\alpha \neq 0$ then $1 \preceq \alpha$.

The **sum** $\alpha + 1$ is defined as that ordinal number which is the type of the ordering $\alpha \cup \{\alpha\}$.

We say that the ordinal number γ is the **successor** of the ordinal number α (or also that α is the immediate **predecessor** of γ) if $\alpha \prec \gamma$ and for each β such that $\alpha \prec \beta$ we have $\gamma \preceq \beta$.

It is easily verified that γ is the successor of the number α just when $\gamma = \alpha+1$.

An ordinal number that has no immediate predecessor is called a **limit** ordinal. Clearly, 0 is a limit number.

An ordinal number that does have an immediate predecessor is referred to as a **successor** ordinal. Clearly, the number 1 and also $2 = 1+1$, $3 = 2+1, \ldots$ are successors.

If we say about some ordinal number α that it exists then we mean that not only α exists but also that the set $[\alpha]$ of all ordinal numbers less that α exists.

2.3 Postulates of Cantor's Set Theory

Ever since Euclid's times, postulates are tasks to create certain mathematical objects. In the ancient Greek geometry they were tasks to create geometrical objects and in Cantor's set theory, tasks to create certain sets. As opposed to the ancient Greek geometry, in Cantor's set theory it is no Greek god let alone man who solves the tasks contained in the postulates but it is the God of the medieval rational theology. Therefore, to understand these tasks, we must strive to appropriate these abilities of his at least in our thinking and via a corresponding mental image.

However, the potentiality of infinity is so crucially connected with the domain of all ordinal numbers (and hence with the domain of all sets) that not even the

God of the medieval and baroque rational theology, enlisted to help in creating infinite sets by Bolzano and Cantor, can push it out of this domain.

Note that the domain of all ordinal numbers is sharply defined, possibly even more so than the domain of natural numbers. The sharp definability of this domain, that can hardly be doubted, has a far reaching consequence.

The domain of all ordinal numbers is not actualisable.

For if there existed a collection of all ordinal numbers then nobody could prevent us from interpreting it as an autonomous individual, and hence as a set.

However, according to Proposition 2.8, for each set it is possible to create an ordinal number that is not an element of it. In fact, the order type of such an assumed collection of all ordinal numbers is an ordinal number that does not belong to this collection.

Hence the catholic philosopher Konstantin Gutberlet (See page 63) was mistaken. He should not have talked about all ordinal numbers but, like Bolzano, only about natural numbers.

Cantor had been aware of the non-actualisability of the domain of all ordinal numbers already during the final years of nineteenth century.[2] This knowledge was, both to Cantor and some other mathematicians who came after him, very unpalatable so that they tried to disregard it in various ways.

On the other hand, the potential infinity present within the sequence of ordinal numbers constituted an irresistible challenge for mathematicians to follow along the suggested path into the mysterious depths of infinity. Therefrom, using set theory, they postulated emergence of ordinal numbers, often with remarkable properties.

Natural numbers, that is, numbers $0, 1, 2, \ldots$ are well orderd by their size. Ordinal natural numbers are the types of the orderings of finite sequences 0,1,2, $\ldots, n$ of natural numbers. Using set theory we can single out ordinal natural numbers from the domain of all ordinal numbers as follows:

We say that an ordinal number α is a **natural ordinal number** if every non-empty cut on the set $[\alpha]$ (in the ordering $\preceq$ of ordinal numbers) has an inner head.

Proposition 2.10. The following properties of an ordinal number α are equivalent:

(a) α is an ordinal natural number.

(b) If $0 \neq \beta \in [\alpha + 1]$ then β is a successor ordinal number.

Proof. If $\alpha = 0$ then α has both properties (a) and (b). Assume $\alpha \neq 0$.

(a) $\Rightarrow$ (b). Let $0 \neq \beta \in [\alpha + 1]$. Clearly $[\beta]$ is a non-empty cut on the set $[\alpha]$ and if γ is its inner head then $\beta = \gamma + 1$.

[2] See Cantor's letter to Dedekind from July 1899.

(b) $\Rightarrow$ (a).[3] Let X be a non-empty cut on the set $[\alpha]$. Clearly X is also a cut on the set $[\alpha+1]$, and since $\alpha \in [\alpha+1]$, but $\alpha \notin X$, $[\alpha+1] \setminus X$ is non-empty and thus has a least element. Hence the cut X has an outer head $\beta \in [\alpha+1]$. Since β is a successor ordinal number, $\beta = \gamma + 1$. Clearly, γ is the inner head of the cut X. □

Proposition 2.11. If α is an ordinal natural number then also any $\beta \in [\alpha]$ and $\alpha + 1$ are ordinal natural numbers.

First Postulate. Create the set N of all natural numbers.

In Section 1.3 we noted that this task had been solved by Bernard Bolzano and that nobody else has ever succeeded to find another solution. The set N thus exists. Its elements are well ordered by their size and Cantor chose the permanent constant ω for the type of this ordering. We can easily see that the following hold:

Proposition 2.12. The permanent constant ω denotes the first ordinal number that is greater than all ordinal natural numbers.

Proposition 2.13. The set $[\omega]$ is the set of all ordinal natural numbers. If δ is an ordinal number, $\delta \notin [\omega]$ then $[\omega]$ is a cut on the set $[\delta]$ (in the ordering of ordinal numbers).

Proposition 2.14.[4] If $A \subseteq [\omega]$ is such that $0 \in A$ and whenever $\alpha \in A$ then $\alpha + 1 \in A$, then $A = [\omega]$.

2.3.1 Cardinal Numbers

Even when the primitive people of our distant past could not count, they still could tell if the total of furs they were offering for trade in exchange for appropriate packets of salt matched, or failed to match, their total. They could do it for example by displaying their furs on the ground and inviting the traders to place one packet on each fur.

The total count of elements of some set, which is what we are looking at here, rests merely on its cardinality (meaning that the nature of its elements and how they relate to each other is irrelevant). Thus the cardinality may be abstracted from the set without a need or even without any means to count its elements. Sets M, N have the same cardinality just when it is possible to arrange a one-to-one correspondence between them.

If we abstract from some set (that is, separate from it) just its cardinality as such and we interpret it as an object, we obtain a **cardinal number**.

This cardinal number gives that basic total count of the elements of the set from which we have abstracted it.

[3] The proof of this proposition in the Czech original is damaged and has been modified. [Ed]

[4] This proposition (principle of mathematical induction) has been added here since it is used below. [Ed]

In this way, albeit in somewhat different words, Cantor introduced the notion of a cardinal number even for infinite sets.

Let $\mathbf{m}$ denote the cardinal number of a set M and let $\mathbf{n}$ denote the cardinal number of a set N. From our definition of cardinal numbers it follows that $\mathbf{m} = \mathbf{n}$ just when it is possible to map M and N onto each other by a one-to-one mapping.

In an obvious way, we can also define when a cardinal number $\mathbf{m}$ is less or equal to a cardinal number $\mathbf{n}$: namely when the set M has the same cardinality as some subset of the set N, in other words, when it is possible to create a one-to-one mapping from M into the set N.

Accordingly, a cardinal number $\mathbf{m}$ is less than $\mathbf{n}$ if $\mathbf{m} \neq \mathbf{n}$ and $\mathbf{m}$ is less or equal to $\mathbf{n}$. That means, if M and N cannot be mapped onto each other by a one-to-one mapping but it is possible to create a one-to-one mapping from M into N.

For the above-defined relation "less or equal" of cardinal numbers to be in some sense a reasonable ordering, this relation must be antisymmetric. That means, if $\mathbf{m}$ is less than $\mathbf{n}$ and at the same time $\mathbf{n}$ is less than $\mathbf{m}$ then we should have $\mathbf{m} = \mathbf{n}$.

If A is a set then $\mathrm{Card}(A)$ denotes its cardinal number.

Thus straight as cardinal numbers are being introduced, we should prove the following assertion:

Proposition 2.15. Let F be a one-to-one mapping from a set M into a set N and let G be a one-to-one mapping from the set N into the set M. Then M and N can be mapped onto each other by a one-to-one mapping.

Cantor was well aware of the need to prove this assertion (which we will prove shortly), and he stated several times that it held. However, he did not publish a proof.

Proposition 2.16. Let α be an ordinal natural number. If $X \subseteq [\alpha]$, $X \neq [\alpha]$ then it is not possible to create a one-to-one mapping of the set $[\alpha]$ onto the set X.

Proof. We shall use the principle of mathematical induction. For $\alpha = 0$ the claim clearly holds. Assume that it holds for number α; we shall prove that it holds also for $\alpha + 1$.

Aiming for a contradiction, suppose that $X \subseteq [\alpha + 1]$, $X \neq [\alpha + 1]$ and let F be a one-to-one mapping of the set $[\alpha + 1]$ onto the set X.

Let $\alpha \notin X$. Then $F(\alpha) \in X \subseteq [\alpha]$, $X \setminus \{F(\alpha)\} \neq [\alpha]$ and $F|[\alpha]$ is a one-to-one mapping of the set $[\alpha]$ onto the set $X \setminus \{F(\alpha)\}$, which is a contradiction.

Let $\alpha \in X$. Then $[\alpha] \cap X \neq [\alpha]$. If $F(\alpha) = \alpha$ then $F|[\alpha]$ is a one-to-one mapping of the set $[\alpha]$ onto the set $[\alpha] \cap X$, which is a contradiction. If $\alpha = F(\gamma)$ where $\gamma \in [\alpha]$ then $F(\alpha) \in [\alpha] \cap X$. We define a function G on the set $[\alpha]$ by setting $G(\gamma) = F(\alpha)$ and $G(\beta) = F(\beta)$ for $\gamma \neq \beta \in [\alpha]$. G is clearly a one-to-one mapping of the set $[\alpha]$ onto the set $X \cap [\alpha]$, which is a contradiction.

□

This means that if α, β are ordinal natural numbers then $\alpha \neq \beta$ just when $\text{Card}([\alpha]) \neq \text{Card}([\beta])$.

2.3.2 Postulate of the Powerset

Second Postulate. If A is a set then create the set of all subsets of A.

Note. The set of all subsets of A is called the powerset of A and denoted $\mathcal{P}(A)$.

If we consider the task set by this postulate to be achievable, as we have just decided to do in accord with many multitudes of mathematicians of the 20th century, we enter the high levels of the hierarchy of actually infinite sets. First evidence of it is the asssertion stated and proved by Cantor in the already mentioned annual report of the German Mathematical Association in 1890–91.[5]

Proposition 2.17. Cantor's Theorem. For any set A, the cardinality of A is less than the cardinality of its powerset $\mathcal{P}(A)$.

Proof. Let G be the function from A to $\mathcal{P}(A)$ such that $G(x) = \{x\}$ for each $x \in A$. Then G is clearly a one-to-one mapping of the set A into the set $\mathcal{P}(A)$. Hence the cardinality of the set A is less or equal to the cardinality of the set $\mathcal{P}(A)$. Assume that F is a one-to one mapping of the set A onto the set $\mathcal{P}(A)$. Let D be the set of all $x \in A$ for which $x \notin F(x)$. Since $D \subseteq A$, there is $d \in A$ such that $D = F(d)$. If $d \in D$ then $d \notin F(d) = D$, contradiction. Hence $d \notin D$, that is $d \notin F(d) = D$ and thus by the definition of the set D, $d \in D$, also contradiction.

□

We remark that the idea of the second (interesting) part of the above proof, called Cantor's diagonal method, found applications later on also in other areas of mathematics.

Proposition 2.18. Let sets A and B have the same cardinality. Then clearly sets $\mathcal{P}(A)$ and $\mathcal{P}(B)$ also have the same cardinality.

We say that a set A is **finite** if there exists an ordinal natural number α such that the sets A and $[\alpha]$ have the same cardinality. We say that the set A is **countable** if it has the same cardinality as the set $[\omega]$. We say that the set A has the **cardinality of the continuum** if it has the same cardinality as the set $\mathcal{P}([\omega])$. We also say that the ordinal natural numbers are finite and the other ordinal numbers are infinite.

When the set A is finite then the existence of $\mathcal{P}(A)$ can be proved by postulates commonly used previously.

However, when the set A is infinite then the act of creating of its powerset requires much larger capabilities, both creative and observational, than were

[5] Georg Cantor, "Uber eine elementare Frage der Mannigfaltigkeitslehre," Jahresbericht der Deutsch. Math. Vereing. 1 (1890–1891): 75–76.

those that we have relied on up to now. This is the case even when A is the most transparent of infinite sets, the set $[\omega]$ of all natural numbers. The creation of the *domain* of all subsets of the set A, just as the insight that this domain is a sharply defined, does not require any special capabilities, which is what undoubtedly contributed to the commonly held view that the Powerset Postulate is not problematic. The difficulty consists in the task of creating the *collection* of all subsets of the set A. The last step, the interpreting of this collection as an object, i.e. a set, is then just a matter of our attitude to this collection and it should not cause us any problems.

As we have pointed out before, for the introduction of cardinal numbers to be possible we need the following assertion:

Proposition 2.19. Cantor-Schröder-Bernstein Theorem. Assume that A,B are sets such that the cardinality of A is less or equal to the cardinality of B and the cardinality of B is less or equal to the cardinality of A. Then the sets A, B have the same cardinality.

Cantor was convinced of the truth of this assertion and he based his conviction on the claim that every set can be well ordered; the above is an easy consequence of this claim. We can also guess the reasons for Cantor's reluctance to publicly talk about a proof of this much stronger claim. It is also why, when in the years 1896 and 1897 the above assertion was proved, independently, by **Ernst Schröder** (1841–1902) and **Felix Bernstein** (1878–1956), Cantor just wrote that he appreciated that they have done it using simple means.

Later on, amongst the controversy surrounding the acceptability of the Axiom of Choice, mathematicians came to value Schröder's and Bernstein's proofs (and also Dedekind's proof, from 1899) for not relying on the validity of this axiom. However, the main contribution of these proofs consisted in offering the first non-trivial general arguments about sets. Thus it transpired that there is no need to understand set theory just as an auxiliary tool for the study of ordinal and cardinal numbers but that it holds a key to obtaining valuable general and far-reaching results concerning more than just these numbers. In other words, it was not until these proofs appeared that set theory began to enter mathematics as a meaningful mathematical theory with an autonomous subject of study, namely the abstract (predominantly infinite) sets. The theory of cardinal and ordinal numbers thus slowly started to be just a part of a general theory of sets, albeit a very important part.

Analyzing some of the proofs of the above theorem, the Polish mathematician **Stefan Banach** (1892–1945) captured more precisely what else they in fact contained. The proof of the theorem stated and proved by Banach then allowed **Alfred Tarski** (1901–1983) to obtain even a more general and more widely applicable theorem.

We shall now state and prove the three above discussed theorems in the reverse order to that in which they were discovered. We do this in order to bring attention to something which is, given its nature, typical for set theory and for all set-theoretical mathematics. To wit, that on the journeys from the general to the particular it is not just their elegance that is remarkable but that in this elegance there resides a power of understanding frequently stronger than the power of understanding of the journeys from the particular to the general dominating elsewhere.

Proposition 2.20. Tarski's Fixed Set Theorem. Let A be a set and let H be a mapping of the set $\mathcal{P}(A)$ into $\mathcal{P}(A)$ such that for each $X, Y \subseteq A$, $X \subseteq Y$ implies $H(X) \subseteq H(Y)$. Then there is a se $C \subseteq A$ such that $H(C) = C$.

Proof. Let $\mathcal{T}$ be the set of sets $X \subseteq A$ such that $X \subseteq H(X)$. Clearly, $\emptyset \in \mathcal{T}$. Let $C = \bigcup \mathcal{T}$. If $X \in \mathcal{T}$ then $X \subseteq C$ and hence $X \subseteq H(X) \subseteq H(C)$. It follows that $C \subseteq H(C)$ and thus $H(C) \subseteq H(H(C))$, that is, $H(C) \in \mathcal{T}$. Consequently, $H(C) \subseteq \bigcup \mathcal{T} = C$. □

Proposition 2.21. Banach's Theorem.[6] Let A, B be sets and let F be a one-to-one mapping of the set A into the set B and G a one-to-one mapping of the set B into the set A. Then there are sets $\bar{A}, \bar{B}$ such that $\bar{A} \subseteq A$, $\bar{B} \subseteq B$ and

$$\bar{B} = \{F(x) : x \in \bar{A}\}, \quad A \setminus \bar{A} = \{G(x) : x \in \ B \setminus \bar{B}\}.$$

Proof. For $Z \subseteq A$ let $\bar{F}(Z) = \{F(x) : x \in Z\}$. Similarly, for $Z \subseteq B$ let $\bar{G}(Z) = \{G(x) : x \in Z\}$. For $X \subseteq A$ let $H(X) = A \setminus \bar{G}(B \setminus \bar{F}(X))$. H is clearly a mapping of the set $\mathcal{P}(A)$ into $\mathcal{P}(A)$. If $X \subseteq Y \subseteq A$ then $\bar{F}(X) \subseteq \bar{F}(Y)$, $B \setminus \bar{F}(Y) \subseteq B \setminus \bar{F}(X)$, $\bar{G}(B \setminus \bar{F}(Y)) \subseteq \bar{G}(B \setminus \bar{F}(X))$ and hence $H(X) \subseteq H(Y)$. By Tarski's Theorem there exists a set $\bar{A} \subseteq A$ such that $\bar{A} = H(\bar{A}) = A \setminus \bar{G}(B \setminus \bar{F}(\bar{A})$. Defining $\bar{B} = \bar{F}(\bar{A})$ yields $A \setminus \bar{A} = \bar{G}(B \setminus \bar{B})$, proving the theorem. □

Proof of Cantor-Schröder-Bernstein Theorem (Proposition 2.19). Let F be a one-to-one mapping of the set A into the set B and G a one-to-one mapping of the set B into the set A. Let $\bar{A} \subseteq A$, $\bar{B} \subseteq B$ be sets guaranteed to exist by Banach's Theorem. Defining $L(x) = F(x)$ for $x \in \bar{A}$ and $L(x) = G^{-1}(x)$ for $x \in A \setminus \bar{A}$ yields a one-to one mapping of the set A onto B.

Proposition 2.22. (About the cardinality of the union of a set of sets.) Let $\mathcal{N} \neq \emptyset$ be a set of sets. Assume that for each set $X \in \mathcal{N}$ there exists a set $Y \in \mathcal{N}$ such that the cardinality of the set X is less than the cardinality of the set Y. Then every set $X \in \mathcal{N}$ has a cardinality less than the cardinality of $\bigcup \mathcal{N}$.

Proof. If $Y \in \mathcal{N}$ then the cardinality of the set Y is less or equal to the cardinality of the set $\bigcup \mathcal{N}$ since $Y \subseteq \bigcup \mathcal{N}$. Aiming for contradiction, assume that $X \in \mathcal{N}$ is a set that has the same cardinality as $\bigcup \mathcal{N}$. Then there is a set $Y \in \mathcal{N}$ with cardinality greater than the cardinality of the set X and hence also greater than the cardinality of $\bigcup \mathcal{N}$, which contradicts the previous assertions. □

The following assertion and its version for ordinal numbers had been known to Cantor already before the year 1900.

[6] In the Czech original, the statement of this theorem, and the proofs of this and the remaining theorems within this section, contain misprints which have been corrected. [Ed]

Proposition 2.23. The domain of all sets is not actualisable – there is no set of all sets.

Proof. Similarly as in the case of ordinal numbers, if the collection $\mathcal{M}$ of all sets existed then nobody could prevent us from interpreting it as an autonomous individual, and hence as a set. We would have $\mathcal{P}(\mathcal{M}) \in \mathcal{M}$ and hence also $\mathcal{P}(\mathcal{M}) \subseteq \mathcal{M}$. The cardinality of the set $\mathcal{P}(\mathcal{M})$ would thus be less or equal to the cardinality of the set $\mathcal{M}$. According to Cantor's Theorem, the set $\mathcal{P}(\mathcal{M})$ has greater cardinality then the set $\mathcal{M}$, contradiction.

□

In 1902 **Bertrand Russel** (1872–1970) published an antinomy concerning the set M such that its elements are just the sets that are not their own element. To wit, if $M \in M$ then according to the definition of M, M must not be its own element and if $M \notin M$ then M must be its own element.

Fast spreading of this antinomy, not only among mathematicians, was aided by the fact that no extraordinary knowledge is needed to express it. In turn, this fact also contributed to the long-prevailing opinion of the wider public that this indeed is a contradiction in the theory of sets.

However, enlightened by Cantor's considerations contained in his last letters to Dedekind (admittedly unknown at the time to anybody but Dedekind) we can see clearly that we are not dealing with a contradiction in set theory but merely with a proof of non-existence of the above set M, defined moreover in breach of the rule that a set can be created only after all its members have been created. Hence this is no paradox, except if somebody chooses to consider it paradoxical that such a set does not exist.

2.3.3 Well-Ordering Postulate

At the beginning of the twentieth century a paper appeared that attracted considerable attention.[7] **Ernst Zermelo** (1871–1953) proved the following assertion in it:

Proposition 2.24. (Zermelo's theorem). Every set can be well ordered.

Third Postulate. The task to well order any set.

Any well ordering R of A is a subset of the set A^2. Hence the third postulate does not ask for creation of some set with cardinality greater than the cardinality of sets that can be created using the previous postulates.

To solve the task demanded by the third postulate, Zermelo used an axiom that came to be called as the **Axiom of Choice** and as such it became widely known. Hence we shall also refer to it by this name.

By a **selector** on a set of sets $\mathcal{A}$ we understand a function F defined on the set $\mathcal{A}$ such that for each $X \in \mathcal{A}$, $X \neq \emptyset$ we have $F(X) \in X$.

[7] Ernst Zermelo, "Beweis, dass jede Menge wohlgeordnet werden kann," in *Mathematische Annalen* 59 (1904).

Axiom of Choice. On every set of sets, a selector can be created.

This axiom does not require a capability to choose an element from any given non-empty set, contrary to its frequent erroneous interpretation, but the capability to choose, whenever some set of sets is given, by a single creative act, one element from each non-empty set belonging to this set of sets. Moreover, it requires doing it in some sharply defined way, remembering which element was thus chosen from which set.

In fact, quite in the reverse, if we manage the task contained in the third postulate, the task required by the Axiom of Choice becomes an easy matter. For if $\mathcal{A}$ is a set of sets and R a well ordering of the set $\bigcup \mathcal{A}$, we can define the function F on $\mathcal{A}$ by setting, for any $\emptyset \neq X \in \mathcal{A}$, $F(X)$ to be the first element of X in the ordering R.

It is significant for the Axiom of Choice that it does not require us to actualise some given sharply defined domain of sets, that is to replace some such domain by a set. The Axiom of Choice is a warning for us not to be lulled by the so-far pervasive impression that our considerations of actual infinity were in fact a redundant luxury. To wit, up to now everything concerned sharply defined domains rather than sets and it sufficed to create this or that object falling in some such domain as and when it became needed. After all, in accepting the Powerset Postulate we came close to such an interpretation. After accepting the Axiom of Choice, the idea based on such an impression collapses: it is no longer possible to return to domains.

From what we have said it is apparent that the Axiom of Choice is the innermost and most characteristic postulate of Cantor's set theory. Not until this postulate had the way opened to new realms of infinity previously inaccessible to the pre-set-theoretical mathematics. It could rightly be expected to lead to extraordinary phenomena. Classical set theory rejecting the Axiom of Choice would be maiming Cantor's vision.

For now we shall formulate just two widely-used simple consequences of the Axiom of Choice.

Proposition 2.25. Let Q be a function on a set $A \neq \emptyset$ such that for each $x \in A$, $Q(x)$ is a non-empty set. Then it is possible to create a function f on the set A such that for each $x \in A$, $f(x) \in Q(x)$.

Proof. Let F be a selector on the set of all sets $Q(x) \times \{x\}$, where $x \in A$. Let f denote the set of all $F(Q(x) \times \{x\})$, where $x \in A$. Clearly f is a function with the required properties. □

Proposition 2.26. Let H be a mapping of a set B onto a set A. Then the set A has cardinality less or equal to that of the set B.

Proof. Let Q be the function on the set A such that for each $x \in A$, $Q(x)$ is the set of all $z \in B$ for which $H(z) = x$. Clearly, $Q(x) \neq \emptyset$ for each $x \in A$. Let f be the function on the set A guaranteed to exist by the previous proposition. Then f is a one-to-one mapping of the set A into the set B. □

2.3.4 Objections of French Mathematicians

The growing number of adherents of Cantor's set theory amongst the German mathematicians at the turn of the nineteenth and twentieth centuries was a challenge for the French to take a stand on this theory. A suitable opportunity arose when Zermelo proved that every set can be well ordered. The *Bulletin de la Société mathématique de France* then published the famous "Five Letters on Set Theory".[8] These letters they were exchanged between four eminent French mathematicians, **Jacques Hadamard** (1865–1963), **Émile Borel** (1871–1956), **René-Louis Baire** (1872–1932) and **Henri Lebesgue** (1875–1941).

Out of the four, only Hadamard defended Cantor's set theory, presented by Zermelo's result in its most challenging form. The remaining three raised objections. At the same time, however, they realized that some considerations carried out within it are useful, as long as they are put in the right way.

We will now present selected statements from these letters,[9] which sketch the opinions of their authors.

Baire's views (from a letter to Hadamard)

> As soon as one speaks of the infinite (even the denumerable, and it is here that I am tempted to be more radical than Borel), the comparison, conscious or unconscious, with a bag of marbles passed from hand to hand must disappear completely. [...]
>
> In particular, when a set is given (we agree to say, for example, that we are given the set of sequences of positive integers), I consider it false to regard the subsets of this set as given. I refuse, a fortiori, to attach any meaning to the act of supposing that a choice has been made in every subset of a set. [...]
>
> Zermelo says: "Let us suppose that to each subset of M there corresponds one of its elements." This supposition is, I grant, in no way contradictory. Hence all that it proves, as far as I am concerned, is that we do not perceive a contradiction in supposing that, in each set which is defined for us, the elements are positionally related to each other in exactly the same way as the elements of a well-ordered set. In order to say, then, that one has established that every set can be put in the form of a well-ordered set, the meaning of these words must be extended in an extraordinary way and, I would add, a fallacious one. [...]
>
> For me, progress in this matter would consist in delimiting the domain of the definable. And, despite appearances, in the last analysis everything must be reduced to the finite.

[8] Jacques Hadamard, "Cinq lettres sur la théorie des ensembles," *Bulletin de la Société mathématique de France* 33 (1905): 261–273.

[9] The translation is from Gregory Moore, *Zermelo's Axiom of Choice: Its Origins, Development and Influence* (Springer-Verlag, 1982).

Lebesgue's views (from a letter to Borel)

Lebesgue writes in this letter that creating a correspondence whereby to each nonempty subset of a set M corresponds an element of it is in fact easy only when this set is well ordered. A general solution of this task however is not possible if we allow the mere specification of a property of its elements to be the definition of a set. For then nothing enables us to distinguish two of its elements, let alone determine one.

> The question comes down to this, which is hardly new: *Can one prove the existence of a mathematical object without defining it?* [...] I believe that we can only build solidly *by granting that it is impossible to demonstrate the existence of an object without defining it.*

Lebesgue doubts justification of some general considerations in set theory and alleges that even the word "existence" assumes a different meaning in them.

> For example, when Cantor's well-known argument is interpreted as saying that there exists a non-denumerable infinity of numbers, no means is given to name such an infinity. It is only shown, as you have said before me, that whenever one has a denumerable infinity of numbers, one can define a number not belonging to this infinity.

Borel's views (from a letter to Hadamard)

> First of all, I would like to call your attention to an interesting remark that Lebesgue made at the meeting of the Society on 4 May: How can Zermelo be certain that in the different parts of his argument he is always speaking of *the same* choice of distinguished elements, since he characterizes them in no way *for himself.* (Here it is not a question that someone may contradict him but rather of his being intelligible to himself.)

Hadamard's views (from a letter to Borel)

> The question appears quite clear to me now, after Lebesgue's letter. More and more plainly, it comes down to the distinction, made in Tannery's article, between what is determined and what can be described.
>
> In this matter Lebesgue, Baire, and you have adopted Kronecker's viewpoint, which until now I believed to be peculiar to him. You answer in the negative the question posed by Lebesgue (above, p. 265): Can one prove the existence of a mathematical object without defining it? I answer it in the affirmative. [...]

> [I]t is clear that the principal question, that of knowing if a set can be well-ordered, does not mean the same thing to Baire (any more than to you or Lebesgue) that it does to me. I would say rather – Is a well-ordering possible? – and not even – Can one well-order a set? – for fear of having to think who this one might be. Baire would say: Can we well-order it? An altogether subjective question, to my way of thinking.

The views of these French mathematicians have not prevailed, even though they were endorsed by **Henri Poincaré** (1854–1912). In fact, he was even more radical.

Of particular note is the fact that the objections raised in the above letters against Cantor's set theory were directed against both the problematic postulates. The Postulate of Powerset was as unacceptable for the positivist mathematicians as the Axiom of Choice, and let us admit that from their point of view this was quite justified.

The Postulate of Powerset was recognized by almost anyone who – to some degree at least – worked in set theory. In the case of the Axiom of Choice, all that could be achieved was that for some time, the assertions with proofs that used the axiom of choice were marked by asterisks.

Cantor's direction in the development of set theory has opened up an enormous area for free yet meaningful mathematical creativity. Any ban on such opportunities was of course ineffective.

The Axiom of Choice was rejected by those mathematicians who wished to keep open the possibility of a return to the pre-set-theoretical conception of infinitary mathematics; and also by those who did not understand who was actually the builder of set theory and whose ability to create and to see they must therefore appropriate to participate in the feast that Cantor had prepared for them.

The above mentioned antinomies (along with some others), together with Zermelo's theorem on well ordering and, last but not least with the fact that Cantor opened the door into mathematics for large infinite sets inimitable in the real world, led the leading figure of French science, Henri Poincaré, to firmly reject the set theory and the timidly emerging set-theoretical mathematics. Poincaré did so in a lecture at the Fourth International Mathematical Congress in Rome in 1908 using the following words:

> Next generations will regard set theory as a disease from which mathematicians have recovered.[10]

[10] There is some dispute amongst the historians of mathematics whether Poincaré actually said this. But it is in the spirit of his views. [Ed]

2.4 Large Cardinalities

2.4.1 Initial Ordinal Numbers

Zermelo's theorem introduced a relationship between cardinal and ordinal numbers. For if A is some set, R its well ordering and γ its type then the structures $\langle A, R\rangle$, $\langle[\gamma], \preceq\rangle$ are isomorphic and consequently the sets $A, [\gamma]$ have the same cardinality. In other words, it is possible to number the set A by ordinal numbers less than γ:

$$A = \{a_\alpha; \alpha \in \gamma\}.$$

We say that an **ordinal number** γ is **initial** if it is infinite and whenever $\beta \prec \gamma$ then the set $[\beta]$ has cardinality less than $[\gamma]$.

We can easily see that the following holds:

Proposition 2.27. (i) Let δ be an infinite ordinal number. Then there exists a unique initial ordinal number γ such that the sets $[\delta], [\gamma]$ have the same cardinality.

(ii) Let A be an infinite set. Then there exists a unique initial ordinal number γ such that the sets $A, [\gamma]$ have the same cardinality.

As apparent from this proposition, when studying cardinal numbers, Zermelo's theorem (proposition 2.24) allows us to limit ourselves to cardinalities of sets $[\gamma]$ such that γ is either a natural number or an initial ordinal number. Thus this theorem elevated the status of the domain of initial ordinal numbers to that of the main pillar of the whole edifice of ordinal and cardinal numbers. Cantor's theorem (proposition 2.17) along with the proposition about the union of a set of sets (proposition 2.22) the made it possible to build this pillar up to a staggering height.

Proposition 2.28. Let M be some set of ordinal numbers. Then there exists (it is possible to create) an initial ordinal number γ such that $M \subseteq [\gamma]$.

Proof. By proposition 2.8 we can find α such that $M \subseteq [\alpha]$. Let γ be an initial ordinal number such that the sets $\gamma, \mathcal{P}([\alpha])$ have the same cardinality. By Cantor's theorem (proposition 2.17) the cardinality of the set $[\gamma]$ is greater than the cardinality of the set $[\alpha]$. Hence $\alpha \prec \gamma$, and hence $M \subseteq [\alpha] \subseteq [\gamma]$. □

If X is some set of ordinal numbers then the type of its ordering by the relation $\preceq$ will be called the **type of the set X**.

We shall surely agree that the domain of all initial ordinal numbers is sharply defined.

We say that an ordinal number α is the **index of the initial ordinal number** γ if α is the type of the set of all initial ordinal numbers less than γ. The initial ordinal number such that its index is the number α is denoted ω_α.

By proposition 2.9, $\alpha \preceq \omega_\alpha$. Clearly $\omega_0 = \omega$ and ω_1 is the first ordinal number such that the set of all smaller ordinal numbers is uncountable. By proposition 2.28, the numbers $\omega_1, \omega_2, \omega_3, \ldots$ exist.

We shall surely agree that the operation that creates from every initial ordinal number its index, is sharply defined, and thus also the domain of all initial ordinal numbers is sharply defined.

Proposition 2.29. For every ordinal number α there exists the number ω_α.

Proof. Aiming for contradiction, assume that α is an ordinal number such that ω_α does not exist. We can assume that α is the least such ordinal number. By proposition 2.28 there exists an initial ordinal number γ such that $\{\omega_\beta : \beta \prec \alpha\} \subseteq [\gamma]$. If γ is the least initial ordinal number with this property then clearly $\gamma = \omega_\alpha$. □

Proposition 2.29 says that the domain of all indices of initial ordinal numbers is identical with the domain of all ordinal numbers.

Lexicographic ordering of the set $[\omega_\upsilon]^2$ is defined as follows: $\langle\alpha,\beta\rangle$ Le $\langle\gamma,\delta\rangle$ just when either $\alpha \prec \gamma$ or $\alpha = \gamma$ and $\beta \preceq \delta$ (where $\alpha,\beta,\gamma,\delta \in [\omega_\upsilon]$).

Maxima-respecting lexicographic ordering of the set $[\omega_\upsilon]^2$ is defined as follows: $\langle\alpha,\beta\rangle$ Ml $\langle\gamma,\delta\rangle$ just when either $\max\{\alpha,\beta\} \prec \max\{\gamma,\delta\}$ or $\max\{\alpha,\beta\} = \max\{\gamma,\delta\}$ and $\langle\alpha,\beta\rangle$ Le $\langle\gamma,\delta\rangle$ (where $\alpha,\beta,\gamma,\delta \in [\omega_\upsilon]$).

It is easy to see that Le and Ml are well orderings of the set $[\omega_\upsilon]^2$.

Proposition 2.30. (i) The type of the set $[\omega_\gamma]^2$ in the ordering Ml is ω_γ.

(ii) Sets $[\omega_\upsilon]$ and $[\omega_\upsilon]^2$ have the same cardinality.

Proof. It suffices to prove (i) because (ii) immediately follows.

Let η be the type of the set $[\omega_\gamma]^2$ in the ordering Ml. It is easy to see that $\omega_\gamma \preceq \eta$. If $\langle\alpha,\beta\rangle \in [\omega_\gamma]^2$ then the set of all ordered pairs of ordinal numbers less than $\langle\alpha,\beta\rangle$ in the ordering Ml is a subset of the set $[\delta]^2$ where $\delta = \max\{\alpha,\beta\}+1$.

If we prove that for every $\delta \prec \omega_\gamma$, the set $[\delta]^2$ has smaller cardinality than the set $[\omega_\gamma]$ then $\eta = \omega_\gamma$ follows. If $\gamma = 0$ then $[\delta]^2$ is a finite set. Aiming for contradiction, assume that $\gamma \neq 0$ is the least ordinal number such that for some $\delta \prec \omega_\gamma$, the cardinality of the set $[\delta]^2$ is not less than the cardinality of $[\omega_\gamma]$. Let $\alpha \prec \gamma$ be an ordinal number such that the set $[\delta]$ has cardinality less or equal to the cardinality of the set $[\omega_\alpha]$. But then the set $[\delta]^2$ has cardinality less or equal to $[\omega_\alpha]^2$ and this means cardinality less or equal to the cardinality of $[\omega_\alpha]$ since $\alpha \prec \gamma$. Thus the set $[\delta]^2$ has cardinality less than $[\omega_\gamma]$, contradiction. □

The following assertion is an almost immediate consequence of proposition 2.30.

Proposition 2.31. Let $A, B \neq \emptyset$ be sets such that $\mathrm{Card}(A)$ is and infinite cardinal number, $\mathrm{Card}(B) \leq \mathrm{Card}(A)$. Then

$$\mathrm{Card}(A \cup B) = \mathrm{Card}(A \times B) = \mathrm{Card}(A).$$

2.4.2 Zorn's Lemma

Considering various proofs employing Zermelo's theorem (Proposition 2.24), Max Zorn (1906–1993) abstracted from them a revealing general assertion that (once it was proved using Zermelo's theorem)[11] in many other cases replaced it. This general assertion is known as **Zorn's lemma**.

We say that the ordering $\leq$ of a set A has **Zorn's property**, if for each set $X \subseteq A$ that is linearly ordered by the relation $\leq$, there exists $x \in A$ such that for each $y \in X$, $y \leq x$.

Clearly, if a partial ordering $\leq$ of a set A has Zorn's property then the set A is non-empty since the empty set $\emptyset \subseteq A$ is linearly ordered by the relation $\leq$, too.

Lemma 2.32. (Zorn's lemma) Let a partial ordering $\leq$ of a set A have Zorn's property. Then there exists a maximal element in the ordering $\leq$.

A sketch of a proof of Zorn's lemma from Zermelo's Theorem. Assume that the ordering $\leq$ has Zorn's property. Let $A = \{d_\alpha; \alpha \in [\gamma]\}$. We define a set D by transfinite induction by letting $d_0 \in D$ and for $0 \neq \alpha \in [\gamma]$ including d_α in D just when for each $\beta \in [\alpha]$ with $d_\beta \in D$ we have $d_\beta < d_\alpha$. The set D is clearly linearly ordered by the relation $\leq$ and has a greatest element which is a maximal element of the set A in ordering $\leq$.

A sketch of a proof of Zermelo's Theorem from Zorn's lemma. Let B be the set of all $\langle X, R\rangle$ where $X \subseteq A$ and R is a well ordering of the set X. For $\langle X, R\rangle$, $\langle Y, S\rangle \in B$ let $\langle X, R\rangle \leq \langle Y, S\rangle$ just when the ordering R of the set X is a cut on the ordering S of the set Y. It is easy to see that the ordering $\leq$ has Zorn's property and a maximal element of the set B in this ordering is a well ordering of the set A.

Zorn's lemma can be strengthened for example in the following way:

Proposition 2.33. Assume that the ordering $\leq$ of a set A has Zorn's property and that a set $C \subseteq A$ is linearly ordered by the relation $\leq$. Then there exists a maximal element b of the set A in ordering $\leq$ such that for all $x \in C$ we have $x \leq b$.

Proof. If B is the set of all $x \in A$ such that for all $y \in C$, $y \leq x$ then the ordering $\leq$ of the set B has Zorn's property and a maximal element $b \in B$ in this ordering is also a maximal element of the set A in the ordering $\leq$ with the required property.

□

Using Zorn's lemma we can easily prove the following assertion known as the **Maximality principle** proved by **Felix Hausdorff**(1868–1942).

[11] Max Zorn, "A Remark on Method in Transfinite Algebra," *Bulletin of the American Mathematical Society* 41, no. 10 (1935): 667–670.

Proposition 2.34. (Hausdorff's maximality principle) Assume that $\leq$ is a partial ordering of a set A. Then there exists a set $C \subseteq A$ linearly ordered by the ordering $\leq$ such that no set $D \neq C$ satisfying $C \subseteq D \subseteq A$ is linearly ordered by the ordering $\leq$. [12]

Proof. Let $\mathcal{A}$ denote the set of all sets $X \subseteq A$ linearly ordered by the relation $\leq$. It is easily checked that $\mathcal{A}$ is partially ordered by the relation $\subseteq$ and that this ordering has Zorn's property. If C is a maximal element of $\mathcal{A}$ in this ordering then C clearly is the required set.

□

We observe that conversely, Zorn's lemma is a straightforward consequence of Hausdorff's maximality principle.

2.5 Developmental Influences

2.5.1 Colonisation of Infinitary Mathematics

In the twentieth century, infinitary mathematics faithfully followed the path outlined for it by Bernard Bolzano in his treatise *The Paradoxien des Unendlichen* (see Section 1.3).

Due to its immense scope, Cantor's set theory opened a convenient field for a free and unrestricted activity to that current within the mathematics of the twentieth century, that sought to identify all mathematics with set theory; that is, to build individual areas and disciplines of mathematics as parts of Cantor's set theory.

This current of thought started to announce itself timidly from the very beginning of set theory, strengthened, justified its existence and confirmed its usefulness all through the turbulent development of the set-theoretical mathematics, until it finally established itself in the consciousness, or at least seeped out of the unconsciousness of almost all mathematicians interested in set theory. It relied on the fact that set-theoretical mathematics created enough tools to transfer the hitherto existing mathematical disciplines into set theory and to make them into its own subdisciplines. Such an interpretation naturally turned the previously more or less autonomous mathematical disciplines into a kind of protectorates or even colonies of the empire represented by set theory.

The loss of independence of individual mathematical disciplines was balanced by the fact that set theory has taken them under its wings in the sense of accepting the responsibility for their consistency. To wit, as long as considerations carried out within some mathematical discipline did not step out of the framework determined by the interpreting this discipline as a part of set theory (a framework which was usually more than sufficient), any contradiction obtain within the discipline would be a contradiction within set theory.

[12] This assertion can be found in the first edition of the book Felix Hausdorff, *Grundzüge der Mengenlehre* (Leipzig: Veit and Company, 1914). It does not appear in the subsequent edition of this book Felix Hausdorff, *Mengenlehre* (Berlin – Leipzig: Walter de Gruyter, 1927).

It should be noted however, that this advantage was afforded to any mathematical discipline only as far as its development obeyed the dictates of set theory.

The change in the position of individual mathematical disciplines did not diminish the contribution that set theory had already made to mathematics. These disciplines continued to enjoy access to the vast spaces and depths of infinity revealed by Cantor's theory. It remained possible to build therein useful extensions of mathematical disciplines such as for example topology for geometry and functional analysis for mathematical analysis.

When carrying out considerations within a given mathematical discipline, it was possible to access its set-theoretical extensions and then come back just as it used to be the case long before this with the theory of complex numbers providing an extension of real numbers. Now however the whole of set theory became an extension for all mathematical disciplines. In it, various mathematical fields influenced each other and linked with each other in a process that led to the birth of new mathematical disciplines such as for example the theory of topological groups.

Still, it should be noted again that only those spaces and depths of infinity were thus made available that were granted by Cantor's set theory.

Naturally, set theory itself profitted greatly from this conception of mathematics. For it could exploit its protectorates and colonies by drawing suitable ideas and stimuli out of them as well as adjusting results gained in them and procedures designed for their development etc.

The interpretation of individual mathematical disciplines as parts of set theory had, beside the already mentioned positive effects, also many negative effects on the development of mathematics. To wit, in transferring some more or less independent mathematical field into set theory, the original contents of the main concepts of such a field were emptied out and replaced by set-theoretical contents. This was not always managed in a fitting and transparent way and then it was often just the names of concepts so violated that dimly recalled their original meanings. Such obfuscation of the original meanings of the main concepts of a mathematical discipline obviously led to obfuscation of the original intentions of its natural development, so that its further development moved in a different direction than it would have done had this discipline resisted set-theoretical reformulation.

Set theory has taken over various mathematical theories mainly by creating models of the subjects of their studies. The subject of study of a mathematical theory is most often formed by a community of some objects. If set theory takes over objects belonging to some such community then it can model the structure of such a community as a relational structure; properties of its objects as sets of objects with this property and relationships between them as relations. In addition, operations performed with these objects can be modeled as functions.

In order for set theory to achieve its goal, it had to model – as far as possible – all that it works with as sets. Relations and functions are interpreted in set theory as sets, but their elements are ordered pairs, triples, etc., that are not sets. Hence it was necessary to replace these objects by suitable sets; in place of the original objects, their set-theoretical substitutes are used.

Probably the most convenient way of replacing ordered pairs (and therefore also triples etc.) by sets was was proposed in 1921 by **Kazimierz Kuratowski** (1896–1980).[13] He replaced the ordered pair $\langle x, y\rangle$ by the set $\{\{x\}, \{x, y\}\}$. The substitute for the ordered triple $\langle x, y, z\rangle$ is then the substitute of the ordered pair $\langle x, \langle y, z\rangle\rangle$ etc.

Objects belonging to the subject of study of some mathematical theory are usually ideal objects and not sets. Hence they must also be replaced by suitable sets. This was accomplished in set theory by selecting some sets from amongst those that could, more or less, play a good role as a substitutes for these ideal phenomena, and by declaring them to be the **canonical substitutes**.

Amongst other things this means that an agreement has been reached, sanctioning, from that point on, the usage of the same names for these canonical substitutes as are those that we use for the phenomena thus substituted.

Numbers, be it natural numbers, integers, rational, real or complex numbers as well as ordinal numbers, are not sets. To grasp them set-theoretically was thus the first task of set theory on its quest to become the world of all mathematics. Replacing numbers with sets was carried out in different ways depending on which numbers they were.

Once the substitutes for natural numbers were created, finding substitutes for integers was a trivial matter. It sufficed to use the signs $+$ and $-$ to denote some two fixed sets and use the ordered pair $\langle +, n\rangle$ as the substitute for the positive integer n, the ordered pair $\langle +, 0\rangle$ as the substitute for 0 and the ordered pair $\langle -, n\rangle$ as the substitute for the negative integer $-n$.

Once the substitutes for real numbers were found, creation of substitutes of the complex numbers was a similarly trivial matter. The complex number $a+ib$ was substituted by the ordered pair $\langle a, b\rangle$.

Numbers can be interpreted as ideal objects abstracted (that is, separated) from their individual occurrences and thus set theory can grasp them as sets of these occurrences. That means, of some suitable occurrences chosen from amongst those that are sets.

For this purpose the following concept was introduced, with signficance far exceeding its usage mentioned here.

A symmetric, reflexive and transitive relation on a set A is called an **equivalence** on the set A.

Every equivalence R on A uniquely determines a partition of the set A into subsets of mutually equivalent elements. By an **equivalence class** of R represented by an element $x \in A$ (denoted $\overline{x}$) we understand the set of all $y \in A$ such that $\langle x, y\rangle \in R$.

If $x, y \in A$ then clearly either $\overline{x} = \overline{y}$ or $\overline{x} \cap \overline{y} = \emptyset$. The set of all equivalence classes (the factor set) is denoted $\overline{A}$, that is, $\overline{A} = \{\overline{x} : x \in A\}$.

Let S be a relation on the set A. We say that the **relation S respects**

[13] Kazimierz Kuratowski, "Sur la notion de l'ordre dans la Théorie des Ensembles," *Fundamenta Mathematicae* 2 (1921): 161–171.

the equivalence R if for eavery $x, y, u, v \in A$ such that $\langle x, y\rangle \in R$, $\langle x, u\rangle \in S$, $\langle y, v\rangle \in S$ we have $\langle u, v\rangle \in R$.

If S is a relation that respects the equivalence R then we define $\langle \overline{x}, \overline{y}\rangle \in \overline{S}$ just when $\langle x, y\rangle \in S$. Clearly, $\overline{S}$ is a relation on the set $\overline{A}$.[14]

Let $F : A \to A$ be a function. We say that the **function** F **respects the equivalence** R if for every $x, y \in A$ such that $\langle x, y\rangle \in R$ we have $\langle F(x), F(y)\rangle \in R$.

If F is a function that respects the equivalence R then we define $\overline{F}(\overline{x}) = \overline{F(x)}$. Clearly, $\overline{F}$ is a function, $\overline{F} : \overline{A} \to \overline{A}$.[15]

We shall use an equivalence on the set A of all ordered pairs $\langle m, n\rangle$ of integers, where $n \neq 0$, to create **substitutes of rational numbers**. The ordered pairs $\langle m, n\rangle, \langle k, l\rangle$ stand for the fractions $\frac{m}{n}, \frac{k}{l}$ and both these fractions capture the same rational number just when $ml = kn$.

The relation R on the set A such that $\langle m, n\rangle R\langle k, l\rangle$ just when $ml = kn$ clearly is an equivalence on the set A. The classes of mutually equivalent elements of the set A in the equivalence R are thus suitable substitutes of rational numbers.

Let B denote the set of all $\langle m, n\rangle \in A$ such that n is a natural number and the numbers m, n are relatively prime. In every class of A partitioned according to R there is a unique element of B. That means that the elements of the set B are also suitable substitutes of rational numbers.

Hence it remains to find suitable substitutes for natural numbers and for real numbers.

Set-theoretical Substitutes for Natural Numbers

In 1919 **Bertrand Russsel** wrote[16]

> The question "What is a number?" is one which has been often asked, but has only been correctly answered in our own time. The answer was given in 1884 by Frege in his Grundlagen der Arithmetik. Although this book is quite short, not difficult, and of the very highest importance, it attracted almost no attention, and the definition of number which it contains remained practically unknown until it was rediscovered by the present author in 1901.

According to the Frege-Russel definition the number 1 is the set of all one-element sets, the number 2 the set of all two-element sets etc. When investigating this definition, Russel realised that already the set of one-element sets is contradictory and thus it cannot exist. This was very sad for Frege, but unlike him, Russel continued research in this direction. He created the type theory,

[14] A similar definition can be introduced for relations of higher arities.

[15] A similar definition can be introduced for functions of more arguments.

[16] Bertrand Russel, *Introduction to Mathematical Philosophy* (London: George Allen & Unwin, 1919).

where objects that are not sets have type 1, sets of these objects are of type 2 and in general, objects of type $n + 1$ are sets such that each of their elements is of type n. Russel thus avoided the problem but paid by having, for example, a number 8 of type 3 and a different number 8 of type 4 etc. This cumbersome definition of natural numbers has not been widely accepted.

A phenomenon that was separated, that is, abstracted from its various bearers can then be with some of them firmly associated again. That means, tied to this chosen bearer by a firmer tie than with the others; a tie such that with some good will the bearer could be interpreted as being the same thing. In other words, abstracted phenomena can be suitably contra-abstracted.

In 1923, **John von Neumann** (1903–1957) demonstrated a text-book example of contra-abstraction of ordinal numbers and cardinal numbers as initial ordinal numbers.[17]

Von Neumann's contra-abstraction of number 0 is the empty set $\emptyset$. The contra-abstraction of number α is the set of all contra-abstractions of numbers less that α. Von Neumann's contra-abstraction covers the natural numbers, amongst others.

Set-theoretical Substitutes for Real Numbers

Up to now, the first postulate of Cantor's set theory was sufficient for the creation of substitutes of various ideal objects. However, it is not enough for the creation of substitutes of real numbers. We cannot do it without the second postulate. In order to have a possibility of creating set-theoretical substitutes of real numbers, it was necessary to complete them first; that means, to add irrational numbers.

Bernard Bolzano demonstrated already in 1817 both of the two common ways of completing real numbers used today.[18] In §9 he completed rational numbers by adding the limits of Bolzano sequences (known also as Cauchy sequences). This was described in more details by George Cantor in a paper that appeared 55 years later.[19] In §12 Bolzano completed rational numbers by adding suprema of non-empty sets of rational numbers, bounded from above. Thanks to this, Richard Dedekind could declare non-trivial cuts without inner or outer head on the ordering of rational numbers to be suitable set-theoretical substitutes of irrational numbers.

To appreciate how set theory took over geometry, it suffices to consult some textbook on geometry informing us that the space is a set of points, the horizon is the set of points at infinity etc. On the other hand, set-theoretical topology

[17] John von Neumann, "Zur Einführung der transfiniten Zahlen," *Acta Scientiarum Mathematicarum* 1, no. 4 (1923): 199–208.

[18] Bernard Bolzano, *Rein analytischer Beweis des Lehrsatzes, dass zwischen je zwey Werthen, die ein entgegengesetzes Resultat gewähren, wenigstens eine reele Wurzel der Gleichung liege* (Prague, 1817).

[19] George Cantor, "Uber die Ausdehnung eines Satzes aus der Theorie der trigonometrischen Reihen," *Mathematische Annalen* 5 (1872): 123–132.

is a show case of spacial abstract shapes that could not be grasped in any other way.

There is one sort of numbers that was ignored, or rather hated by Cantor's set theory, in keeping with Cantor's own attitude. To wit, they were the infinitesimal numbers underlying differential and integral calculus. This debt of set-theoretical mathematics had not been paid until the arrival of non-standard models of analysis.

2.5.2 Corpuses of Sets

The non-existence of the set of all sets forced a reduction of the subject of study for set theory. Still, the non-actualisability of the domain of all sets does not mean that various important parts of it also fail to be actualisable. After all, no postulate accepted so far requires the creation of the set of all sets. For example, the set defined below are for set-theoretical studies sufficient.

We say that a set K is a **corpus** of sets if the following hold:

(a) $[\omega] \in K$.

(b) If $X \in K$ then $X \subseteq K$.

(c) If $X, Y \in K$ then $X \setminus Y \in K$.

(d) If $X, Y \in K$ then $\{X, Y\} \in K$.

(e) If $X \in K$ then $\mathcal{P}(X) \in K$.

(f) If $X \in K$ then $\bigcup X \in K$.

(g) If F is a function, $F \subseteq K$, $\mathrm{dom}(F) \in K$ then $F \in K$, $\mathrm{rng}(F) \in K$.

(h) On the corpus of sets K there exists a selector.

It is clear from the above conditions that no corpus of sets can be created by using the three postulates introduced so far. Its cardinality cannot be reached by these postulates. Its existence, as long as we care for it and believe in it, has to be guaranteed by another postulate.

Fourth postulate To create at least one corpus of sets.

Let us denote a corpus created using the fourth postulate by K_0. If we so dare, we can postulate the creation of a corpus K_1 such that $K_0 \in K_1$ or even, for any ordinal number α, the creation of a corpus K_α such that for all $\beta \prec \alpha$ we have $K_\beta \in K_\alpha$.

This repetitious process may be disrupted by a qualitatively new (that means, not accessible from below) **Ramsey** inaccessible initial ordinal (cardinal) number. From there we can continue using the above algorithm to the next inaccessible number, a **measurable** one. This tower of postulates is illustrated on the title page of the book *The Higher Infinite*, see page 101. The line 0=1 at its very top warns of the danger of logical contradiction hanging over the process of acceptance of ever stronger postulates.

2.5.3 Introduction of Mathematical Formalism in Set Theory

If we model individual objects belonging to some community under investigation as abstract ur-objects (that means as objects emptied of their contents), we create a formal structure of the community under investigation. Independent study of such formal structures then relies merely on logical proofs (usually in predicate calculus) of various formal assertions from previously chosen formal axioms.

It was basically in this way that **David Hilbert** (1862–1943) approached Euclidean geometry in his book *Grundlagen der Geometrie*, that appeared in 1903.[20] Then he formally captured the structure of real numbers in similar way.[21]

In both these cases, Hilbert could not do without infinite sets, in particular Bolzano's theorem about suprema. This blemish on the beauty of formal mathematics (only artificially repaired with the help of second-order logic) notwithstanding, in his lecture *Grundlagen der Logic und der Arithmetik*[22] Hilbert emphasised the necessity of a strict mathematical formulation of the language of mathematics and logic.

Hilbert's emphasis on axiomatisation of mathematical theories was just as important. Results in this or that mathematical theory must be obtained by purely logical arguments based on previously chosen axioms, that is, in a purely formal way, without reliance on intuition. He thus initiated the mathematical-philosophical approach called mathematical formalism.

The obligation to axiomatise mathematical theories advocated by Hilbert naturally concerns set theory too. On account of the non-actualisability of the set of all sets, only corpuses of sets can be axiomatised (using predicate calculus). If that can be done, the above-mentioned blemish upon axiomatisation of real numbers could be removed. **Ernst Zermelo** (1871–1953) undertook this task and he published such an axiomatization in *Mathematische Annalen.*[23] However, he neglected to include an axiom corresponding to the condition (g) required of corpuses of sets. This was remedied by **Dimmitrij Mirimanov** (1861–1925) in a paper published in 1917.[24] The same was also achieved by **Adolf Fraenkel** (1891–1965) in *Mathematische Annalen*[25] and **Thoralf**

[20] Vopěnka refers to the second, extended edition, David Hilbert, *Grundlagen der Geometrie* (Leipzig: Teubner, 1903). *Grundlagen der Geometrie* first appeared in 1899. [Ed]

[21] In the Czech original, Vopěnka refers here to David Hilbert, "Die Theorie der algebraischen Zahlkörper," *Jahresbericht der DMV* 4 (1894/95): 175–546. The usual reference, used by Vopěnka in Chapter 12, is David Hilbert, "Über den Zahlbegriff," *Jahresbericht der DMV* 8 (1900): 180–194. [Ed]

[22] David Hilbert gave this lecture at the third international mathematical congress in 1904 in Heidelberg.

[23] Ernst Zermelo, "Untersuchungen über die Grundlagen der Mengenlehre," *Mathematische Annalen* 65 (1908): 261–281

[24] Dimitrij Mirimanov, "Les antinomies de Russell et de Burali-Forti et le probléme fondamental de la théorie des ensembles," *L'Enseigment Mathématique* 19 (1917): 37–52.

[25] Adolf Fraenkel, "Zu den Grundlagen der Cantor-Zermeloschen Mengenlehre," *Mathematische Annalen* 5, no. 86 (1922): 230–237.

Skolem (1887–1963) at the congress of Scandinavian mathematicians.[26]

Axiomatisation of not only corpusses of sets but also of their subsets makes it possible to eliminate axiom schemata from formalisation of set theory This was the path taken by **John von Neumann** (1903–1957) who created and developed such an axiomatisation in the paper "Eine Axiomatisierung der Mengenlehre".[27]

A more natural form was given to it by **Paul Bernays** (1888–1977) in a series of articles that started to appear in 1937 in *Journal of Symbolic Logic* under a shared name "A system of axiomatic set theory"[28] and by **Kurt Gödel** (1906–1978) in a slender volume from 1940.[29] This axiomatisation came to be called **Gödel-Bernays axiomatisation.**

If set theory is to guarantee the consistency of mathematical theories that are modelled in it, then obviously this theory itself needs to be consistent. That means at least some axiomatisations of it need to be consistent. The question of consistency of various axiomatisations of set theory has been forsaken by mathematicians. After all, they could not do anything else. They had to content themselves with Cantor's insight into the world of actually infinite sets not being a deceptive and collapsing mirage.

Even though mathematical formalism failed to reach the Hilbert's goal of guaranteeing the consistency of axioms of set theory, or at least of axioms for natural numbers, its entrance into set theory was an extraordinary event nevertheless. To wit, it was the second grand encounter and blending of two most important currents in mathematics: that of intuition and that of calculations.

In the mathematics of intuition the task is to make results evident. To evidence something means to see it (it the widest sense of this word) and to know that we are seeing it. This current of mathematics was born in the geometry of Greek antiquity, and in the twentieth century it worked upon Cantor's set theory although the agent who evidenced the results was not Zeus but the God of medieval and scholastic theology.

In the mathematics of calculations the task is to gain the results by applying correct calculations with signs, that means calculations carried out according to various previously established fixed rules. This current emerged from ancient India in creating the arithmetical and algebraic calculus.

Their first grand encounter occurred when **René Descartes** (1596–1650) introduced algebra into geometry. Their second encounter and blending occured when David Hilbert brought mathematical formalism into set theory.

The entrance of mathematical formalism into set theory had similar impact on mathematics as in its own time the entrance of algebra into geometry. A

[26] Thoralf Skolem, lecture "Einige Bemerkungen zur axiomatischen Begründung der Mengenlehre," published in *Wissenschaftliche Vorträge auf dem Fünften Kongress der Skandinavischen Mathematiker in Helsingfors* (1922).

[27] John von Neumann, "Eine Axiomatisierung der Mengenlehre," *Journal für die reine und angewandte Mathematik* 154 (1925): 219–240.

[28] Paul Bernays, "A system of axiomatic set theory – Part I–III," Journal of Symbolic Logic, Part I (1937): 2 (1): 65–77; Part II (1941): 6 (1): 1–17; Part III (1942): 7 (2): 49–104.

[29] Kurt Gödel, *Consistency of the Axion of Choice and of the Generalized Continuum Hypothesis with the Axioms of Set Theory* (Princeton University Press, 1940).

whole new and extraordinarily stimulating body of problems arose in set theory. Up to that point, mathematicians looked for assertions that were true in set theory, but now their interest transferred to assertions such that could not be proved from the axioms of set theory. Some such assertions, or possibly also their negations, could be added as further axioms of set theory. They were usually identified by creating models (interpretations) within axiomatic set theory of theories in which they did not hold. The question whether these assertions are true in set theory does not concern the mathematics of calculations. The answer should be provided by the mathematics of intuition.

All that has been achieved using the axiomatic approach in set theory could fill many volumes. Let us just recall, for the interest it holds, that it is possible to define a certain subset of the set of all natural numbers such that its existence is provable from the axiom of existence of an inaccessible cardinal number. However, it is not provable from the axioms of set theory without this additional axiom! In other words, the existence of this set of natural numbers excludes the existence of inaccessible cardinal numbers.

In his work *Consistency of the Axiom of Choice and of the Generalized Continuum Hypothesis with the Axioms of Set Theory*, Kurt Gödel defined constructible sets and proved that they form a model of axiomatic set theory in which the axiom of choice and continuum hypothesis hold.[30] Moreover, in this model the ordinal numbers are absolute.

[30] Cantor's continuum hypothesis is the assertion according to which the set of all subsets of natural numbers has the same cardinality as the set of all ordinal numbers smaller that the first uncountable initial ordinal number.

Title page of the book The Higher Infinite: Large Cardinals in Set Theory from Their Beginnings by Akihiro Kanamori (2001). Berlin – Heidelberg – New York: Springer-Verlag.

Chapter 3

Explication of the Problem

This chapter explicitly states what many mathematicians have subconsciously suspected since the 1960s, but feared to bring to light. To wit, the fact that the domain of all natural numbers is non-actualisable and consequently there exists no set of all natural numbers. This fact relegates amongst mere illusions the whole world of classical infinitary set-theoretical mathematics, which has been based precisely on the existence of the set of all natural numbers. This is because:

> Almost all infinitary mathematics of the twentieth century stands and falls with the existence of the set of all natural numbers.

All the same, many beautiful and ingenious things were accomplished in this illusory world of the twentieth century mathematics and its subversion brought about by refuting its cornerstone may be seen as a truly Barbarian act.

Hence the stakes are also for what from the infinitary mathematics of the twentieth century can still be salvaged.

3.1 Warnings

> There is no actual infinity. The Cantorians forgot this, and so fell into contradiction. It is true that Cantorism has been useful, but that was when it was applied to a real problem, whose terms were clearly defined, and then it was possible to advance without danger.

These words were written by Jules-Henri Poincaré (1854–1912) in the article "Les dernieres efforts des logisticiens" published in the book *Science et méthode.*[1] Should the above quoted words concern just the transfinite ordinal numbers or the domain of all sets etc., then it could be said that the mainstream development of set mathematics was directed by them, in the sense that mathematicians simply accepted the fact that the domain of all ordinal numbers is

[1] See Henri Poincaré, *Science et méthode*, here in the translation of Francis Maitland, *Science and Method* (London: T. Nelson & sons, 1914): 195.

not actualisable and limited their interest to those ordinal numbers that belong to some conveniently chosen cut On on the domain of all ordinal numbers. In accordance with this, the domain of all sets was limited to a set V which strictly speaking only contained regular sets whose type is an ordinal number belonging to the set On. Obviously On and V do not belong to V. The cut On can be expanded as needed, but never to such an extent that it would contain all ordinal numbers.

Restricting the subject of enquiry of set-theoretical mathematics in this way has made it possible to involve the predicate calculus to explore sets and ordinal numbers, and therefore also to axiomatise the thus-limited subject of enquiry of set theory. In the Gödel–Bernays axiomatic system, sets V, On (called classes therein) appear explicitly. The Zermelo–Fraenkel axiomatic system does not mention them explicitly but implicitly presupposes their existence. Otherwise the existential quantifier used in definitions of relationships and properties and also in propositions would have no clearly defined scope.

However, Poincaré also had in mind natural numbers. He recognized only potential infinity, and did so without any exception. In his article "Réflexions sur les deux notes précédentes" he wrote

> There is no actual infinity; what we call the infinite is only the possibility of constantly adding new objects, however numerous are the objects already created.[2]

3.2 Two Further Emphatic Warnings

At the time when mathematicians had to cope with the non-existence of the set of all ordinal numbers, the question whether something similar does not occur also in the case of natural numbers evaded attention. Something similar occurring means that when we assume the existence of a set of all natural numbers then we find that other natural numbers can be added to it. Naturally, such an extension of natural numbers needs to be carried out in a different manner than that used by Cantor when he added ordinal numbers to natural numbers. At the start of 1930s two eminent mathematicians published results that should have been understood by attentive readers already at the time as very strong warnings against the existence of the set of all natural numbers.

First warning. In 1930–1931, the journal *Monatshefte für Mathematik und Physik* published two crucially important mathematical-logical works:

"Die Vollständigkeit der Axiome des logischen Funktionenkalküls."[3]

---

[2] See Henri Poincaré, "Réflexions sur les deux notes précédentes", *Acta mathematica* 32 (1909): 195–200. "Il n'y a pas d'infini actuel; ce que nous appelons l'infini, c'est uniquement la possibilité de crées sans cesse de nouveaux objets, quelque nombreux que soient les objets déja créés."

[3] See Kurt Gödel, "Die Vollständigkeit der Axiome des logischen Funktionenkalküls," *Monatshefte für Mathematik und Physik* 37 (1930): 349–360.

"Über formal unentscheidbare Sätze der Principia Mathematica und verwandter Systeme I."[4]

The author of these works was Kurt Gödel (1906–1978). The latter of the two cited articles showed that even if we added to the Peano axiomatic system a finite number of other axioms (recorded in the corresponding predicate calculus, possibly supplemented with other primitive predicates representing some relations between natural numbers or properties of these numbers) and the thus-extended Peano's axiomatic system were consistent, it would be possible to find a proposition (which can be formulated in this language) which can be neither proved nor disproved from these axioms.

Cantor's set theory was able to cope with this unexpected and remarkable piece of knowledge very easily. Gödel simply proved that even an extended Peano's axiomatic system must fail to contain all the true propositions about natural numbers. If we want to axiomatize natural numbers (which was more of interest to mathematical logic than to Cantor's set theory), we place such an undecidable proposition or its negation among the axioms depending on which of these two cases is true for the numbers belonging to the set of all natural numbers. Cantor's set theory would only become interested in this if a proposition were to be found which was transparently true as regards its arithmetic or at least set-theoretical meaning, but which was also unprovable using the Peano axioms. However, this would not happen for some time.

Mathematicians thus paid little attention to the case of the Peano axiomatic system supplemented with a consistent axiom which however fails to be true for the set of all natural numbers. This was despite the fact that results presented in the first of Gödel's two works cited above shows that this axiomatic theory has a model. Let us note that it must necessarily be a non-standard model, that is, one in which the original natural numbers have been extended.

It was these models which held the first warning.

A second serious warning to those who believed in the existence of a set of all natural numbers and based their work on it, that is, to almost everybody who used Cantor's set theory, came from Bergen, Norway in 1933. It was contained in a short article "Über die Unmöglichkeit einer vollständingen Charakterisiernug der Zahlenreihe mittels eines endlichen Axiomensystems" which was written by **Thoralf Skolem** (1887–1963).[5]

Skolem's result is present in the very title of the article. The point is that no matter how the Peano axiomatic system is supplemented as discussed above, it is never a complete description of the sequence of all natural numbers. From a

[4] See Kurt Gödel, "Über formal unentscheidbare Sätze der Principia Mathematica und verwandter Systeme I," *Monatshefte für Mathematik und Physik* 38 (1931): 173–198.

[5] Thoralf Skolem, "Über die Unmöglichkeit einer vollständingen Charakterisiernug der Zahlenreihe mittels eines endlichen Axiomensystems," *Norks matematisk fornings skriften* 2 (10) (1933): 73–82. It was subsequently also published in *Fundamenta Mathematicae* in a somewhat modified and extended version in 1934: Thoralf Skolem, "Über die Nichtcharakterisierbarkeit der Zahlenreihe mittels endlich oder abzählbar unendlich vieler Aussagen mit ausschliesslich Zahlenvariablen," *Fundamenta Mathematicae* 23 (1934): 150–161.

narrowly formalistic perspective, this was not a revolutionary result. Formalists are not interested in what the language describes, but whether it adheres to the laws of logic. They are not interested in whether a proposition is true but whether it can be proved from certain axioms.

On the other hand, mathematicians who are not formalists and do not underestimate the role of intuition in mathematical research were (or should have been) especially attracted to the way in which Skolem obtained his results. For apparently it was in this work that a **non-standard model of natural numbers** has been created in set theory for the very first time.

Skolem extended the sequence of natural numbers with further objects (later referred to as non-standard natural numbers) and he also extended arithmetical operations and relations to them. He carried all this out in such a way that the resulting structure (later referred to as a non-standard model of natural numbers) satisfied all of Peano's axioms (and even those that might have been added).

The original (standard) natural numbers constitute a cut on all (standard and non-standard) natural numbers in this model, which naturally has no last element. Moreover if $\varphi(x)$ were a formula of the language of Peano arithmetic (even the supplemented one) such that for every x, $\varphi(x)$ if and only if x is a standard (that is, original) natural number, then according to the scheme of induction $\varphi(x)$ would hold for every standard and non-standard natural number.

Standard natural numbers can only be separated out in this model using the means which Skolem employed for its creation, for example with the help of set theory in which this model was constructed.

There are good grounds to believe that Thoralf Skolem knew, or at least suspected, that the set of all natural numbers does not exist for reasons similar to those which feature in the case of infinite ordinal numbers. It is suggested not only by his discovery of a nonstandard model of Peano arithmetic of natural numbers, but in a way by all his work.

Skolem's extension of the set of all natural numbers alone does not justify an announcement of the non-existence of this set. For that purpose it is necessary to extend not only the (assumed) set of all natural numbers but also every infinite set since the original (standard) natural numbers must not be in any way identifiable inside this extended world of set-theoretical mathematics.

Such an extension of the whole universe of sets can be created using the ultrapower. That however had not been discovered until shortly before Skolem's death, and it is essentially a generalization of the original Skolem's model.

3.3 Ultrapower

Having become acquainted with Skolem's construction, I managed to construct a non-standard model of the entire set theory (or more precisely of the Gödel–Bernays axiomatic theory within this theory) at the end of the year 1960. This construction later became known as the **ultrapower**. I sent a brief description of the model to the Muscovite academician P. S. Alexandrov, who recommended the content of this letter for publication on 12th October 1961. This work was then published in *Doklady Akademii Nauk SSSR*.[6]

At roughly the same time, the same model was also constructed in the USA on the other side of the impenetrable "iron curtain" with the only difference being that the Americans used **ultrafilters**, while I used **maximal ideals**. It was the use of ultrafilters that won out shortly afterwards.

The ultrapower does not have to be perceived only from a formal point of view. It is not just a model (an interpretation) of the axiomatic system of the set theory in which it was constructed. The ultrapower models Cantor's entire original set theory and at any moment it can be adapted to become a model of that part of Sensorium Dei[7] which has just been actualised. The only prerequisite for the ultrapower construction is the existence of the set of all natural numbers and of a non-trivial ultrafilter on this set. Moreover, the latter condition may be weakened in various ways; if general collapse[8] is used, meeting it becomes almost trivial.

The ultrapower is a faithful model of that part of Sensorium Dei in the actualisation of which it was constructed. In looser terms, this means that, if we find ourselves inside the ultrapower, then there is no way to check (without stepping out of it) whether we are inside or outside it.

The ultrapower (on the set of all natural numbers) contains non-standard natural numbers, similarly to Skolem's aforementioned model. The difference is that once inside the ultrapower it is impossible to define the original (standard) natural numbers by any means of Cantor's set theory whatsoever.

In the ultrapower, just as in Cantor's set theory, each non-empty set of natural numbers has a first element. The non-standard (and consequently also the standard) natural numbers therefore do not form a set in the ultrapower on the set of all natural numbers.

3.4 There Exists No Set of All Natural Numbers

Only a small step was needed to realise that for reasons similar to those applying in the case of ordinal numbers, the set of all natural numbers does not

[6] Petr Vopěnka, "Odin metod postrojenija nestandardnoj modeli aksiomatičeskoj teorii množestv Bernaysa Gödela (A method of constructing a non-standard model in Bernays-Gödel axiomatic set theory)," *Doklady Akademii Nauk SSSR* 143, no. 1 (1962): 11–12.

[7] In the early mathematics, the question whether or not a geometric object could be actualised was decided by the classical geometric space, namely so that the space either did or did not provide a place for this object. During the first stage of modern mathematical natural science, the whole real world was embedded in this space and **Isaac Newton** used the name *Sensorium Dei* for it. For several more or less obvious reasons we will now be using this term in the following new sense. Sensorium Dei is that which decides actualisability of finite and infinite sets by providing or not providing a place for these sets.

[8] **Postulate of general collapse** concerns creating a bijective mapping of a given infinite set onto the set of all natural numbers. The general collapse, which can first be glimpsed in Bolzano's *Paradoxien des Unendlichen* (1851) and more conspicuously in the work of Leopold Löwenheim, "Über Möglichkeiten im Relativkalkül," *Math. Ann.* 76 (1915), emerged backed by mathematical results of the 1960s as a valid principle for creating actually infinite sets.

exist either. All that remained was to acknowledge what the ultrapower directly showed. Ever since 1960, anybody familiar with the ultrapower could have formulated this message; no further non-trivial mathematical results were required.

The only problem with taking this step was that a lot of courage was required to face its consequences, because it results in devastation of the amazing abstract construct created in the twentieth century by the set-theoretical mathematics. And it could not have been disclosed by an innocent like in Andersen's famous fairy tale where a child simply exclaimed that "the emperor is naked!"

The question as to why mathematicians shut their eyes to the legacy of the ultrapower is a question for psychologists.

The existence of a set N of all natural numbers is the fundamental axiom of classical set theory in the sense that it enables us to build the entire edifice of Cantor's set theory on this set, step by step. As we have already stated, ultrapower (on the set N) contains non-standard natural numbers much like Skolem's model mentioned above. The difference is that inside the ultrapower the original natural numbers (that is, the numbers belonging to N) cannot be defined by any means of the entire Cantor's set theory, rather than just arithmetical means.

It is also possible to build the entire edifice of Cantor's set theory on set N^* (that is, on the set of the natural numbers in the ultrapower), step by step in the same way as the Cantor's set theory was built from the set N. This results in a second exemplar of Cantor's set theory which is indistinguishable from the one in which it was created. This means that set N^* also satisfies the fundamental axiom of classical set theory, but it is an extension of the set N.

Therefore, if we believe that a set N is the set of all natural numbers and we base this belief for example on the fact that the entire Cantor's set theory can be built on it, the ultrapower enables us to find easily a set N^* which has the same properties, but is also longer than N. This therefore means that N is not the set of all natural numbers.[9]

Any set N that we consider to be the set of all natural numbers is only a cut on the domain of all natural numbers, with no inner or outer head.

This implies that the set of all natural numbers whatsoever does not exist. In other words, **the domain of all natural numbers is not actualisable**. Precisely this is the main legacy of the ultrapower.

The non-actualisability of the domain of all natural numbers therefore does not amount to the fact that nobody is able to conceive all the actual (meaning metamathematical) natural numbers one by one, as was claimed by the opponents of the actual infinity, and which was probably also Poincaré's position. This non-actualisability is based on the fact that even if somebody was able to reach this far (or even farther), they would still not reach all natural numbers.

[9] At a symposium in Berkeley in 1963 I demonstrated an expansion of the whole universe that has more natural numbers than any external set.

Scholion for phenomenologists

The answer to the question of whether the domain of all natural numbers is actualisable can only be sought in the mysterious understanding we have for the passing of natural numbers towards infinity. For this purpose, we have to cleanse this understanding as far as possible from all additional interpretations which obscure this passing of natural numbers towards infinity.

If we carry out such a purification, or the phenomenological $\epsilon\pi o\chi\acute{\eta}$ as Edmund Husserl would call it, we have no option but to state that the purified understanding of the passing of natural numbers towards infinity does not give us the slightest cause to shorten the domain of all natural numbers. On the contrary, the raw infinity present in this domain **a priori** excludes all the obstacles which could hinder natural numbers as they flee towards infinity.

3.5 Unfortunate Consequences for All Infinitary Mathematics Based on Cantor's Set Theory

A mathematician who works with a set N believing it to be the set of all natural numbers, is wrong, because the domain $\mathcal{N}$ of all natural numbers is not actualisable. This supposed set N is a mere collection which constitutes a convenient headless cut on the domain $\mathcal{N}$ and in no way exhausts the whole domain $\mathcal{N}$. Since no non-empty headless cut on the domain $\mathcal{N}$ is sharply defined, the collection N cannot be interpreted as a set. Hence the supposed set N is not even a set; it is merely an indistinctly defined collection interpreted as an autonomous object.

Choosing some convenient cut to stand for the supposed set of all natural numbers however means that all concepts and results of the 20th century infinitary mathematics based on Cantor set theory are relativised, they lose their certainty and clear comprehensibility and sometimes they even cease to make sense.

Even if we accepted a suitable cut as a representative of the supposed set of all natural numbers, we would find it much harder to accept representatives of the supposed subsets of this supposed set, let alone a representative of the supposed set of all the supposed subsets of this supposed set etc.

Predicate calculus (unlike propositional calculus) loses its firm ground in infinitary set mathematics. The general (let alone the existential) quantifier used for natural numbers, or for elements of any supposedly infinite set, has an indefinite scope and therefore the results obtained are not entirely convincing.

Topology, mathematical analysis, functional analysis, topological-algebraic structures, ..., all this breaks down and needs to be revised or replaced.

To put it briefly, mathematical enquiry must start again from the position where it finished at the beginning of the 20th century, as in fact Henri Poincaré clairvoyantly recommended.

It could appear that saving the set of all natural numbers is a simple matter. There has always been a conviction within the collective unconscious as regards the "sanctity" of the set of all natural numbers. **Leopold Kronecker** (1823–1891) pointed it out when he stated:

> The integer numbers were made by our beloved God, all the rest is the work of man.[10]

Kronecker said this when Cantor declared the ordinal and cardinal numbers to be a transcendental reality deposited in God's absolute mind.

In other words: God created the set of natural numbers. Even if no man, not even an Olympian god living in some expansion of the world of actually infinite sets, has the capability to identify the standard, that is the actual natural numbers created by God, God does have this capability. For He knows what numbers he created and what numbers are added by men.

However, in the twenty first century this argument for the existence of the set of all natural numbers is untenable.

[10] "Die ganzen Zahlen hat der liebe Gott gemacht, alles andere ist Menschenwerk." Quoted in H. Weber, "Leopold Kronecker," *Jahresbericht der DMV* 2 (1893): 19.

Chapter 4

Summit and Fall

4.1 Ultrafilters

In a lecture at the fourth International Congress of Mathematics in Rome in 1908, **Frigyes Riesz** (1880–1956) introduced the concept of an ultrafilter. How very useful this notion was had gone unnoticed for almost thirty years until it was pointed out by **Henri Cartan** (1904–2008). Since then, the notion of an ultrafilter has become one of the most remarkable and most fruitful in set theory.

We say that a set of sets[1] $\mathcal{D}$ is a **filter** on a set A, if the following holds:

(a) $\emptyset \neq \mathcal{D} \subseteq \mathcal{P}(A)$; $\emptyset \notin \mathcal{D}$.

(b) If $X, Y \in \mathcal{D}$ then $X \cap Y \in \mathcal{D}$.

(c) If $X \in \mathcal{D}$ and $X \subseteq Y \subseteq A$ then $Y \in \mathcal{D}$.

We say that a filter $\mathcal{D}$ is an **ultrafilter** (on the set A) if, moreover,

(d) If $X \cup Y \in \mathcal{D}$ then $X \in \mathcal{D}$ or $Y \in \mathcal{D}$.

If A is a non-empty set, $a \in A$ then the set of all $X \subseteq A$ such that $a \in X$ clearly is an ultrafilter on the set A. We say that such an ultrafilter is **principal**.

If A is a non-empty finite set then every ultrafilter on A is principal.
If A is an infinite set then clearly the set of all $X \subseteq A$ such that the set $A \setminus X$ has cardinality smaller than the set A is a filter on the set A. We shall denote this filter $\mathcal{D}_A$.

We say that the filter $\mathcal{D}$ on an infinite set A is **uniform**, if $\mathcal{D}_A \subseteq \mathcal{D}$.

A uniform ultrafilter $\mathcal{D}$ on an infinite set A is obviously non-principal.

[1] Vopěnka introduces a special word (soubor) for "set of sets", see page 69 for the definition of this notion. Recall that sets of sets are denoted by calligraphic capital letters. In the English translation, no new word was introduced, and if clear from the context, in this chapter"set of sets" is sometimes translated as "set". [Ed]

We say that a set of sets $\mathcal{M}$ of sets has the **finite intersection property** if $\mathcal{M} \neq \emptyset$ and if for every finite nonempty $\mathcal{N} \subseteq \mathcal{M}$ we have $\bigcap \mathcal{N} \neq \emptyset$.

Every filter clearly has the finite intersection property.

Proposition 4.1. Let a set $\mathcal{M}$ of subsets of a set A have the finite intersection property. Let $\mathcal{D}$ denote the set of all $X \subseteq A$ for which there exists a nonempty finite set $\mathcal{N} \subseteq \mathcal{M}$ such that $\bigcap \mathcal{N} \subseteq X$. Then $\mathcal{D}$ is a filter on the set A and $\mathcal{M} \subseteq \mathcal{D}$.

Proof. It is obvious that $\mathcal{D}$ satisfies the conditions (a) and (c) from the definition of a filter. The inclusion $\mathcal{M} \subseteq \mathcal{D}$ is also obvious. We shall show that $\mathcal{D}$ satisfies also the condition (b). Let $X, Y \in \mathcal{D}$ and let $\mathcal{N}_1, \mathcal{N}_2$ be nonempty finite subsets of $\mathcal{M}$ such that $\bigcap \mathcal{N}_1 \subseteq X$, $\bigcap \mathcal{N}_2 \subseteq Y$. Then $\bigcap(\mathcal{N}_1 \cup \mathcal{N}_2) \subseteq X \cap Y$ and hence $X \cap Y \in \mathcal{D}$.

□

Proposition 4.2. Let A be a non-empty set and let $\mathcal{D}$ be a filter on the set A. Then the following conditions on $\mathcal{D}$ are equivalent:

(a) $\mathcal{D}$ is an ultrafilter.

(b) $\mathcal{D}$ is maximal in the ordering of all filters on the set A by the relation of inclusion.

Proof. (a) $\Rightarrow$ (b). For contradiction, assume that $\mathcal{C}$ is a filter on the set A such that $\mathcal{D} \subseteq \mathcal{C}$, $\mathcal{D} \neq \mathcal{C}$. Let $X \in \mathcal{C}$ be a set such that $X \in \mathcal{C}$, $X \notin \mathcal{D}$. Since $X \cup (A \setminus X) = A \in \mathcal{D}$, we have $A \setminus X \in \mathcal{D}$ and hence also $A \setminus X \in \mathcal{C}$. Consequently, $\emptyset = X \cap (A \setminus X) \in \mathcal{C}$, contradiction.
(b) $\Rightarrow$ (a). Again, for contradiction, let $X, Y \subseteq A$ be such that $X \cup Y \in \mathcal{D}$, $X \notin \mathcal{D}$, $Y \notin \mathcal{D}$. Then there are no sets $Z_1, Z_2 \in \mathcal{D}$ such that $Z_1 \cap X = \emptyset$, $Z_2 \cap Y = \emptyset$, since if that was the case then $(Z_1 \cap Z_2) \cap (X \cup Y) = \emptyset$ and also $Z_1 \cap Z_2 \in \mathcal{D}$, $X \cup Y \in \mathcal{D}$. It means that either $\mathcal{D} \cup \{X\}$ or $\mathcal{D} \cup \{Y\}$ has the finite intersection property and by Proposition 4.1 there exists a filter $\mathcal{C}$ on the set A such that either $\mathcal{D} \cup \{X\} \subseteq \mathcal{C}$ or $\mathcal{D} \cup \{Y\} \subseteq \mathcal{C}$. Consequently the filter $\mathcal{D}$ is not maximal, contradiction. □

Proposition 4.3. Let A be a non-empty set and let $\mathcal{D}$ be a filter on the set A. Then there exists an ultrafilter $\mathcal{C}$ on the set A such that $\mathcal{D} \subseteq \mathcal{C}$.

Proof. Let $\mathcal{F} \neq \emptyset$ be a set of filters on the set A which is linearly ordered by the relation $\subseteq$. We can easily check that $\bigcup \mathcal{F}$ is a filter on the set A and since every filter $\mathcal{A} \in \mathcal{F}$ satisfies $\mathcal{A} \subseteq \bigcup \mathcal{F}$, the ordering $\subseteq$ of the set of all filters on the set A has Zorn's property. Hence there exists a filter $\mathcal{C}$ maximal in this ordering and satisfying $\mathcal{D} \subseteq \mathcal{C}$. By the assertion 4.2, $\mathcal{C}$ is an ultrafilter. □

Finally, we give an important example of an ultrafilter. Let R be an infinite set and let A be the set of all finite subsets of the set R. For $x \in A$, we denote by $|x|$ the set of all $y \in A$ such that $x \subseteq y$. The set $\mathcal{M}$ of all $|x|$ where $x \in A$ has the finite intersection property since for each $x, y \in A$ we have $|x \cup y| \subseteq |x| \cap |y|$.

Let $\mathcal{B}$ be an ultrafilter on the set A, $\mathcal{M} \subseteq \mathcal{B}$ guaranteed by Propositions 4.1 and 4.3. The ultrafilter $\mathcal{B}$ is uniform on the set A: if $X \subseteq A$ is a set such that the set $A \setminus X$ has smaller cardinality than the set A, then there is $a \in R$ such that for each $u \in A \setminus X$ we have $a \notin u$. Hence $(A \setminus X) \cap |\{a\}| = \emptyset$ and consequently $X \in \mathcal{B}$.

4.2 Basic Language of Set Theory

The basic language of set theory uses the following metamathematical notation:

(a) Variables $x_1, x_2, \ldots$ and constants $c_1, c_2, \ldots$ for sets. The subscripts denote metamathematical natural numbers.

(b) Signs $=, \in$ for equality and the relation of being an element of sets.

(c) Logical connectives: $\wedge$ for conjunction, $\neg$ for negation.

(d) Existential quantifier $\exists$.

(e) Signs φ, ψ for expressions composed of the above signs.

(f) Auxiliary signs: brackets, commas, ...

By **set formulae** we mean expressions that can be obtained as follows:

(a) $x_i \in x_j$, $x_i \in c_j$, $c_i \in x_j$, $c_i \in c_j$, $x_i = x_j$, $x_i = c_j$, $c_i = x_j$, $c_i = c_j$ are set formulae (where $x_1, x_2, \ldots$ are variables, $c_1, c_2, \ldots$ are constants for sets).

(b) If φ, ψ are set formulae then also $\varphi \wedge \psi$, $\neg\varphi$ are set formulae.

When saying that $\varphi(x_1, \ldots, x_n)$ is a set formula, we mean that φ is a set formula in which none of the variables $x_1, \ldots, x_n$ is quantified (that is, it does not appear behind the existential quantifier $\exists$) and any variable that occurs non-quantified in φ is listed amongst the variables $x_1, \ldots, x_n$.

(c) If $\varphi(x_1, \ldots, x_{n+1})$ is a set formula then also $(\exists x_{n+1})\varphi(x_1, \ldots, x_{n+1})$ is a set formula.

4.3 Ultrapower Over a Covering Structure

Let A denote some infinite set. Recall that the set C is a **disjoint covering** of the set A if $A = \bigcup C$ and $(\forall u, v \in C)(u \neq v \Rightarrow u \cap v = \emptyset)$.

If C, D are disjoint coverings of the set A then we define:

(a) $C \leq D \stackrel{\text{df}}{\equiv} (\forall u \in C)(\exists v \in D)(u \subseteq v)$.

(b) $C \wedge D \overset{\text{df}}{\equiv} \{u \cap v; u \in C \wedge v \in D\}$.

Clearly $C \wedge D$ is a disjoint covering of the set A and $C \wedge D \leq C, D$.

We say that a set $\mathcal{C}$ is a **covering structure on the set** A if the following conditions hold:

(a) $\mathcal{C} \neq \emptyset$ and every element of the set $\mathcal{C}$ is a disjoint covering of the set A.

(b) If $C, D \in \mathcal{C}$ then also $C \wedge D \in \mathcal{C}$.

(c) If $C \in \mathcal{C}$ and D is a disjoint covering of the set A such that $C \leq D$ then also $D \in \mathcal{C}$.

By the **complete covering structure on a set** A we understand the set of all disjoint coverings of the set A.

Let $\mathcal{C}$ be a covering structure on a set A. We say that an **ultrafilter** $\mathcal{D}$ on the set A is **non-principal on the structure** $\mathcal{C}$ if there exists $C \in \mathcal{C}$ such that $C \cap \mathcal{D} = \emptyset$ (in other words, no set from the covering C is an element of the ultrafilter $\mathcal{D}$).

In what follows, $\mathcal{C}$ denotes some given covering structure on the set A and $\mathcal{D}$ denotes some given ultrafilter on the set A, which is non-principal on the covering structure $\mathcal{C}$.

If f is a function such that $\text{dom}(f) = A$ then we define

$$C(f(x)) = \{y; y \in \text{dom}(f) \wedge f(y) = f(x)\}.$$

$C(f) = \{C(f(x)); x \in A\}$ is clearly a disjoint covering of the set A and if $u \in C(f)$ then the function f is constant on the set u.

$\mathcal{G}(\mathcal{C})$ denotes the domain of all functions f such that $\text{dom}(f) = A$ and $C(f) \in \mathcal{C}$.

When there is no danger of confusion, we will denote $\mathcal{G}(\mathcal{C})$ simply just by the letter $\mathcal{G}$. In what follows we will use the letter z exclusively as a variable for the elements of the set A.

Let k_x denote such a function belonging to $\mathcal{G}$ for which $k_x(z) = x$ for all z. Clearly for each x, $C(k_x)$ contains just A.

For $f, g \in \mathcal{G}$ we define:

(a) $f \overset{*}{=} g \overset{\text{df}}{\equiv} \{z; f(z) = g(z)\} \in \mathcal{D}$.

(b) $f \overset{*}{\in} g \overset{\text{df}}{\equiv} \{z; f(z) \in g(z)\} \in \mathcal{D}$.

We can easily check that $\stackrel{*}{=}$ is an equivalence on the collection $\mathcal{G}$; $k_x \stackrel{*}{=} k_y$ just when $x = y$.

The relation $\stackrel{*}{\in}$ on the collection $\mathcal{G}$ is called the **ultrapower** of the relation $\in$ **over the covering structure** $\mathcal{C}$ (or simply the ultrapower of the relation $\in$ when $\mathcal{C}$ is complete on the set A).

Letters $g_1, g_2, \ldots$ will be used as variables for functions belonging to $\mathcal{G}$.

If φ is a set formula then φ^* denotes the formula which obtains from φ when we replace each constant c occurring in φ by the constant k_c and each variable x_i by the variable g_i, and when we add stars to the signs $\in$, $=$.

Proposition Schema. Let $\varphi(x_1, \ldots, x_n)$ be a set formula. Then

$$(\forall g_1, \ldots, g_n)(\varphi^*(g_1, \ldots, g_n) \Leftrightarrow \{z; \varphi(g_1(z), \ldots, g_n(z))\} \in \mathcal{D}).$$

Proof. For $\varphi(x_1, \ldots, x_n)$ of the form $x_i \in x_j$, $x_i \in c_j$, $c_i \in x_j$, $c_i \in c_j$ it follows directly from the definition of the relation $\stackrel{*}{\in}$. The cases $x_i = x_j$, ..., are also obvious.

Let φ be of the form $\psi_1 \wedge \psi_2$ and assume that for each $g_1, \ldots, g_n$ we have

$$\psi_i^*(g_1, \ldots, g_n) \Leftrightarrow \{z; \psi_i(g_1(z), \ldots, g_n(z))\} \in \mathcal{D}; \;\; i = 1, 2.$$

Then for each $g_1, \ldots, g_n$ also

$$\psi_1^*(g_1, \ldots, g_n) \wedge \psi_2^*(g_1, \ldots, g_n) \Leftrightarrow$$

$$\Leftrightarrow \{z; \psi_1(g_1(z), \ldots, g_n(z))\} \in \mathcal{D} \wedge \{z; \psi_2(g_1(z), \ldots, g_n(z))\} \in \mathcal{D} \Leftrightarrow$$

$$\Leftrightarrow \{z; \psi_1(g_1(z), \ldots, g_n(z)) \wedge \psi_2(g_1(z), \ldots, g_n(z))\} \in \mathcal{D}.$$

Let φ be of the form $\neg\psi$ and assume that for any $g_1, \ldots, g_n$ we have

$$\psi^*(g_1, \ldots, g_n) \Leftrightarrow \{z; \psi(g_1(z), \ldots, g_n(z))\} \in \mathcal{D}.$$

Then

$$\neg\psi^*(g_1, \ldots, g_n) \Leftrightarrow \{z; \psi(g_1(z), \ldots, g_n(z))\} \notin \mathcal{D} \Leftrightarrow \{z; \neg\psi(g_1(z), \ldots, g_n(z))\} \in \mathcal{D}.$$

Let $\varphi(x_1, \ldots, x_n)$ be of the form $(\exists x_{n+1})\psi(x_1, \ldots, x_{n+1})$ and assume that for each $g_1, \ldots, g_{n+1}$ we have

$$\psi^*(g_1, \ldots, g_{n+1}) \Leftrightarrow \{z; \psi(g_1(z), \ldots, g_{n+1}(z))\} \in \mathcal{D}$$

.

Fix $g_1, \ldots, g_n$. First assume that $\varphi^*(g_1, \ldots, g_n)$. Then there is g_{n+1} such that $\psi^*(g_1, \ldots, g_{n+1})$, and hence $B' = \{z; \psi(g_1(z), \ldots, g_{n+1}(z))\} \in \mathcal{D}$. Moreover $B' \subseteq \{z; (\exists x)\psi(g_1(z), \ldots, g_n(z), x)\} = \{z; \varphi(g_1(z), \ldots, g_n(z))\} \in \mathcal{D}$.

Now assume that $B = \{z; \varphi(g_1(z), \ldots, g_n(z))\} \in \mathcal{D}$. Let $C = C(g_1) \wedge \ldots \wedge C(g_n)$. If $z, z' \in u \in C$ then $g_i(z) = g_i(z')$ for every $i = 1, 2, \ldots, n$ and hence

$$\varphi(g_1(z), \ldots, g_n(z)) \Leftrightarrow \varphi(g_1(z'), \ldots, g_n(z')).$$

Consequently, either $u \subseteq B$ or $u \cap B = \emptyset$. By the Axiom of Choice there is a set $D \subseteq A$ such that for each $\emptyset \neq u \in C$, the set $D \cap u$ is a singleton. We define the function $f \in \mathcal{G}$ as follows. If $z \in u \in C$, where $u \cap B = \emptyset$, then we set $f(z) = \emptyset$. Let $\emptyset \neq u \in C$, $u \subseteq B$, $z' \in u \cap D$. We choose one set – denoted $x(u)$ – such that $\psi(g_1(z'), \ldots, g_n(z'), x(u))$. For $z \in u$ we then set $f(z) = x(u)$. For $z \in u$ we thus have $\psi(g_1(z), \ldots, g_n(z), f(z))$ and hence

$$B = \{z; \psi(g_1(z), \ldots, g_n(z), f(z))\}.$$

Let $g_{n+1} = f$. Then $\{z; \psi(g_1(z), \ldots, g_{n+1}(z))\} \in \mathcal{D}$, and hence by the inductive assumption $\psi^*(g_1, \ldots, g_{n+1})$; consequently also $\varphi^*(g_1, \ldots, g_n)$. □

Since the relation $\overset{*}{\in}$ on the collection $\mathcal{G}$ satisfies all the Zermelo Fraenkel axioms, the ultrapower is a model of this axiomatic theory. However, more than that. In the ultrapower, every assertion holds that can be expressed by a set formula and that holds in the set theory in which the ultrapower was formed. In other words, the ultrapower is a faithful model of the set theory. Since also the property of "being a natural number" (meaning von-Neuman natural number) can be expressed by a set formula, a function f belonging to the collection $\mathcal{G}$ is a natural number of the ultrapower just when the set of all $x \in A$ for which $f(x)$ is a natural number belongs to the ultrafilter $\mathcal{D}$.

4.4 Ultraextension of the Domain of All Sets

$\mathcal{D}$ denotes some fixed non-principal ultrafilter on the set N of all natural numbers, $\mathcal{V}$ denotes the domain of all sets and $\mathcal{G}$ denotes the domain of all functions defined on the set N.

By an **ultraextension operator** we understand an operator ∇ that creates a set $\nabla(f)$ from each function f belonging to the domain $\mathcal{G}$, so that for each f, g belonging to the domain $\mathcal{G}$ the following holds:
$\nabla(f) = \nabla(g)$ just when $f \overset{*}{=} g$, that is $\{n; f(n) = g(n)\} \in \mathcal{D}$
$\nabla(f) \in \nabla(g)$ just when$f \overset{*}{\in} g$, that is $\{n; f(n) \in g(n)\} \in \mathcal{D}$

By **ultraextension** of the domain $\mathcal{V}$ we understand the domain $\mathcal{W}$ of all sets $\nabla(f)$ where f belongs to the domain $\mathcal{G}$.

If y is a set belonging to the domain $\mathcal{V}$ then, to simplify our expressions, we write $\bar{y}$ for the set $\nabla(k_y)$. For the same reasons we understand natural (or ordinal) numbers to be von-Neumann natural (or ordinal) numbers. Thus $\mathrm{N} = \omega_0$.

From the Proposition Schema on page 115 we have the following:

Proposition 4.4. Let $\varphi(x_1, \ldots, x_n)$ be a set formula and let $y_1, \ldots, y_n$ be sets belonging to the domain $\mathcal{V}$. Then $\varphi(y_1, \ldots, y_n)$ holds in the domain $\mathcal{V}$ just when $\varphi(\bar{y}_1, \ldots, \bar{y}_n)$ holds in the domain $\mathcal{W}$.

Hence if $\varphi(x_1)$ is a set formula then the set y belonging to the domain $\mathcal{V}$ has the property φ in the domain $\mathcal{V}$ just when the set $\bar{y}$ has the property φ in the domain $\mathcal{W}$.

Proposition 4.5. Let y be a finite set belonging to the domain $\mathcal{V}$, $y = \{y_1, \ldots, y_i\}$. Then $\bar{y} = \{\bar{y}_1, \ldots, \bar{y}_i\}$.

Proof. Let $1 \leq j \leq i$. Then by the definition of the ultraextension operator we have $\bar{y}_j \in \bar{y}$ since $y_j \in y$. Conversely, let z be a set belonging to the domain $\mathcal{W}$ such that $z \in \bar{y}$. Let $z = \nabla(f)$, where $f \in \mathcal{G}$. Since $\nabla(f) \in \nabla(k_y)$, we have $\{n; f(n) \in y\} \in \mathcal{D}$. Let $v_j = \{n; f(n) = y_j\}$. Since $v_1 \cup \cdots \cup v_i \in \mathcal{D}$, there is j such that $v_j \in \mathcal{D}$ and hence $\{n; f(n) = y_j\} \in \mathcal{D}$. Consequently, $z = \nabla(f) = \nabla(k_{y_j}) = \bar{y}_j$. □

Proposition 4.6. Let y be a finite set belonging to the domain $\mathcal{V}$.

(a) The set $\bar{y}$ is also finite and it has the same number of elements as the set y.

(b) If $\bar{z} = z$ for each $z \in y$ then $\bar{y} = y$.

(c) $\bar{\emptyset} = \emptyset$.

(d) For every natural number n belonging to the domain $\mathcal{V}$ we have $\bar{n} = n$.

Proof. The first three assertions follow from the concept of ultraextension and from Proposition 4.4. The last assertion (d) is proved by induction: $\bar{0} = 0$ by (c); let $\bar{n} = n$, then $\overline{n+1} = \overline{n \cup \{n\}} = \bar{n} \cup \overline{\{n\}} = n \cup \{\bar{n}\} = n \cup \{n\} = n + 1$. □

Proposition 4.7. Let y be an infinite set belonging to the domain $\mathcal{V}$. Then in the domain $\mathcal{W}$ there exists $g \in \bar{y}$ such that for each $z \in y$ we have $g \neq \bar{z}$.

Proof. Let $y_1, y_2, \ldots$ be an infinite sequence in the domain $\mathcal{V}$ of distinct elements of the set y. We define a function f belonging to $\mathcal{G}$ so that for $n \in \mathrm{N}$, $f(n) = y_n$. Let $g = \nabla(f)$. Since $\bar{y} = \nabla(k_y)$ and $\{n; f(n) \in y\} = \mathrm{N} \in \mathcal{D}$, also $g = \nabla(f) \in \nabla(k_y) = \bar{y}$. Let $z \in y$ and assume first that for some i we have $z = y_i$. Then $\bar{z} = \nabla(k_{y_i}) \neq \nabla(f)$ since $\{n; y_i = f(n)\} = \{i\} \notin \mathcal{D}$. Assume that for each i we have $z \neq y_i$. Then $\bar{z} = \nabla(k_z) \neq \nabla(f) = g$ since $\{n; z = f(n)\} = \emptyset \notin \mathcal{D}$. □

Proposition 4.8. The set $\bar{\omega}_0$ of all natural numbers in the domain of sets $\mathcal{W}$ is an extension of the set ω_0 of all natural numbers in the domain of sets $\mathcal{V}$.

Proof. By Proposition 4.6(d) for any natural number $n \in \omega_0$ also $n \in \bar{\omega}_0$ and if $m \in n$ in the domain $\mathcal{W}$ then also $m \in n$ in the domain $\mathcal{V}$. The set ω_0 is thus a cut on the set $\bar{\omega}_0$ and by Proposition 4.7, $\omega_0 \neq \bar{\omega}_0$. □

By induction in the domain of sets $\mathcal{V}$ we define regular sets of rank n, where $n \in \omega_0$, as follows.
A regular set of rank zero is the empty set.
A regular set of rank at most $n+1$ is a set such that each element of it is a regular set of rank at most n.

Proposition 4.9.

(a) Every regular set of rank at most n is finite.

(b) If x is a regular set of rank at most n then $\bar{x} = x$

Proof. (b) follows from Proposition 4.6 (b), (c). □

The domain of sets $\mathcal{W}$ thus arose by an expansion of infinite sets belonging to the domain of sets $\mathcal{V}$. Each infinite set belonging to the domain $\mathcal{V}$ gained many previously non-existent elements, sets that have newly appeared. That is why no infinite set belonging to the domain $\mathcal{V}$ appears in the domain $\mathcal{W}$.

Inside the domain $\mathcal{V}$, it is – thanks to the ultrapower – merely possible to describe the structure of sets belonging to the domain $\mathcal{W}$, not to create this domain of sets.

In the domain of sets $\mathcal{W}$, $\bar{\omega}_0$ is the set of all natural numbers. Natural numbers belonging to the domain of sets $\mathcal{V}$ are also natural numbers belonging to the domain $\mathcal{W}$ (cf. Proposition 4.8), but they do not exhaust natural numbers from $\mathcal{W}$ (cf. Proposition 4.7). After expansion into the set $\bar{\omega}_0$, the set ω_0 disappeared; it ceased to exist.

In the domain $\mathcal{W}$, the set ω_0 is not even identifiable. The only thing that can be said is that ω_0 should be some proper cut on the set $\bar{\omega}_0$, a cut which however no longer exists.

This knowledge of the expired existence of a non-existent set induces a suspicion that even the domain of sets $\mathcal{V}$ is an ultraextension of some non-existent domain of sets $\mathcal{V}'$, similarly as the domain of sets $\mathcal{W}$ is an ultraextension of the domain $\mathcal{V}$. This suspicion is supported by our knowledge of not knowing which of the domains $\mathcal{V}$, $\mathcal{W}$ we mean when we talk about set theory.

4.5 Ultraextension Operator

The creating of the domain $\mathcal{W}$ of new sets by the ultraextension operator ∇ is a turn of thought which breaks the hitherto prevailing customs of mathematics stated below.

(a) Newly created infinite sets are created later than all their elements, not simultaneously with them.

(b) Newly created infinite sets do not cancel the existence of sets from which they were created.

To (a) we add that violating this rule brings a danger of contradiction, as transpired on the example of the set of all sets almost immediately after the birth of Cantor's set theory. All the same, we do violate it whenever we talk about some infinite set, for example about the set of all natural numbers. For we subconsciously assume that there is somebody who can create first all natural numbers and then their set. In our case the danger of contradiction is warded off through the ultrapower. Thanks to the ultrapower any contradiction in the domain $\mathcal{W}$ would be transferred into a contradiction in the domain $\mathcal{V}$.

As regards (b), the incompatibility of concurrent existence of the domains of sets $\mathcal{V}$, $\mathcal{W}$ is not a logical contradiction, but a new principle that significantly enhances Cantor's set theory. To defend against objections regarding a justification of this principle, a simile can be used, of a newly sowed wheat field (domain $\mathcal{V}$) and the field of young wheat blades simultaneously grown from the seeds (domain $\mathcal{W}$).

4.6 Widening the Scope of Ultraextension Operator

Proposition 4.10. The set of all ultrafilters on an infinite set A has the same cardinality as the set $\mathcal{P}(\mathcal{P}(A))$.[2]

Proposition 4.11. An ultraextension operator ∇ is determined by a fixed non-principal ultrafilter $\mathcal{D}$ on the set N of all natural numbers.

There are many such ultrafilters and hence many corresponding mutually incompatible ultraextensions of the domain $\mathcal{V}$. In order to distinguish them, we shall denote the ultraxtension of the domain $\mathcal{V}$ determined by the ultrafilter $\mathcal{D}$ by $\mathcal{W}(\mathcal{D})$.

Let us replace the set N by some infinite set A and assume that $\mathcal{D}$ is a non-principal ultrafilter on the set A and that $\mathcal{G}$ is the domain of all functions defined on the set A. If we copy exactly the definition $\nabla(f)$ (page 116) we obtain a definition of ultraextension of the domain of sets $\mathcal{V}$ which we will also denote $\mathcal{W}(\mathcal{D})$. Further we copy all up to Proposition 4.7.

We say that an ultrafilter $\mathcal{D}$ is $\aleph_1$-complete if it is non-principal and if for each countable disjoint covering C of the set A there exists $u \in C$ such that $u \in \mathcal{D}$.

Proposition 4.12. Assume that the ultrafilter $\mathcal{D}$ is not $\aleph_1$-complete. Then if y is an infinite set belonging to the domain $\mathcal{V}$ then in the domain of sets $\mathcal{W}(\mathcal{D})$ there exists $g \in \bar{y}$ such that for each $z \in y$ we have $g \neq \bar{z}$.

Proof. Let $\{C_1, C_2, \dots\}$ be a covering of the set A such that $C_n \in \mathcal{D}$ holds for no n. Let $y_1, y_2, \dots$ be an infinite sequence of distinct elements of the set y in the domain $\mathcal{V}$. We define a function f belonging to $\mathcal{G}$ so that $f(x) = y_n$ for $x \in C_n$. Let $g = \nabla(f)$. Since $\bar{y} = \nabla(k_y)$ and $\{x; f(x) \in y\} = A \in \mathcal{D}$, we have

[2] Bedřich Pospíšil, "Remark on bicompact spaces," *Annals of Mathematics* 38 (1937): 845–846.

$g = \nabla(f) \in \nabla(k_y) = \bar{y}$. Let $z \in y$ and assume that for some i we have $z = y_i$. Then $\bar{z} = \nabla(k_{y_i}) \neq \nabla(f)$ since $\{x; y_i = f(x)\} = C_i \notin \mathcal{D}$. Assume that for each i we have $z \neq y_i$. Then $\bar{z} = \nabla(k_z) \neq \nabla(f) = g$ since $\{x; z = f(x)\} = \emptyset \notin \mathcal{D}$. □

When $\mathcal{D}$ is not $\aleph_1$-complete then clearly also Proposition 4.8 holds (that is, when we write everywhere in its formulation and in its proof $\mathcal{W}(\mathcal{D})$ in place of $\mathcal{W}$).

Proposition 4.13. When $\mathcal{D}$ is $\aleph_1$-complete then the domain of sets $\mathcal{W}(\mathcal{D})$ is a subdomain of the domain of all sets $\mathcal{V}$ and $\bar{\omega}_0 = \omega_0$.

This case was studied in detail in the 1960s and hence we will not discuss it. For our purposes is is not interesting. Unless stated otherwise, we will always assume that the ultrafilter $\mathcal{D}$ is not $\aleph_1$-complete.

We can widen the scope of the ultraextension operator in an obvious way also by replacing "ultrapower" in the definition of the operator ∇ (page 116) by "ultrapower over a non-trivial covering structure $\mathcal{C}$." In such case we denote the ultraextension of the domain of all sets $\mathcal{V}$ by $\mathcal{W}(\mathcal{D}, \mathcal{C})$.

4.7 Non-existence of the Set of All Natural Numbers

It is clear from our considerations above that Bolzano's proof of the existence of an infinite set of all truths in themselves, and of the set of all natural numbers derived from it, cannot be used. If we should still appeal to the God of medieval and baroque rational theology then it would be appropriate to concede to him a power of perceiving in some special way all simultaneously non-existent domains of all natural numbers in all possible domains of all sets.

The notion of the set of all natural numbers is relative. If we talk about this set, we should say which domain of all sets we have in mind. However, this cannot be done since the notion of the domain of all sets is also relative.

It might appear that such a fundamental domain of all sets is the domain $\mathcal{V}$. However, that is fundamental only in the sense of us having declared it to be fundamental. We could just as well denote any other domain of all sets by the letter $\mathcal{V}$.

If there exists one set of all natural numbers then there exist many different sets of all natural numbers. Consequently, the set of all natural numbers does not exists, for the same reasons as the set of all ordinal numbers.

Note. The expected objection that the non-existence of the set of all natural numbers was proved using the second and third postulate of Cantor's set theory (Postulate of Powerset and Axiom of Choice) is answered as follows: we could have chosen the domain $\mathcal{L}$ of all Gödel constructible sets as the initial domain of all sets.

Note. The genuine natural numbers are the natural numbers called the meta-mathematical numbers by mathematical logicians. They are, so to speak, the numbers that can be reached from below, realised, etc. The collection FN of these genuine natural numbers is a headless cut on the domain of all natural numbers. This means that if $n \in \mathrm{FN}$ and $m < n$ then also $m,\ n+1 \in \mathrm{FN}$. These numbers are invariant with respect to the ultraextension operator since if $n \in \mathrm{FN}$ then for each such operator ∇ we have $n = \nabla(k_n) = \bar{n}$. The collection FN is not sharply defined and hence it is not a set. It cannot be identified with the set of all natural numbers in a domain of all sets since clearly for each ultraextension operator ∇ we have $\mathrm{FN} \neq \nabla(k_{\omega_0}) = \bar{\omega}_0$.

4.8 Extendable Domains of Sets

Let $\mathcal{T}$ be a compact Hausdorf extremally disconnected non-empty topological space[3] without isolated points. A denotes the set of all points of the space $\mathcal{T}$. The **closure** of a set $X \subseteq A$ is denoted $\overline{X}$; the **interior** of a set $X \subseteq A$ is defined as $\mathrm{Int}(X) = A \setminus \overline{(A \setminus X)}$.

We say that a set $X \subseteq A$ is:

(a) **closed** if $X = \overline{X}$

(b) **open** if $A \setminus X$ is closed

(c) **clopen** if it is both open and closed

(d) **dense** if $\overline{X} = A$

(e) **nowhere dense** if $A \setminus \overline{X}$ is dense.

The set $\overline{X}$ is closed, the set $\mathrm{Int}(X)$ is open.

In the topological space $\mathcal{T}$ which we are considering the closure of each open set is clopen. Hence the interior of each closed set is clopen. For if X is a closed set then $A \setminus X$ is open, $\overline{A \setminus X}$ closed, and hence the set $\mathrm{Int}(X)$ is clopen.

Assertions below without proof are either common knowledge or they are trivial consequences of basic properties of the space $\mathcal{T}$.

Proposition 4.14.

(a) If X is an open set then the set $\overline{X} \setminus X$ is nowhere dense.

(b) If X is a closed set then the set $X \setminus \mathrm{Int}(X)$ is nowhere dense.

(c) The set of all open dense sets has the finite intersection property.

Proposition 4.15.

[3] A topological space $\mathcal{T}$ is **Hausdorff** if for any two distinct points x and y from $\mathcal{T}$ there exists such neighbourhoods U of x and V of y such that $U \cap V = \emptyset$. A compact Hausdorff space is **extremally disconnected** when the closure of every open set is clopen. See Petr Vopěnka, "The limits of sheaves over extremally disconnected compact Hausdorff spaces," *Bull. Acad. Polon. Sci. Sér. Sci. Math. Astronom. Phys.* 15 (1967): 1–4.

(a) For each point x there exists a sequence $\{u_n\}$ of clopen sets such that for each n we have $x \in u_n$, $u_{n+1} \neq u_n$, $u_{n+1} \subseteq u_n$.

(b) For each point x there exists a sequence $\{v_n\}$ of distinct clopen sets such that for each n we have $x \notin v_n$ and if $n \neq m$ then $v_n \cap v_m = \emptyset$.

Proof. (a) follows from the normality of the space $\mathcal{T}$.
(b) Let $v_n = u_n \setminus u_{n+1}$. □

B denotes the set of all clopen sets of the space $\mathcal{T}$.

By a (B)-**covering** of the set A we understand a set C of pairwise disjoint sets such that the set $\bigcup(B \cap C)$ is dense.

$\mathcal{C}$ denotes the set of all (B)-coverings.

Proposition 4.16. $\mathcal{C}$ is a covering structure.

Proposition 4.17. There exists a (B)-covering C of the set A such that the set $B \cap C$ is infinite.

Proof. Let v denote the closure of the set $\bigcup\{v_n\}$, where $\{v_n\}$ is a sequence guaranteed to exist by Proposition 4.15(b). If $v = A$ then $\{v_n\}$ is the required (B)-covering. If $v \neq A$ then we add to the set $\{v_n\}$ also the set $A \setminus v$. □

Proposition 4.18. On the set A there exists an ultrafilter, non-principal on the covering structure $\mathcal{C}$, which contains every open dense set.

Proof. Let C be a (B)-covering guaranteed to exist by Proposition 4.17. Let H denote the set of all open dense sets and H' the set of all $A \setminus \bigcup S$, where S is a finite subset of the set C. The set $H \cup H'$ clearly has the finite intersection property and an ultrafilter containing it as a subset has the required property. □

We define the meet and join of a non-empty set $Q \subseteq B$ as follows:

$$\bigwedge Q = \operatorname{Int}\left(\bigcap Q\right); \quad \bigvee Q = \overline{\bigcup Q}.$$

Clearly $\bigwedge Q, \bigvee Q \in B$ and if $u, v \in B$ then

$$\bigwedge\{u, v\} = u \cap v; \quad \bigvee\{u, v\} = u \cup v.$$

Proposition 4.19. B is a complete Boolean algebra with operations $\bigvee, \bigwedge$ and complement.

Let $\mathcal{D}$ be some fixed ultrafilter guaranteed to exist by Proposition 4.18.
By Proposition 4.14(a)(b) the sets $\bigcap Q \setminus \bigwedge Q$, $\bigvee Q \setminus \bigcup Q$ are nowhere dense and hence

Proposition 4.20.

$$\bigvee Q \in \mathcal{D} \Leftrightarrow \bigcup Q \in \mathcal{D}$$

$$\bigwedge Q \in \mathcal{D} \Leftrightarrow \bigcap Q \in \mathcal{D}$$

If $x \in A$ then $|x|$ denotes the set of all $u \in B$ such that $x \in u$.

Proposition 4.21. A set $Q \subseteq \mathcal{P}(B)$ is an ultrafilter on the complete Boolean algebra B just when there exists a unique $x \in A$ such that $Q = |x|$.

Proof. If Q is an ultrafilter then Q is a set of clopen sets with the finite intersection property. Since the space $\mathcal{T}$ is compact, we have $\bigcap Q \neq \emptyset$. Since the space $\mathcal{T}$ is Hausdorf, the intersection of the set Q is a singleton. The converse implication is trivial. □

S denotes the set of all functions $f \in \mathcal{G}(\mathcal{C})$ such that $\{x;\ f(x) \in |x|\} \in \mathcal{D}$.

Proposition 4.22. Let $f, g \in \mathcal{G}(\mathcal{C})$, $C(f) = C(g) = C$.

(a) Let $\{x;\ f(x) \subseteq g(x)\} \in \mathcal{D}$. If $f \in S$ then also $g \in S$.

(b) Let $\{x;\ f(x) \cup g(x) \in |x|\} \in \mathcal{D}$. Then $\{x;\ f(x) \in |x|\} \in \mathcal{D}$ or $\{x;\ g(x) \in |x|\} \in \mathcal{D}$; that is, $f \in S$ or $g \in S$.

Proposition 4.23. Let $g \in \mathcal{G}(\mathcal{C})$ be a function such that for all $x \in A$, $g(x) \subseteq |x|$. Then the function g_0 such that $g_0(x) = \bigwedge g(x)$ for each $x \in A$, belongs to the set S.

Proof. Let $u \in C(g)$, $x \in u$. Then for each $w \in g(x)$ we have $x \in w$ and hence $x \in \bigcap g(x)$. Since the function g is constant on the set u, for each $y \in u$ we have also $y \in \bigcap g(y) = \bigcap g(x)$. Hence $u \subseteq \bigcap g(x)$ a thus $u \subseteq \operatorname{Int} \bigcap g(x) = \bigwedge g(x) = g_0(x)$, $x \in g_0(x) \in |x|$. □

Let $\mathcal{W}$ denote the domain of all sets $\nabla(f)$ where $f \in \mathcal{G}(\mathcal{C})$. In the domain of sets $\mathcal{W}$, $\overline{B}$ is a complete Boolean algebra.

Let $Z = \{\nabla(f); f \in S\}$. Although $Z \subseteq \overline{B}$, Z is a set which does not belong to the domain $\mathcal{W}$. In 1967 I have called such sets semisets.

Using Proposition 4.22 we can easily see that Z is a non-trivial ultrafilter on the complete Boolean algebra $\overline{B}$. By Proposition 4.23, if $X \subseteq Z$ is a set belonging to the domain $\mathcal{W}$ then $\bigwedge X \in Z$. This already shows, amongst other things, that the set Z does not belong to the domain $\mathcal{W}$.

The set (more precisely, semiset) Z is a support on the domain of sets $\mathcal{W}$. That means that even though it does not belong to the domain $\mathcal{W}$, it is possible to add it to the domain $\mathcal{W}$ extending it to a larger domain $\overline{\mathcal{W}}$ so that $\mathcal{W}$ is a subdomain of the domain $\overline{\mathcal{W}}$.

Such extensions of the domain of all sets with supports and sets formed from them are described in detail in the book *The Theory of Semisets.*[4] There it is

[4] Petr Vopěnka and Petr Hájek, *The Theory of Semisets* (Amsterdam – London: North-Holland Publishing Company, 1972).

also proved that the above described ultrafilter Z is a support and conversely, that every support can be modified into the form of such an ultrafilter on a complete Boolean algebra.

Consequently, for each assertion such that its consistency with the axioms of set theory has been proved using the so-called Boolean models, it is possible to create via an ultraextension operator ∇ a domain of all sets in which this assertion is true.

Scholion

In 1940, the famous work of Kurt Gödel appeared, *The Consistency of the Axiom of Choice and the Generalized Continuum Hypothesis with the Axiom of Set Theory.*[5] Without using the axiom of choice, the author created in this book the class L of constructible sets. If we restrict set-theoretical investigations to elements of this class, we obtain a model $\triangle$ of set theory where all axioms of this theory hold and moreover also the Axiom of Choice and the Generalised Cntinuum Hypothesis. The assertion that no other than constructible sets exist came to be called the Axiom of Constructibility.

The question whether in set theory with the Axiom of Choice there exist also non-constructible sets, haunted mathematicians for over twenty years. At that point the development of set theory began to stagnate since obtaining results based on the Axiom of Constructibility, or even more so, on its negation, involved a risk of doing useless work.

For those mathematicians who did not believe that the Axiom of Constructibility was true, the class L represented a cage with no apparent way of escaping from it. Their task was to create a non-constructible set from the sets of the class L, or at least to prove that the existence of such a set is consistent.

Only two ways to this goal offered themselves at the time. The first one amounted to extending a countable model of the entire set theory by other sets that were constructible, but not constructible within this model. The other one appeared after the discovery of the ultrapower. It amounted to extending some submodel of the ultrapower. Namely the ultrapower over some suitable covering structure by some functions that did not belong to it.

It was Paul J. Cohen who reached the goal by the first of the above ways in the article "The Independence of the Continuum Hypothesis"[6], breaking thus the bars of the cage that held set theory prisoner.

I had been trying to reach the goal by the other way since 1961. I preferred it to the first way since it did not require accepting an axiom of the existence of a countable model of set theory. Somehow it appeared natural that the most suitable covering structure was formed by all countable coverings of the set of all real numbers, since the corresponding ultrapower had the same natural numbers as the ultrapower over the complete covering structure; it however had fewer

[5] Kurt Gödel, *The Consistency of the Axiom of Choice and the Generalized Continuum Hypothesis with the Axiom of Set Theory* (Princeton: Princeton University Press, 1937).

[6] Paul Cohen, "The Independence of the Continuum Hypothesis," *Proc. Nat. Acad. Sci. U.S.A.* 50 (1963): 1143–1148.

subsets of the set of all natural numbers. This choice was wrong and it did not lead to the goal.

Only after having studied the above mentioned work of Cohen did I realise that the correct covering structure was formed by all open coverings of some topological space. Success followed almost immediately so I could publish a model of Gödel-Bernays axioms of set theory where Continuum Hypothesis did not hold, without assuming the axiom of the existence of a countable model of this theory.[7]

In the same year 1964 I developed a general method for creating models of set theory by extending ultrapowers over covering structures formed by open coverings of topological spaces. In contrast to Gödel's model $\triangle$ I used the name ∇-models[8] for these models. When constructing them, I introduced valuations of formulae by open sets of topological spaces, analogous to that introduced in the so-called Boolean models created later on.

In autumn of 1964 I lectured about the thus-developed ∇-models in Warsaw and Wroclaw. Subsequently I sent four transcripts of these lectures to professor Jan Mostowski who presented them successively in the Bulletin of Polish Academy of Sciences in January, February, March and May 1965.[9]

In spring of 1966 I realised that in creating ∇-models, open sets can be replaced by regular open sets. M. H. Stone then proved that the algebra of these sets is isomorphic to the complete Boolean algebra of clopen sets of a compact extremely disconnected Hausdorff topological space.

Therefore I sent to Warsaw a further article "The limits of sheaves over extremally disconnected compact Hausdorff spaces", which was presented by J. Mostowski in the Bulletin of Polish Academy of Sciences in September 1966.[10]

I lectured on this new form of ∇-models in 1966 at a summer seminar organised by my collaborators and students. After the summer vacations my student Karel Příkrý left for the USA where he met with Robert Solovay and Dana Scott.

The search for qualitatively new models of set theory led me to the theory of semisets since any such model is determined by subsets of the class of all ordinal numbers. I lectured about this theory in Amsterdam at the International Congress for Philosophy of Science in 1967.

[7] Petr Vopěnka, "Nezavisimost kontinuum-gipotezy," *Comment. Math. Univ. Carolinae* 5 (1964). English translation Petr Vopěnka, "The Independence of Continuum Hypothesis," *American Mathematical Society Translations*, Series 2, 57, no. 2 (1964): 85–112.

[8] Naturally the usage of ∇ here is different from its usage for an ultraextension operator.

[9] Petr Vopěnka, "The limits of sheaves and applications on construction of models," *Bull. Acad. Polon. Sci. Sér. Sci. Math. Astronom. Phys.* 13 (1965): 189–192.
Petr Vopěnka, "On ∇-model of Set Theory," *Bull. Acad. Polon. Sci. Sér. Sci. Math. Astronom. Phys.* 13 (1965): 267–272.
Petr Vopěnka, P. "Properties of ∇-model," *Bull. Acad. Polon. Sci. Sér. Sci. Math. Astronom. Phys.* 13 (1965): 441–444.
Petr Vopěnka, "∇-models in which the generalized continuum hypothesis does not hold," *Bull. Acad. Polon. Sci. Sér. Sci. Math. Astronom. Phys.* 14 (1966): 95–99.

[10] Petr Vopěnka, "The limits of sheaves over extremally disconnected compact Hausdorff spaces," *Bull. Acad. Polon. Sci. Sér. Sci. Math. Astronom. Phys.* 15 (1967): 1–4.

In 1968 Czechoslovakia was occupied by the Soviet army and I disappeared from the international mathematical scene.

Finally I confess that I had known about the collapse of Cantor's set theory ever since the discovery of ultrapower in 1960. Absorbed in the whirl of events happening in this theory at the time, I suppressed this knowledge into the depths of my subconscious, from where it eventually guided me to the alternative set theory.

4.9 The Problem of Infinity

Results we have reached so far affect the existing interpretations of infinity so significantly that they require us to change our understanding of this phenomenon. We note that we are largely dealing with the infinity associated with multitudes, since in the 20th century it was the investigation of just such infinity that most dramatically inspired the radical change in our interpretation of infinity.

The fundamental theory of infinity associated with multitudes (and, according to Bolzano, of infinity in general), against which all theories of infinity have been judged, has been Cantor's set theory. The subject of its study, the domain $\mathcal{V}$ of all sets, which cannot be actualised because of its incomprehensible voluminousness, is not a set. Nevertheless, the actualisation of the domain of all natural numbers has been generally accepted and the set formed from them has been admitted into the domain of all sets.

Although the domain $\mathcal{V}$ went through being restricted or extended in various ways, ordinal numbers, in particular natural numbers, remained untouched through all these modifications. They were seen as sharing in the absolute.

The ultraextension operator ∇ exploded this idyll. It is possible to create many different domains of all sets. However, only one can exist; we denote it $\mathcal{V}$. All the other ones incompatible with $\mathcal{V}$ perish. But if some other of these domains is activated, that is, brought to existence, then the domain $\mathcal{V}$ perishes.

If domains of all sets differed only in the lengths of the sets of all natural numbers then it would not matter from the viewpoint of mathematics which of these domains had been activated. However, some domains of all sets differ in the truth of various so-called undecidable propositions of set theory. This was discussed in Section 4.8. A mathematician working in set theory and more generally, in set-theoretical mathematics, can thus activate for example a domain where the Continuum Hypothesis holds just as well as he or she can activate a domain of all sets in which the cardinality of the set of all real numbers is $\aleph_{22}$. However, they cannot activate them both at the same time.

If we accept that the set of all natural numbers is not absolute – and after all, we have no other option – and continue to insist on the exclusive position of Cantor's set theory, we can conclude our considerations by noting that in many important cases the problem of truth in Cantor's set theory has been solved.

For if some assertion of Cantor's set theory is undecidable then it cannot be proved nor disproved from the generally accepted axioms of set theory. This sorts out the problem for formalists. On the other hand, realists ask whether this assertion is true in the domain of all sets, or not. Since no universal domain of all sets exists, the question has to be rephrased as asking whether or not the assertion is true in this or that domain of all sets.

Much more interesting and pressing than searching for propositions undecidable in the collapsing Cantor's set theory is the surprising malleability of infinity. It was brought to light by the ultraextension operator ∇ when working with the original rigid domain of all cantorian sets.

> The malleability we have in mind here concerns infinity as such, not as an accompanying phenomenon pertaining to multitudes but infinity as an autonomous individual. As that, from which the domains of all sets take their rise and then pass away through mutual incompatibility of their existences.

This paraphrasing of Anaximander's statement is intended to point out that the infinity has become Anaximander's apeiron. Pure apeiron, all-encompassing and all-governing.

We shall not discuss it any further. Interpreting it any further would deform it.

Part II

New Theory of Sets and Semisets

Introduction

The non-actualisability of the domain of al natural numbers, that is, the non-existence of the set of all these numbers, has far-reaching consequences for the infinitary mathematics of the 20th century. The guarantor of the existence of this set, namely the God of the medieval and baroque rational theology, who was identified as such by Bolzano and for whom no substitute has ever been found, no longer fulfills his role. And the problem is not just the non-existence of this particular set; even more importantly, most mathematicians working in theories based on Cantor's set theory and dealing with various subjects of mathematical inquiry, rely on capabilities of this divine observer, which they appropriated.

Thus the attitude of mathematicians to infinity has been as if it lay before them in its actual form; as Adam lies before God on Michelangelo's picture in the Sistine chapel.

Infinity remains even after its divine observer and maker has disappeared. We have been thrown into pre-set-theoretical mathematics. We are not above the world but within it. We are not looking for infinity in God's mind but within the world.

We are seeking that infinity which has been always present in our journeys to the boundaries of our world (or even a little beyond these boundaries). We encounter it in all directions, not just travelling to a distance but also on the way to the micro-world. In our search we can draw on mathematical knowledge and experience from the past millennia.

We shall not enter the pre-set-theoretical mathematics as it was when we left it; we shall enter it with an answer to the following question, that should be asked by every philosophically thinking mathematician.

> How is it possible that some results of classical infinitary mathematics can be applied in the natural real world[11] and others have no use even when interpreting the real world?

Contemporary infinitary mathematics is based on Cantor's theory of classical, that is, classically interpreted, (actually) infinite sets. Such sets do not exist in the natural real world and hence it is not them that infinitary mathematics applies its results to.

If some results of classical infinitary mathematics are somehow applicable in the natural real world, then this involves interpreting indistinct[12,13] phenomena

[11] By the *natural real world* we basically mean the environment whose phenomena we perceive. By the *real world*, we mean an extension of the natural real world presupposed by science.

[12] *Indistinctness* is a phenomenon of the natural real world and it should be built also into the real world, if we want to avoid Descartes' dualism. If we interpret the real world as sharply defined, then we cannot avoid declaring indistinctness to be a subjective phenomenon that has no place in the objective real world. Consequently we must grant a soul to man, which – albeit not necessarily immortal – is separated from the real world. On the other hand, if man along with his perception is a part of the real world then the phenomenon of indistinctness is is also a part of it. Moreover, indistinctness is the primary phenomenon; it is no poorly captured sharpness. On the contrary, in most cases sharpness is idealized indistinctness.

[13] "Indistinct" and "indistinctness" translates the Czech words "neostrý" and "neostrost"

of this world. For example, many results about the classical continuum are applicable when interpreting the shape of a table standing in front of us. The shape of the table is also a continuum, but a different continuum than that studied in topology by classical infinitary mathematics. Similarly, results from classical infinitary probability theory or statistics can be applied to study the indistinctly defined multitude of birds on the territory of our country at the end of summer, and so on.

If we say that these applications of classical infinitary mathematics in the natural real world are only approximate (in other words that these phenomena of the natural real world are only more or less accurately modeled in classical infinitary mathematics), we have not answered the question posed above. We have only reformulated it (or more precisely, we have reformulated its first part) as follows: How is it possible that some phenomena of the natural real world, namely the indistinct phenomena, can be modeled in classical infinitary mathematics, and sometimes even very accurately?

If infinity is applicable when interpreting indistinct phenomena of the natural real world, it must already be present in the indistinctness of these phenomena in some form. If it were not there, we could not use it in these interpretations or, if you like, these phenomena could not be modeled in classical infinitary mathematics.

The presence of infinity in indistinctness can be illustrated by the following narrative. In classical mathematics, using the non-fading effect of the principle of mathematical induction we can prove that, if we add one extra element to a finite set A, we get a set B which cannot be bijectively mapped onto the set A.[14] This proposition does not hold for infinite sets and Richard Dedekind even tried to characterize infinite sets by its invalidity. An author of a popular book demonstrated the invalidity of this proposition for sets whose elements can be numbered by all the natural numbers by the following story.

Imagine a hotel with infinitely many rooms numbered by all the natural numbers, with all the rooms occupied, each by one guest only. Let A be the set of all the hotel guests. It is still possible to accommodate another guest who has arrived to request a room, without placing two guests in one room. We can do that by putting up this guest in the room number 1 and for each n, moving the guest from the room number n to the room $n + 1$. It is obvious that we map the set B which has been created by adding one extra guest to the set A, bijectively onto the set A. Let us now imagine that the hotel has only one thousand rooms, all occupied. We can nevertheless proceed in the same way os outlined above. The newcomer is accommodated in the room number 1, the guest from the room number 1 moves to the room number 2, etc. As the whole procedure of moving guests is realised gradually, it will surely not be finished by morning (undoubtedly we will not have time to move the guest out of the room number 1000). And every guest will still have been accommodated for

respectively, which is literally "non-sharp" and "non-sharpness". [Ed]

[14] That is, we cannot pair elements of A with elements of B so that each element of A is paired with a unique element from B and each element from B is paired with a unique element from A.

the best part of the night. In this case the set of one thousand rooms has an indistinctly defined part, or semiset, which contains all the rooms where guests change. This semiset behaves in a similar way as Cantor's classical set of all natural numbers.

To be consistent, we have to admit that infinity shows up in some form or other in the indistinctness of phenomena of the natural real world. For example, on semiset parts of large – from the classical point of view, finite – sets, not somewhere behind these large sets. We shall call this form of infinity **natural infinity**.

Of course, we could object that in the latter of the above stories, the moving of guests could be accelerated, at least theoretically. In that case we would increase the number of rooms in the hotel to match it. We would not remove the semiset. We can only make it sharper, more distinct.

The infinity we are now talking about is a sharpening of the natural infinity,[15] just as (ideal) geometric objects are sharpened shapes of objects of the natural real world.

Bolzano's programme for infinitary mathematics[16] can be expressed using the following imperative:

> Wherever infinity occurs in some form, look for a multitude on whose structure this form of infinity shows!

Bolzano's programme plays an essential role in mathematics based on Cantor's set theory. By contrast, in mathematics based on the new theory of sets and semisets the role of Bolzano's programme is only auxiliary, albeit very important. The programme of the new infinitary mathematics can be characterized by the following imperative:

> Wherever there is indistinctness, look for a horizon and natural infinity that has caused this indistinctness; then idealise this situation!

Natural real world is thus a source of immediate inspiration for the new infinitary mathematics and also the place for almost immediate applications of its results.

[15] By going back to natural infinity, we are emphasizing a crucial even if suppressed cross-reference between infinity and indistinctness.

[16] See Section 1.3. [Ed]

Chapter 5

Basic Notions

5.1 Classes, Sets and Semisets

In this introductory chapter we remind the readers of familiar notions related to classes, and we summarise some concepts important for the the new infinitary mathematics.

If we single out some previously formed objects, we will obtain a **collection** of these singled-out objects.

A **domain** is not a totality of some existing objects; it is a source or a container into which the suitable emerging or created objects fall.

Every collection of objects can be interpreted as a domain, albeit an exhausted one.

By **actualisation** of a domain we mean its exhaustion, that is, substitution of this domain by the collection of all objects that fall or can fall within it.

By a **class** we mean any collection of given objects (its elements), such that we interpret this collection as being an autonomous entity, or a single object.

By a **set** we mean a class that is sharply defined.[1] Every set is finite from the classical point of view.

By a **semiset** we mean an indistinctly defined class which is a part (that is, a subclass) of a set.

Symbols or groups of symbols used to denote objects are called **terms**.

Letters $A, B, C, \ldots$ and $a, b, c, \ldots$ will be used as particularly simple terms, namely as unspecified constants for objects. This means that if we say within an argument that A is an object, then A denotes the same object, no matter which one, to the very end of this argument. Consequently, the argument carried out with A applies to every object, since any object could have been denoted by this letter.

If we say that A is a constant then we mean the symbol A; if we say that A is an object then we mean the object denoted by the symbol A. One object may be denoted by various terms.

[1] Sharpness refers to distinctness, clarity and definiteness, ..., in brief, the ancient perfection of geometric objects studied in Euclid's *Elements*.

The fact that two terms, for example the constants A and B, denote the same object is recorded as $A = B$ and we say that "A equals B". Even though the notation $A = B$ is sometimes read as "object A equals object B" or "object A is identical to object B", strictly speaking we always mean that symbols A and B denote the same object. The fact that two terms, for example constants A and B, denote different objects is recorded as $A \neq B$ and read "A is different from B" or "A does not equal B" etc.

The notation $A \in B$ means that the object B is a class and the object A is an element of (or belongs to) the class B.

The notation $A \notin B$ means that the object B is a class and the object A is not an element of (or does not belong to) the class B.

We say that a class A is a subclass (or a part) of a class B (notation $A \subseteq B$), if every element of the class A is also an element of the class B.

If A and B are classes for which it holds that both $A \subseteq B$ and $B \subseteq A$ then $A = B$; we create only one class from the given collection of objects (the so-called **extensionality of classes**).

An **empty class** is a class which has no elements. Extensionality implies that there is only one such class. Obviously, an empty class is a set. We will continue using the symbol $\emptyset$ for it.

The notation and definitions for classes that are unordered or ordered pairs, triples etc. of objects, or are intersections, unions etc. of other classes are as in Section 2.1. Similarly, Cartesian products of classes, definitions and properties of relations and functions along with orderings and their properties are as they were described for sets in sections 2.1.1 and 2.1.2.

5.2 Horizon

Every look[2] we cast, no matter in what direction, is limited. Either there is a firm boundary which disrupts (or deflects) it sharply, or it is limited by a **horizon** in whose direction clarity decreases and sharpness blunts.

We interpret horizon as a border separating the illuminated part of an observed object from the unilluminated part, that is, from the part not perceived by our look. In other words, by a horizon we understand only the border itself, not the illuminated part of the observed object nor all that we can capture by our look as is sometimes the case.

A look at some comprehensive object of interest is therefore connected to three distinct phenomena: the **illuminated** and the **unilluminated parts** of the object (in other words its parts lying before or beyond the horizon) and the **horizon** separating these two parts. We shall call this triple the **fundamental triad** since it will be the basis, the starting point and the reference point for all our considerations.

[2] *A look* here does not refer merely to a look by sight (with physical eyes); we understand it in a broad sense of regarding something that has been encountered.

The illuminated and unilluminated parts of the observed object cover it entirely and do not intersect; the unilluminated part of an object is defined as that from the object which does not lie within the illuminated part. The horizon belongs to none of these two parts and unlike them it is not a part of the observed object (after all, we left there no place for it to occupy).

The horizon itself, as such, is a sharp, definite and constant phenomenon.

The form of the horizon depends on the nature of a particular look. An example of a horizon that is evident[3] merely to our physical sight is the sky. In this case, the horizon shows up in an almost tangible form; this led to the sky having sometimes been interpreted as a firm and sharp boundary of the world (as a crystal surface encircling the world), and not as a horizon.

Although the horizon itself is a sharp and definite phenomenon, it does not sharply separate the illuminated part of an object from its unilluminated part. (Otherwise it would be a sharp and fixed boundary sharply and clearly delimiting the illuminated part.) The illuminated (and therefore also the unilluminated) part of an observed object is therefore delimited by the horizon indistinctly. The closer to the horizon an observed phenomenon lies, the vaguer it is. In other words, the closer we get to the horizon, the more indefiniteness there is on the illuminated part of the object. In this sense, the horizon is the place where we meet the phenomenon of indefiniteness and indistinctness in one of its purest forms.

The sharpness and definiteness of the horizon thus enable us to grasp the phenomenon of indefiniteness more definitely.

Principle of continuous transition. The illuminated part (of an object) passes continuously into the unilluminated part.

Our knowledge of this continuity does not come from our experience but, on the contrary, it precedes experience. This continuity is the very thing which distinguishes a horizon from a fixed boundary. The horizon is not some line drawn on an observed object, it is not a part of it and it is not attached to it. Therefore no phenomena belonging to the observed object mark the place where the horizon lies. Either the horizon does not touch these phenomena at all, or it separates them, too, into an illuminated and an unilluminated part. In other words, there is nothing on the observed object that would disrupt its passing from the illuminated to the unilluminated part. Naturally this does not mean that somewhere in the unilluminated part of the object (not immediately behind the horizon) there could not be a totally unexpected or even unpredictable phenomenon.

The observed object thus smoothly continues beyond the horizon until there is another phenomenon in its way which disrupts this continuity. As a consequence, the part of the observed object known to us – this **known land (terra cognita)** – goes beyond the illuminated part of the object, everywhere where

[3] To see something as *evident*, to evidence it, means to see it – in a wide sense of the word, not necessarily by our physical sight – and to be aware that we are seeing it.

this part is being bounded by the horizon. Naturally, this consideration does not apply to sharp and fixed boundaries of the illuminated part. What lies beyond the known land is **terra incognita**.

By the **overflow of the known land** (on an observed object, possibly also in a given direction) we mean the part by which the known land extends the illuminated part of the object; in other words, what lies simultaneously in the known land and in the unilluminated part.

The overflow of the known land therefore contains no phenomenon we would not be somehow familiar with from the illuminated part. Our far from matter-of-course confidence that our knowledge of the observed object reaches into the unilluminated part is based on this fact.

The closer to the horizon particular phenomena from the illuminated part lie, the vaguer they are and the more indistinct our knowledge of these phenomena is. Moreover, our active attention that responds to illuminated phenomena falters in the direction of the horizon. In the overflow of the known land vagueness clouds phenomena to such an extent that it is only our knowledge of their existence that remains and our awareness of these phenomena slips into passively not expecting anything unexpected.

Principle of Backward Projection. Phenomena lying on the horizon challenge us to interpret them as traces left by phenomena that have fallen beyond (or under) the horizon. More precisely: to interpret them as backward projections of these phenomena onto the places on the horizon through which there leads a way to them.

An exemplary case of this are again stars in the sky.

Phenomena lying on the horizon interpreted in this way often have no resemblance to the phenomena whose backward projections they are. They cannot be deduced from them by pure reason, not even by our wildest imagination. For example, colour is a backward projection of something that has no colour; sound is a backward projection of some waves that themselves have no sound, as witnessed by deaf people. It is the horizon that makes it possible to interpret something as a backward projection of something else; more precisely, it is the position of the horizon in the tangle of phenomena which we have been thrown into.

Therefore it is necessary to pay at least as much attention to phenomena that arise (lie) on the horizon as we pay to phenomena lying before or beyond the horizon, or to those divided by the horizon.

Scholion

Let us imagine that we could see, for example, a table standing in front of us more and more sharply. For instance, that we could observe it using an increasingly strong magnifying glass and then an increasingly strong microscope, and thus move the horizon further and further backwards. In that case we would probably soon begin to see bulges and depressions that would eventually turn into holes running through the table. The shape of the table would keep changing and eventually we would feel reluctant to call what we see a table. And it would be the same if our sight remained equally powerful but we grew smaller and smaller. In that case the observed table would gradually grow beyond the field of our vision and, should we be suddenly placed in this situation, we would not be able to recognise that we were inside some table. The observed table would disappear together with its observed shape. The preceding shapes of the table thus disappear with the horizons of our preceding looks. In other words, the shape of the observed table – and other things of course – is a phenomenon showing itself only on the horizon of a particular look.

A table without a shape is no table; it is a table only because it has that shape which shows itself to our ordinary looks into the tangle of phenomena around us. And it is the same in the case of other phenomena of the sensorily perceptible world. To put it briefly, the horizon forms our natural real world. Were it not for the horizon, our looks would bounce off rigorous boundaries or we would stare into emptiness. Our natural real world would be a cage open to nothingness in some directions; we would not inhabit a world, but a coffin. The horizon is thus the most important element of the fundamental triad; it is not an auxiliary phenomenon, but one of the fundamental footholds in the tangle of phenomena that we are thrown into.

The modern natural scientist regards phenomena on the horizon merely as a kind of a gateway into the objective real world; these phenomena are merely subjective (or inter-subjective, if they are perceived by several people). The scientist believes that objective phenomena may only be obtained if he or she manages to go beyond the horizon on which the subjective phenomena lie.

Modern natural science, which aims to explore the objective real world, thus regards phenomena on the horizon as inferior. Their role is merely auxiliary and sometimes even misleading.

However, the preceding considerations must naturally call into question whether there are any grounds for speaking about some objective shapes of various things, that is, about shapes which are independent of any looks. This would in fact only be possible if they were shapes showing on the horizon of a sharpest look (that is, a look cast by God). However, even if the aforementioned table had such an objective shape, it would not be the shape that we see now. Neither would it be a shape that we see through a microscope etc.

Moreover, there is no reason to believe that if our sight were sharper or if we grew so much smaller that we would get insight into some great depths of the microcosm, we would continue to see shapes as we are able to see them now. Perhaps we would find ourselves in a four-dimensional world and if we submerged deeper, in a ten-dimensional world, and even deeper again in a two-dimensional world twisted as a Möbius strip etc. The assumption of objective shapes is nothing other than an assumption that these changing shapes converge to some limit, or at least that different ways of sharpening our looks and submerging into the depths of the microcosm lead to limits that are in some way mutually coherent etc. However, we have no cogent argument supporting this hypothesis.

These were only a few random examples from the inexhaustible multitude of sometimes even fantastic shapes which could conceivably appear on the horizon in case that our eyesight or sense of touch continued to sharpen or if we kept growing considerably smaller; in other words, if we could cast some unnatural looks into the sensorily perceptible world. But we cannot cast such looks into this world. Even when looking through a microscope, we use our natural (that is, innate) sight, whose horizon may be brought closer or pushed back to a larger distance; but this shifting is relatively strictly limited. We do not move closer to what we observe through the microscope, we move it closer to us; we place it on the horizon of our natural look into the sensorily perceptible world, and as a consequence we see familiar shapes. We deduce from this that even in the depths of the microcosm there are the same shapes as on the horizon of our natural looks so we explain the convergence of changes in the shapes as their sharpening. There is a considerable danger of error here since everything might be completely different in the depths of the microcosm. On the other hand, we are not able to see it differently than here, and in this sense it is the same there as it is here.

From this point of view any scientific cognition of the real world is merely an organic supplementation and re-creation of phenomena appearing on a variously positioned horizon.

The famous narration about a cave that opens the seventh book of Plato's dialogue *The Republic* is usually interpreted (probably correctly) as an allegory of Plato's teaching regarding the relationship between the world of ideas and the sensorily perceptible world. Our above considerations however offer an almost realistic interpretation of Plato's enigmatic narration.

So let us recall an extract from Plato (translated by Benjamin Jowett) which will serve as the basis for the interpretation promised above.[4]

> And now, I said, let me show in a figure how far our nature is enlightened or unenlightened: – Behold! Human beings living in an underground den, which has a mouth open towards the light and reaching all along the den; here they have been from their childhood, and have their legs and necks chained so that they cannot move, and can only see before them, being prevented by the chains from turning round their heads. Above and behind them a fire is blazing at a distance, and between the fire and the prisoners there is a raised way; and you will see, if you look, a low wall built along the way, like the screen which marionette players have in front of them, over which they show the puppets.
>
> I see.
>
> And do you see, I said, men passing along the wall carrying all sorts of vessels, and statues and figures of animals made of wood and stone

[4] Plato, *The Republic*, trans. B. Jowett, 7.514a1–515c2 (Project Gutenberg, August 2008).

> and various materials, which appear over the wall? Some of them are talking, others silent.
>
> You have shown me a strange image and they are strange prisoners.
>
> Like ourselves, I replied; and they see only their own shadows, or the shadows of one another which the fire throws on the opposite wall of the cave?
>
> True, he said; how could they see anything but the shadows if they were never allowed to move their heads?
>
> And of the objects which are being carried in like manner they would only see the shadows?
>
> Yes, he said.
>
> And, if they were able to converse with one another, would they not suppose that they were naming what was actually before them?
>
> Very true.
>
> And suppose further that the prison had an echo which came from the other side, would they not be sure to fancy when one of the passers-by spoke that the voice which they heard came from the passing shadow?
>
> No question, he replied.
>
> To them, I said, the truth would be literally nothing but the shadows of the images.
>
> That is certain.

To put it shortly, scientific cognition does not depart from the opposite wall of Plato's cave.

5.3 Geometric Horizon

The picture below shows a well-known construction of the tangents of a given circle k (with centre S) passing through a given point A lying outside the circle.

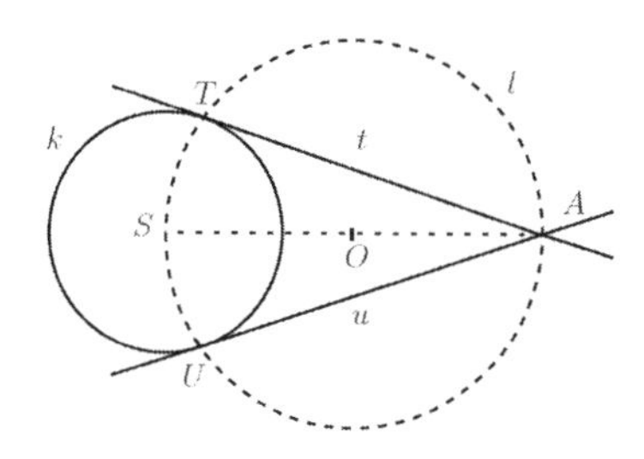

At the point when we say this, when we become aware of it, our original perception of the picture changes into geometric perception. The small dot marked by the letter T becomes an irreducible geometric point, the straight lines t, u become revealed segments of ideally straight and irreducible geometric lines, the line k – and even the broken line l – become ideal geometric circles. The line t touches the circle k at a unique point T (the point T is their only common point) and this point is simultaneously one of the two intersection points (that is, common points) of circles k, l etc. To put it briefly, our understanding of the picture now is that we are looking through it into the geometric world.

However, we will never reach the geometric world using our bodily senses. No matter how hard we try, we will never find an ideal circle through haptic reception and neither will we see it with our very own eyes. We still do look into this world, namely using our **geometric sight**, which is a kind of a spiritual sight. Using it, we can see as far as individual points and ideally straight and irreducible lines, and looking at a square we grow perfectly sure that both its diagonals are equally long. We could not pursue geometry if we were not capable of this. We could never agree on the fact that a line touches a circle at a unique point. Our imagination would fail to guide us through the geometric world when uncovering its less obvious phenomena and laws. We do not choose axioms of geometry arbitrarily but so that they are true in the geometric world which we can see by our geometric sight.

We call this world, discovered in Greek antiquity, the **intuitive** or also the **ancient geometric world.**

The lines t and k in the picture above do have some indeterminate blurred section in common. If we make them thinner and thinner, if we straighten t and smoothly round k, that is, if we abrade them smoothly, the section they have in common will grow smaller and smaller. Eventually – that is, after seeing through to the geometric world, once the line t becomes a true geometric line and the line k becomes a true circle – their common section will turn into an irreducible geometric point.

This interpretation makes it possible to place ideal geometric objects (and as a consequence the whole ancient geometric world) on the very horizon lying beyond (or below) all horizons limiting the human ability of cognition through bodily senses. To see geometrically is a special ability to see as far as this horizon by means of some spiritual sight. We call this horizon **geometric horizon**.

The horizon on which we have placed the geometric world is of a peculiar nature since it lacks some properties characteristic for a horizon. It is rigid and definite. This has enabled the pure and in the ancient Greek sense ideal science – geometry – to be set on it. Rigidity here refers to the fact that, broadly speaking, this horizon is always at the same distance from us; it cannot be pushed further away. In this respect it quite resembles the sky which is always above us, at the same distance from us, no matter whether we are standing or moving. However, in contrast to the sky, the geometric horizon is definite, namely because we can safely perceive it by means of our geometric sense. The geometric horizon could consequently become a stage for the characteristic phenomena of the geometric world situated on it. (Let us note here that as long as the sky had been interpreted in the same manner, the science of celestial phenomena had been perceived to be of equal worth to geometry in the ancient Greek sense. In other words, also this science used to have more of a claim to be knowledge of permanent and unchangeable truths than natural sciences.)

Summarizing, ancient geometry is not pursued by people, but by Olympian gods. A human can practice it only when he or she replaces his or her human outlook by that of Zeus and his or her human abilities by the abilities of a superhuman. This substitution of sensorium humanum with sensorium deorum has been made possible by the miracle of the ancient discovery of the intuitive geometric world. Geometric horizon is the horizon limiting looks cast by the Olympian gods.

The ancient geometric world and geometric perception is a gift of the Olympian gods to European humankind. Or rather, geometry is knowledge stolen from the Olympian gods. It is necessary to note here that the Europeans were punished for this theft by being bound to the intuitive geometric world with manacles even stronger than those that bound Prometheus to the rock in the Caucasus.

5.4 Finite Natural Numbers

Since the domain of all natural numbers is not actualisable, natural numbers along with their progress to the horizon, beyond the horizon and possibly as far as the depths of terra incognita, have to be approached – so to speak – from below. It means that we shall break into this domain starting from individual natural numbers and we shall actualise only those parts of it where it appears to serve some purpose. Naturally we do not intend to give up advantages obtained by actualisation of at least some parts of the domain of all natural numbers.

By natural numbers we understand numbers 0,1,2,3,.... Obviously, those which we have written down form a negligible part of the domain of all natural numbers, and this would be the case even if we wrote down the first hundred, million or trillion natural numbers. It is so even in the case of numbers smaller than 10^{421} which, according to the ancient Indian manuscript Lalitavistara, Buddha could name.

The remaining infinitely many natural numbers, that is, those which even in the wildest imagination sensorium humanum cannot comprehend, hide in those three dots (or in the words "and so on") which follow the writen-down or uttered numbers whenever we wish to acquaint somebody with natural numbers.

The remarkable thing is that these three dots (or the words "and so on") represent sufficiently, and in a way very clearly, our idea of natural numbers and invoke understanding for their progress to infinity. Since we are unable to represent natural numbers in any better way, we are undoubtedly dealing here with an archetypal idea that belongs amongst the mysterious outgrowths from the collective unconscious of – at least – Indo-European mankind (in the sense of C. G. Jung).

Similar archetypes of the collective unconscious formed those European gods which, thanks to their immortality and the unlimited penetration of their sight, are capable of reaching successively natural numbers beyond sensorium humanum.

By **finitely large** (briefly just finite) **natural numbers** we understand those natural numbers that Zeus can reach. In other words, those natural numbers from the sequence $0, 1, 2, 3, \ldots$ that lie before the geometric horizon.

If α is a finite natural number then any smaller natural number is clearly also finite. All these smaller natural numbers form a set which we denote $[\alpha]$.

Since even the geometric horizon is a horizon, the sequence of natural numbers also enters the overflow of its known land. That allows us to consider the following postulate to be at least theoretically performable.

Postulate of weak actualisability of the domain of natural numbers. For every natural number α belonging to the known land adjacent to the geometric horizon it is possible to create the set $[\alpha]$ of all natural numbers smaller than α.

The existence of at least one infinite natural number ϑ along with the fact that every finite natural number is smaller than ϑ implies the existence of the collection FN of all finite natural numbers; for the collection FN is a part of the set $[\vartheta]$. In other words, the actualisability of the domain of all finite natural numbers is a consequence of the principle of continuous transition and the postulate of weak actualisability of the domain of natural numbers.

The collection FN cannot be interpreted as a set. If FN were a set, as a subset of some $[\vartheta]$, where ϑ is an infinite natural number, it would have to have a maximum element. However, this is not possible as it is straightforward to check that if n is finite, $n + 1$ is also a finite natural number.

There is no reason why we should not view the collection FN as an independent individual, that is, to interpret this collection as an object. On the contrary, such an interpretation is undoubtedly useful.

Consequently we must conclude that the collection FN is not sharply defined. Hence the class FN is a semiset.

This means that we can use predicate calculus when studying elements of the semiset FN, their properties and relations. However, we must be very careful with the quantifiers because the semiset FN is not sharply defined in the direction of the horizon limiting the size of finite natural numbers. By contrast, predicate calculus can be used without hesitation when studying the set $[\vartheta]$, where $\mathrm{FN} \subseteq [\vartheta]$.

Chapter 6

Extension of Finite Natural Numbers

6.1 Natural Numbers within the Known Land of the Geometric Horizon

Let ϑ denote some fixed infinite natural number lying in the known land of the geometric horizon. Let $\alpha, \beta, \gamma, \dots$ denote elements of the set $[\vartheta]$ and $m, n, k \dots$ elements of the semiset FN. Obviously $\mathrm{FN} \subseteq [\vartheta]$.

The transition from finite natural numbers to infinite numbers is so continuous that it is impossible to determine the place where it actually happens. Thanks to the continuity of this transition, various properties of natural numbers and also many propositions and their proofs pass almost automatically from the semiset FN into the set $[\vartheta]$. For example:

(a) Natural numbers belonging to the set $[\vartheta]$ are linearly ordered according to their magnitude.

(b) Every non-empty subset of the set $[\vartheta]$ has a least and a greatest element.

(c) Arithmetical laws of natural numbers hold in the entire known land of the geometric horizon and so on.

We say that sets u, v have the same number of elements if there is a bijective function f such that f is a set and

$$\mathrm{dom}(f) = u, \quad \mathrm{rng}(f) = v.$$

Proposition 6.1. If $\alpha \in [\vartheta]$, $u \subseteq [\alpha]$ and $u \neq [\alpha]$ then the sets u and $[\alpha]$ do not have the same number of elements.

Proof.[1] Assume that $\alpha + 1 \in [\vartheta]$ is the smallest natural number for which there is $u \subseteq [\alpha + 1]$, $u \neq [\alpha + 1]$ and a bijective function f such that f is a set and

$$\mathrm{dom}(f) = [\alpha + 1], \qquad \mathrm{rng}(f) = u.$$

[1] In this proof by contradiction, we can focus on non-zero numbers since the proposition clearly holds for $\alpha = 0$. [Ed]

First let $\alpha \notin u$. Then $f(\alpha) \in u \subseteq [\alpha]$, $u \setminus \{f(\alpha)\} \neq [\alpha]$ a $f|[\alpha]$ is a bijective function mapping the set $[\alpha]$ onto the set $u \setminus \{f(\alpha)\}$, which is a contradiction.

So let $\alpha \in u$. Then $[\alpha] \cap u \neq [\alpha]$. If $f(\alpha) = \alpha$ then $f|[\alpha]$ is a bijective function mapping the set $[\alpha]$ onto the set $[\alpha] \cap u$, which is a contradiction. If $\alpha = f(\beta)$, where $\beta \in [\alpha]$, then $f(\alpha) \in [\alpha] \cap u$. We define a function g on the set $[\alpha]$ by setting $g(\beta) = f(\alpha)$ and $g(\gamma) = f(\gamma)$ for $\beta \neq \gamma \in [\alpha]$. Obviously g is a bijective function, $\text{dom}(g) = [\alpha]$, $\text{rng}(g) = u \cap [\alpha] \neq [\alpha]$, which is a contradiction. □

Proposition 6.2. For every set u there is at most one $\alpha \in [\vartheta]$ such that the sets u and $[\alpha]$ have the same number of elements.

Proof. Let sets $[\alpha], [\beta]$, where $\alpha, \beta \in [\vartheta]$, $\alpha \neq \beta$ have the same number of elements as a set u. Let for example $\beta \in [\alpha]$. Then also the sets $[\beta], [\alpha]$ have the same number of elements, which contradicts Proposition 6.1. □

If $\alpha \in [\vartheta]$ and sets $u, [\alpha]$ have the same number of elements then the number α is called the **number of elements of the set** u and denoted $\alpha = \text{Card}(u)$.

Proposition 6.3. Let $\alpha \in [\vartheta]$, $u \subseteq [\alpha]$, $u \neq [\alpha]$. Then there exists $\beta \in [\alpha]$ such that $\beta = \text{Card}(u)$.

Proof. Assume that $\alpha + 1 \in [\vartheta]$ is the smallest natural number for which there is a set $u \subseteq [\alpha + 1]$, $u \neq [\alpha + 1]$ such that $\beta = \text{Card}(u)$ for no $\beta \in [\alpha]$. Let δ be the greatest element of set u (obviously $u \neq \emptyset$). Then $u \setminus \{\delta\} \subseteq [\alpha]$ and so there is some $\gamma \in [\alpha]$ such that $\gamma = \text{Card}(u \setminus \{\delta\})$. Hence $\gamma + 1 = \text{Card}(u)$, which is a contradiction. □

We say that a **set** u is **finite**, if there exists $m \in \text{FN}$ such that $m = \text{Card}(u)$. We leave it to the reader to prove the following proposition.

Proposition 6.4. (i) Let sets u, v be finite. Then also $u \cup v$ and $u \times v$, are finite sets.

(ii) Let u be a finite set whose all elements are finite sets. Then also $\bigcup u$ is a finite set.

(iii) Let u be a finite set and let $v \subseteq u$. Then also the set v is finite.

(iv) Let a relation r be a finite set. Then also the sets $\text{dom}(r)$ and $\text{rng}(r)$ are finite.

If we observe some set which has, for example, a million elements then various **semiset parts** of it begin to appear; for no human being has the capability to clearly see this set along with each of its element, as evident. Zeus however does have this capability. In the ancient geometric world there are no semisets, only phenomena sharpened to the limit. Since we are claiming Zeus' capabilities, we have the following:

Proposition 6.5.

(i) If α is a finite natural number and X a class such that $X \subset [\alpha]$ then X is a set.

(ii) If however $\alpha \notin \mathrm{FN}$ then FN is a semiset part of the set $[\alpha]$.

This allows us to formulate the following equivalent definition of the class of all finite natural numbers.

FN is the class of all natural numbers α such that for each $X \subset [\alpha]$, X is a set.

Scholion

The above definition of the semiset FN of all finite natural numbers can be used also when applying our theory in the real world. In such case the class FN will be much shorter than it is in the ancient geometric world. After all, line segments in the ancient geometric world are much thinner than those drawn on a piece of paper, too.

6.2 Axiom of Prolongation

Hard sets are sets in whose interior structure there is no semiset. We define them as follows.

Hard sets of type 1 are:

(a) sets of abstract ur-objects;[2]

(b) sets of natural numbers;

(c) sets of other kinds of numbers that will be discussed later.

Hard sets of type n, $n \neq 1$, are sets whose elements are only sets of smaller types than n, or abstract ur-objects, natural numbers and numbers introduced later. (For the considerations that we will be carrying out we will make do with hard sets of very small types.)

We say that a **function** F is **stable on the semiset** FN (in the following text just **stable**) if $\mathrm{dom}(F) = \mathrm{FN}$ and for every $n \in \mathrm{FN}$, the restriction of F to the set $[n]$ (that is, the function $F|[n]$) is a hard set.

Not only the transition from finite natural numbers to infinite natural numbers is so smooth that it is impossible to pinpoint the place where it actually happens. It is the same in case of various stable functions on the semiset FN. This phenomenon can be captured as follows:

Axiom of Prolongation. Let F be a stable function (on the semiset FN). Then there exists a function f such that f is a set and for every $n \in \mathrm{FN}$,

$$f(n) = F(n) \qquad \text{(that is, } F = f|\mathrm{FN}\text{)}.$$

[2] In other words, objects whose contents have been emptied; they are mere objects and nothing more.

Note. Without loss of generality we can assume about f that $\operatorname{dom} f = [\gamma]$ where γ is an infinite natural number, $\gamma \in [\vartheta]$. If this is not the case, we set γ as the greatest element of the set of all $\beta \in dom(f)$, for which $\operatorname{dom}(f|[\beta]) = [\beta]$. Obviously $\gamma \notin \mathrm{FN}$.

It is often more convenient to use the Axiom of Prolongation in the following form:

Axiom of Prolongation. Let $\{a_n\}_{n \in \mathrm{FN}}$ be a stable sequence. Then this sequence has a (set) extension $\{a_\alpha\}_{\alpha<\gamma}$, where γ is an infinite natural number, $\gamma \in [\vartheta]$.

In mathematics based on the new (formerly alternative) theory of sets and semisets we use the Axiom of Prolongation to replace all possible forms of the notion of *limit.*

6.3 Some Consequences of the Axiom of Prolongation

Proposition 6.6. Let F be a stable function such that $F(n)$ is a set for every n (or also $F(n) \subseteq w$ respectively, where w is some given set). Then there exists an extension f of the function F such that for every $\alpha \in \operatorname{dom}(f)$, $f(\alpha)$ is a set (or also $f(\alpha) \subseteq w$ respectively) .

Proof. Let g be an extension of a function F. Let γ be the largest natural number such that $\gamma \in \operatorname{dom}(g)$ and $g(\beta)$ is a set (or also $g(\beta) \subseteq w$ respectively) for every $\beta \leq \gamma$. Evidently $\gamma \notin \mathrm{FN}$. So it suffices to set $f = g|[\gamma]$. □

Proposition 6.7. Let F be a stable function such that $F(n)$ is a set for every n. Let $F(n) \subseteq F(m)$ (or $F(m) \subseteq F(n)$ respectively) for $n \leq m$.

Then there exists an extension f of the function F such that $f(\alpha)$ is a set for every $\alpha \in \operatorname{dom}(f)$ and $f(\alpha) \subseteq f(\beta)$ (or $f(\beta) \subseteq f(\alpha)$ respectively) for $\alpha \leq \beta$, $\alpha, \beta \in \operatorname{dom}(f)$.

Proof. Let g be an extension of the function F such that $g(\alpha)$ is a set for every $\alpha \in \operatorname{dom}(g)$ (see Proposition 6.6). Let γ be the largest natural number such that $\gamma \in \operatorname{dom}(g)$ and $g(\alpha) \subseteq g(\beta)$ (or $g(\beta) \subseteq g(\alpha)$ respectively) for every $\alpha < \beta < \gamma$. Evidently $\mathrm{FN} \subseteq [\gamma]$. It therefore suffices to set $f = g|[\gamma]$. □

Proposition 6.8. Let F be a stable function such that $F(n)$ is a set for every n. Assume that $\bigcup \operatorname{rng}(F)$ is a set; let us denote it w. Then there exists $m \in \mathrm{FN}$ such that $w = \bigcup \operatorname{rng}(F|[m])$.

Proof. Let f be an extension of the function F such that $f(\alpha)$ is a set, $f(\alpha) \subseteq w$ for every $\alpha \in \operatorname{dom}(f)$. If $\gamma \in \operatorname{dom}(f) \setminus \mathrm{FN}$, then $w = \bigcup \operatorname{rng}(f|[\gamma])$. The smallest natural number γ such that $w = \bigcup \operatorname{rng}(f|[\gamma])$ is therefore an element of the semiset FN. □

Proposition 6.9. Let F, G be stable functions such that for every m and n, $F(m)$ and $G(n)$ are sets satisfying $F(m) \cap G(n) = \emptyset$. Then there exist sets u, v such that $u \cap v = \emptyset$ and for every n

$$F(n) \subseteq u, \quad G(n) \subseteq v.$$

Proof. Let f, g be extensions of the functions F, G such that $f(\alpha)$, $g(\alpha)$ are sets for every $\alpha \in \mathrm{dom}(f) \cap \mathrm{dom}(g)$ (see Proposition 6.6). Let γ be the largest natural number such that $\gamma \in \mathrm{dom}(f) \cap \mathrm{dom}(g)$ and $f(\alpha) \cap g(\beta) = \emptyset$ for every $\alpha, \beta \in [\gamma]$. Evidently $\gamma \notin \mathrm{FN}$. If we set

$$u = \bigcup \mathrm{rng}(f|[\gamma]), \qquad v = \bigcup \mathrm{rng}(g|[\gamma]),$$

then u, v are sets with the desired properties. □

Proposition 6.10. Let F, G be stable functions such that $F(n)$, $G(n)$ are sets satisfying $F(n) \cap G(n) \neq \emptyset$ for every n and $F(m) \subseteq F(n)$, $G(m) \subseteq G(n)$ for every $n \leq m$. Then there exists x such that $x \in F(n) \cap G(n)$ for every n.

Proof. Let f, g be extensions of functions F, G. Let γ be the largest natural number such that $[\gamma + 1] \subseteq \mathrm{dom}(f) \cap \mathrm{dom}(g)$ and $f(\alpha)$, $g(\alpha)$ are sets satisfying $f(\alpha) \cap g(\alpha) \neq \emptyset$ for every $\alpha \leq \gamma$ and for every $\alpha < \beta \leq \gamma$ we have $f(\beta) \subseteq f(\alpha)$, $g(\beta) \subseteq g(\alpha)$. Evidently $\gamma \notin \mathrm{FN}$. If $x \in f(\gamma) \cap g(\gamma)$, then $x \in f(n) \cap g(n) = F(n) \cap G(n)$ for every n. □

Proposition 6.11. Let F be a stable function such that $F(n)$ is a non-empty set for every n. Let $F(m) \subseteq F(n)$ for every $n \leq m$. Then there exists x such that $x \in F(n)$ for every n.

Proof. This proposition is a special case of Proposition 6.10, namely for $F = G$.
□

If $\gamma \in [\vartheta]$ is an infinite number then evidently also for every n, $\gamma - n$ is an infinite number.

Proposition 6.12. Let $\{\gamma_n\}$ be a sequence of infinite numbers smaller than ϑ such that $\gamma_{n+1} \leq \gamma_n$ for every n. Then there exists an infinite number β such that $\beta < \gamma_n$ for every n.

Proof. Evidently $[n] \cap ([\vartheta] \setminus [\gamma_n]) = \emptyset$ for every n. According to Proposition 6.9 there exists a set v such that $\mathrm{FN} \subseteq v$ and $v \cap ([\vartheta] \setminus [\gamma_n]) = \emptyset$ for every n. Let us choose $\beta \in v \setminus \mathrm{FN}$. Then $\beta \notin [\vartheta] \setminus [\gamma_n]$ for every n and so $\beta \in [\gamma_n]$, in other words $\beta < \gamma_n$. □

6.4 Revealed Classes

It this section, we consider only classes whose every element is either an abstract ur-object or a hard set. A class of such classes will be called a **class cluster**.

We remark that the Axiom of Prolongation may not apply to a class cluster (sequence) Z. For example the class cluster $\{\mathrm{FN} \setminus [n]\}_{n\in\mathrm{FN}}$ has no suitable extension.

Similarly to previous sections, ϑ denotes a fixed infinite number lying in the known land of the geometric horizon.

Whenever a set is considered in this section, it means a set X such that $\mathrm{Card}(X) \in [\vartheta]$.

Let F be a stable function, $\mathrm{rng}(F) \subseteq X$. We say that f is a **set extension of the function** F in the class X if

(i) f is a set and a function,

(ii) $F \subseteq f$,

(iii) $\mathrm{rng}(f) \subseteq X$,

(iv) $\mathrm{dom}(f) = [\gamma]$ for some $\gamma \notin \mathrm{FN}$.

We say that a **class** X is **revealed** if every stable function F such that $\mathrm{rng}(F) \subseteq X$ has a set extension in the class X.

In other words, a class X is revealed if every sequence $\{a_n\}_{n\in\mathrm{FN}}$ such that $a_n \in X$ for every n has a set extension $\{a_\alpha\}_{\alpha\leq\gamma}$ where γ is an infinite natural number and $a_\alpha \in X$ for every $\alpha \leq \gamma$.

In very broad terms, a revealed class is such a class from which there is no way out (not even for Zeus).

The following proposition is an immediate consequence of Proposition 6.6.

Proposition 6.13. (i) A class X is revealed if and only if for every sequence $\{a_n\}_{n\in\mathrm{FN}}$ such that $a_n \in X$ for every n there exists a set $v \subseteq X$ such that $a_n \in v$ for every n.

(ii) Every set X is revealed.

Proposition 6.14. Let classes X, Y be revealed. Then also the class $X \times Y$ is revealed.

Proof. Let $\{\langle x_n, y_n\rangle\}_{n\in\mathrm{FN}}$ be a sequence of elements of the class $X \times Y$. Then obviously there are sets u, v such that $u \subseteq X$, $v \subseteq Y$ and $x_n \in u$, $y_n \in v$ for every n. It means that $\langle x_n, y_n\rangle \in u \times v \subseteq X \times Y$ for every n. □

Proposition 6.15. Let classes X, Y be revealed. Then also the class $X \cup Y$ is revealed.

Proof. Let $\{a_n\}_{n\in\mathrm{FN}}$ be a sequence of elements of the class $X \cup Y$. If the entire sequence lies in the class X (or the entire sequence lies in the class Y) then there is nothing to prove. Let $a_k \in X$, $a_l \in Y$ and define

$$a'_n = \begin{cases} a_n & \text{for } a_n \in X, \\ a_k & \text{for } a_n \notin X, \end{cases} \qquad a''_n = \begin{cases} a_n & \text{for } a_n \in Y, \\ a_l & \text{for } a_n \notin Y. \end{cases}$$

Let u, v be sets such that $u \subseteq X$, $v \subseteq Y$ and $a'_n \in u$, $a''_n \in v$ for every n. Then obviously $a_n \in u \cup v \subseteq X \cup Y$ for every n. □

Proposition 6.16. Let classes X, Y be revealed. Then also the class $X \cap Y$ is revealed.

Proof. Let $\{a_n\}_{n \in \mathrm{FN}}$ be a sequence of elements of the class $X \cap Y$. Let $\{a_\alpha\}_{\alpha \leq \gamma_1}$ where $\gamma_1 \notin \mathrm{FN}$ (or $\{a_\alpha\}_{\alpha \leq \gamma_2}$ where $\gamma_2 \notin \mathrm{FN}$ respectively) be its set extensions in the class X (or Y respectively). Let β be the greatest natural number for which the two extensions agree up to β. Then obviously $\beta \notin \mathrm{FN}$ and $\{a_\alpha\}_{\alpha \leq \beta}$ is a set extension of the sequence $\{a_n\}_{n \in \mathrm{FN}}$ in the class $X \cap Y$. □

Silent Version of the Axiom of Choice. Let F be a function such that $\mathrm{dom}(F) = \mathrm{FN}$ and $F(n)$ is a non-empty class for every n. Then there exists a function G such that $\mathrm{dom}(G) = \mathrm{FN}$ and $G(n) \in F(n)$ for every n.

This is a very weak version of the Axiom of Choice implicitly used by mathematicians since long before the birth of set theory without it being seen as problematic. In fact, it was accepted by some opponents of the general Axiom of Choice even later on; for example, in the proof that two different definitions of continuity of a real function at a given point are equivalent. That is why we too will be using this axiom without mentioning it; in fact we might have implicitly used it already.

By induction for finite natural numbers we can easily prove that the above postulated **choice function** G is stable. This enables us to state the silent Axiom of Choice as follows.

Proposition 6.17. Let F be a function, $\mathrm{dom}(F) = \mathrm{FN}$ and let $F(n)$ be a non-empty class for every n.

Then there exists a function f such that f is a set, $\mathrm{dom}(f) = [\gamma]$ where $\gamma \notin \mathrm{FN}$, $\gamma \in [\vartheta]$ and $f(n) \in F(n)$ for every n.

Proposition 6.18. Let G be a function such that $\mathrm{dom}(G) = \mathrm{FN}$ and $G(n)$ is a revealed class for every n. Then the class

$$X = \bigcap\{G(n); n \in \mathrm{FN}\}$$

(the intersection of all the classes $G(n)$) is revealed.

Proof. Let $\{a_n\}_{n \in \mathrm{FN}}$ be a sequence of elements of the class X. Let $\{a_\alpha\}_{\alpha \leq \gamma}$ where $\gamma \notin \mathrm{FN}$ be a set extension of it. Let $H(n)$ denote the class of all $\alpha \leq \gamma$ such that $a_\alpha \in G(n)$. If $\{\overline{a}_\alpha\}_{\alpha \leq \delta}$ is a set extension of the sequence $\{a_n\}_{n \in \mathrm{FN}}$ in the class $G(n)$ then the smallest β such that $\overline{a}_\beta \neq a_\beta$ satisfies $\beta \notin \mathrm{FN}$. Hence $H(n) \setminus \mathrm{FN} \neq \emptyset$.

By the Silent Version of the Axiom of Choice there exists a sequence $\{\beta_n\}_{n \in \mathrm{FN}}$ of natural numbers such that $\beta_n \in H(n) \setminus \mathrm{FN}$ for every n. By Proposition 6.12 there exists β such that $\beta \notin \mathrm{FN}$ and $\beta \leq \beta_n$ for every n. In consequence, the sequence $\{a_\alpha\}_{\alpha \leq \beta}$ is a set extension of the sequence $\{a_n\}_{n \in \mathrm{FN}}$ in every class $G(n)$, and so also in the class X. □

Proposition 6.19. Let G be a function such that $\text{dom}(G) = \text{FN}$ and $G(n)$ is a non-empty revealed class for every n. Let $G(n) \subseteq G(m)$ for $m \leq n$. Then the class

$$X = \bigcap\{G(n); n \in \text{FN}\}$$

is non-empty.

Proof. By the Silent Version of the Axiom of Choice there exists a sequence $\{a_n\}_{n\in\text{FN}}$ such that $a_n \in G(n)$ for every n. If $n \leq m$, $a_m \in G(n)$. Let $\{a_\alpha\}_{\alpha\leq\gamma}$ be a set extension of this sequence. The class $G(n)$ is revealed so there exists $\beta_n \notin \text{FN}$ such that $a_\alpha \in G(n)$ for $n \leq \alpha \leq \beta_n$. Again let $\beta \notin \text{FN}$ be such that $\beta \leq \beta_n$ for every n. Then $a_\beta \in G(n)$ for every n, and so $a_\beta \in X$. □

Proposition 6.20. Let relation R be a revealed class. Then also the classes $\text{dom}(R)$, $\text{rng}(R)$ are revealed.

Proof. Let $\{a_n\}_{n\in\text{FN}}$ be a sequence of elements of the class $\text{dom}(R)$. By the Silent Version of the Axiom of Choice there exists a sequence $\{b_n\}_{n\in\text{FN}}$ such that $\langle b_n, a_n\rangle \in R$ for every n. Let $\{\langle b_\alpha, a_\alpha\rangle\}_{\alpha\leq\gamma}$ be a set extension of the sequence $\{\langle b_n, a_n\rangle\}_{n\in\text{FN}}$ in the class R. Then obviously $\{a_\alpha\}_{\alpha\leq\gamma}$ is a set extension of the sequence $\{a_n\}_{n\in\text{FN}}$ in the class $\text{dom}(R)$.

The case of $\text{rng}(R)$ is analogical. □

Proposition 6.21. Let G be a function such that $\text{dom}(G) = \text{FN}$. Let $G(n)$ be a revealed class for every n. Assume that $G(0)$ is a relation and $G(m) \subseteq G(n)$ for $n \leq m$. Let $X = \bigcap\{G(n); n \in \text{FN}\}$. Then

$$\text{dom}(X) = \bigcap\{\text{dom}\, G(n); n \in \text{FN}\}.$$

Proof. Obviously $\text{dom}(X) \subseteq \text{dom}\, G(n)$ for every n. Conversely let x be such that $x \in \text{dom}\, G(n)$ for every n. Let $S(n) = \{y; \langle y, x\rangle \in G(n)\}$. $S(n)$ is a non-empty revealed class for every n and $S(n) \subseteq S(m)$ whenever $m \leq n$. Hence there is some y such that for every n, $y \in S(n)$ so $\langle y, x\rangle \in G(n)$ and consequently $\langle y, x\rangle \in X$. Hence $x \in \text{dom}(X)$. □

6.5 Forming Countable Classes

We say that a **class** X is **countable** if there exists a bijective function F such that

$$\text{dom}(F) = \text{FN} \qquad \text{and} \qquad \text{rng}(F) = X.$$

We say that a **class** X is **finite** if $n \in \text{FN}$ and there exists a bijective function F such that

$$\text{dom}(F) = [n] \qquad \text{and} \qquad \text{rng}(F) = X.$$

If X is a finite class then X is a finite set since the collection of its elements is sharply defined by the function F.

We say that a **class** X is **uncountable** if it is not finite nor countable.

We say that a **class** X is **at most countable** if it is either countable or finite.

Proposition 6.22. Let X be a subclass of the class FN. Then the following holds:

(i) Let $X \neq \emptyset$. Then the class X has a first element.

(ii) Let $0 \in X$ and let $n+1 \in X$ for every $n \in X$. Then $X = \mathrm{FN}$.

Proof. (i) Choose $n \in X$. Then $[n] \cap X$ is a set. If this set is empty, then n is the first element of the class X; if this set is non-empty, then its first element is obviously also the first element of the class X.

(ii) Assume that $X \neq \mathrm{FN}$. Let n be the first element of the class $\mathrm{FN} \setminus X$ (it exists according to (i)). Since $n \neq 0$, there exists m such that $n = m + 1$. Obviously $m \in X$ and therefore also $n \in X$, which is a contradiction. □

Proposition 6.23. Let $X \subseteq \mathrm{FN}$ and assume that for ever $n \in X$ there exists $m \in X$ such that $n < m$. Then the following is true:

(i) For every $n \in X$ there exists $m \in X$ such that $\mathrm{Card}(X \cap [m]) = n$.

(ii) There exists a bijective function F for which

$$\mathrm{dom}(F) = \mathrm{FN}, \qquad \mathrm{rng}(F) = X$$

and if $n < m$ then $F(n) < F(m)$; hence the class X is countable.

Proof. (i) For a proof by contradiction, assume that n is the smallest natural number for which there exists no $m \in X$ such that $\mathrm{Card}(X \cap [m]) = n$. Obviously $n \neq 0$ and therefore $n = n_0 + 1$. Let $m_0 \in X$ be a number such that $\mathrm{Card}(X \cap [m_0]) = n_0$. Let $m \in X$ be the smallest number such that $m_0 < m$. Then $\mathrm{Card}(X \cap [m]) = n$, which is a contradiction.

(ii) For $n \in X$, let us define $G(n) = \mathrm{Card}(X \cap [n])$. Obviously G is a bijective function and $\mathrm{dom}(G) = X$. By (i), $\mathrm{rng}(G) = \mathrm{FN}$ and for $m, n \in X$, $m < n$ if and only if $G(m) < G(n)$. The function F, which is inverse to the function G, has all the desired properties. □

It is straightforward to check that the following holds.

Proposition 6.24. Let G be a bijective function,

$$X = \mathrm{dom}(G), \qquad Y = \mathrm{rng}(G).$$

If one of the classes X, Y or G is countable, the remaining two are also countable. Similarly, if one of the classes X, Y or G is finite, the remaining two are also finite.

If X is a countable class, $Y \subseteq X$, then the class Y is at most countable.

Proposition 6.25. Let F be a function such that $\mathrm{dom}(F) = \mathrm{FN}$ and $F(n)$ is a hard set for every $n \in \mathrm{FN}$. Then the function F is stable.

Proof. Let X be a class of all n for which $F|[n]$ is a set. Obviously $0 \in X$, since $F|\emptyset = \emptyset$. Let $n \in X$. Then

$$F|[n+1] = F|[n] \cup \{\langle F(n), n\rangle\}; \qquad \text{hence} \quad n+1 \in X.$$

By Proposition 6.22, $X = \mathrm{FN}$. □

We clearly have

Proposition 6.26. Let Z be a countable class. Then Z is not a set.

Proposition 6.27. (i) Let X be a countable class, Y at most countable. Then the class $X \cup Y$ is countable.

(ii) Let A be a finite cluster of countable classes. Then $\bigcup A$ is a countable class.

Proof. (i) Let F_1 and F_2 be bijective functions,

$$\begin{aligned} \mathrm{dom}(F_1) &= X, & \mathrm{rng}(F_1) &= \mathrm{FN}, \\ \mathrm{dom}(F_2) &= Y, & \mathrm{rng}(F_2) &\subseteq \mathrm{FN}. \end{aligned}$$

Let

$$F(x) = \begin{cases} 2F_1(x) & \text{for } x \in X, \\ 2F_2(x) + 1 & \text{for } x \in Y \setminus X. \end{cases}$$

Then F is obviously a bijective function,

$$\mathrm{dom}(F) = X \cup Y, \qquad \mathrm{rng}(F) \subseteq \mathrm{FN}.$$

Since the class $\mathrm{rng}(F)$ is countable, the class $\mathrm{dom}(F)$ is also countable.

(ii) For a proof by contradiction, let n be the smallest natural number such that there is a cluster A of countable classes such that $\mathrm{Card}(A) = n$ but $\bigcup A$ is not a countable class.

Obviously $A \neq \emptyset$; let us choose $Y \in A$. Let us set $B = A \setminus \{Y\}$. Since $\mathrm{Card}(B) < n$, the class $\bigcup B$ is countable (or empty). Hence by (i), $\bigcup A = \bigcup B \cup Y$ is a countable class, which is a contradiction. □

Proposition 6.28. Let F be a function such that $\mathrm{dom}(F)$ is a countable class. Then the class $\mathrm{rng}(F)$ is at most countable.

Proof. Without loss of generality, we can assume that $\mathrm{dom}(F) = \mathrm{FN}$. Let X be a class of all n for which the following holds: if $m < n$ then $F(m) \neq F(n)$. Obviously $F|X$ is a bijective function, $\mathrm{rng}(F) = \mathrm{rng}(F|X)$. Since the class X is either countable or finite, the class $\mathrm{rng}(F)$ is also at most countable. □

Proposition 6.29. (i) The class $\mathrm{FN} \times \mathrm{FN}$ is countable.

(ii) Let X be a countable class, $Y \neq \emptyset$ at most countable. Then $X \times Y$ is a countable class.

Proof. (i) Define $G(\langle m,n\rangle) = 2^m \cdot 3^n$. It is straightforward to check that G is a bijective function, $\mathrm{dom}(G) = \mathrm{FN} \times \mathrm{FN}$, $\mathrm{rng}(G) \subseteq \mathrm{FN}$. The class $\mathrm{rng}(G)$ is countable because it is not finite, and therefore the class $\mathrm{dom}(G)$ is also countable.

(ii) Let F_1 and F_2 be bijective functions,

$$\begin{aligned} \mathrm{dom}(F_1) &= X, & \mathrm{rng}(F_1) &= \mathrm{FN}, \\ \mathrm{dom}(F_2) &= Y, & \mathrm{rng}(F_2) &\subseteq \mathrm{FN}. \end{aligned}$$

For $\langle x,y\rangle \in X \times Y$ let

$$F(\langle x,y\rangle) = \langle F_1(x), F_2(y)\rangle.$$

Obviously F is a bijective function, $\mathrm{dom}(F) = X \times Y$, $\mathrm{rng}(F) \subseteq \mathrm{FN} \times \mathrm{FN}$. The class $\mathrm{rng}(F)$ is countable because it is not finite. As a consequence, the class $\mathrm{dom}(F)$ is also countable. □

Various functions defined on the semiset FN (and therefore also various countable classes) can be generated by means of the so-called construction by mathematical induction. One relatively general version of this principle is contained in the following proposition.

Proposition 6.30. (On construction of a sequence by induction) Let $\mathcal{O}$ be some unary sharply defined operation that turns any finite sequence $\{z_i\}_{i<n}$ into an object $\mathcal{O}(\{z_i\}_{i<n})$. Then there exists a unique sequence $\{v_n\}_{n\in\mathrm{FN}}$ such that for each n, $v_n = \mathcal{O}(\{v_i\}_{i<n})$.

Proof. First we will prove that there exists at most one sequence with the desired property. Assume that $\{v_n\}_{n\in\mathrm{FN}}$, $\{u_n\}_{n\in\mathrm{FN}}$ are two such different sequences. Let n be the smallest natural number such that $u_n \neq v_n$. Then $v_n = \mathcal{O}(\{v_i\}_{i<n}) = \mathcal{O}(\{u_i\}_{i<n}) = u_n$, which is a contradiction.

Now we shall prove that at least one such sequence does exist. Let A be the class of all n for which there exists a finite sequence $\{u_i\}_{i<n}$ such that $u_k = \mathcal{O}(\{u_i\}_{i<k})$ for every $k < n$. Analogously to the first case, we can now prove that for every $n \in A$ there exists only one such finite sequence. If $n \in A$ then, if we add one more term $u_n = \mathcal{O}(\{u_i\}_{i<n})$ to the sequence $\{u_i\}_{i<n}$ corresponding to number n, we get a desired sequence for $n+1$ again. Hence $n+1 \in A$. Consequently $A = \mathrm{FN}$.

It suffices to set $v_n = u_n$, where $\{u_i\}_{i<n+1}$ is the sequence corresponding to the number $n+1$. Obviously $\{v_n\}_{n\in\mathrm{FN}}$ is as required. □

Although the construction by induction is a very useful tool for creating countable classes, it is far from being a sufficient tool for this purpose. We have witnessed this when circumstances forced us to accept the so-called Silent Version of the Axiom of Choice.

Those variants of the Axiom of Choice that are sufficient for the new (alternative) theory of sets and semisets and are both sufficiently weak and transparent that they would have been acceptable in mathematics even before set theory, can be summarized in the axiom of horizon attainability stated below.

As it is in the case of the Axiom of Prolongation, this axiom is also an axiom in the original sense of the word. This means an axiom as something that should be accepted, not an axiom as the word is used today.

We say that a **function** f is **short** if there exists $n \in \mathrm{FN}$ so that

$$\mathrm{dom}(f) = [n].$$

Axiom of Horizon Attainability. Let $Z \neq \emptyset$ be a class whose every element is a short function, and such that if $f \in Z$, $\mathrm{dom}(f) = [n]$ then there exists $g \in Z$ satisfying $f \subseteq g$ and $\mathrm{dom}(g) = [n+1]$. Then there exists a function F such that $\mathrm{dom}(F) = \mathrm{FN}$ and $F|[n] \in Z$ for every n.

Scholion

One of the incentives for the formulation and acceptance of the Axiom of Horizon Attainability is the following consideration.

The ability to make a decision, even if we value the options from which we choose almost equally, is one of the characteristic human features. In other words, no remotely sensible person would behave as the legendary donkey that perished on a dusty road because he could not make up his mind whether to assuage his unbearable hunger grazing on the field to the right or on the field to the left.

Naturally, a super-human cannot be denied this ability either, not even in the case where s/he must choose from possibilities that are of entirely equal worth as regards some abstract object of study. This would, for example, be the case in the following story.

Assume that a super-human wants to walk to the horizon through an environment where after each step s/he needs to select a place to set foot from many possible places. S/he selects one of them and on taking the step s/he is faced with a similar decision again. Such an environment is passable for the super-human. And usually there exists more than one direct way leading through as far as the horizon.

Proposition 6.31. Let Y be an infinite class. Then there exists a countable class X such that $X \subseteq Y$.

Proof. Let Z be the class of all bijective short functions f such that $\mathrm{rng}(f) \subseteq Y$. If $f \in Z$, $\mathrm{dom}(f) = [n]$, then the class $Y \setminus \mathrm{rng}(f) \neq \emptyset$. Let us choose $y \in Y \setminus \mathrm{rng}(f)$ and set $g = f \cup \{(y, n)\}$. Obviously $g \in Z$. The class Z satisfies the requirements of the Axiom of Horizon Attainability. Let F be a function guaranteed to exist by this axiom. If we set $X = \mathrm{rng}(F)$ then X is as required. □

Now we have the means to prove the Silent Version of the Axiom of Choice from the Axiom of Horizon Attainability.

Proposition 6.32. Let G be a function such that $\mathrm{dom}(G) = \mathrm{FN}$ and $G(n)$ is a non-empty class for every n. Then there exists a function F such that $\mathrm{dom}(F) = \mathrm{FN}$ and $F(n) \in G(n)$ for every n.

Proof. Let Z be the class of all short functions f such that if $\mathrm{dom}(f) = [n]$, then $f(m) \in G(m)$ for each $m < n$. It is straightforward to check that the class Z meets the requirements of the Axiom of Horizon Attainability; if F is a function guaranteed to exist by this axiom then F has the required properties. □

Proposition 6.33. Let A be a countable class cluster. Assume that for every $X \in A$, X is at most countable. Then the class $\bigcup A$ is at most countable.

Proof. Obviously we can suppose that $X \neq \emptyset$ for every $X \in A$. Let $A = \{X_n\}_{n\in\mathrm{FN}}$ and let us choose $\gamma \in \mathrm{N} \setminus \mathrm{FN}$. Define $B = \{Y_n\}_{n\in\mathrm{FN}}$, where Y_n is a class of all functions f such that $\mathrm{dom}(f) = [\gamma]$ and $X_n = \mathrm{rng}(f|\mathrm{FN})$. According to the Silent Version of the Axiom of Choice there exists a function G such that $\mathrm{dom}(G) = \mathrm{FN}$ and $G(n) \in Y_n$ for every n. Let $g_n = G(n)$. Let F be a function such that

$$\mathrm{dom}(F) = \mathrm{FN} \times \mathrm{FN} \qquad \text{and} \qquad F(\langle n, m\rangle) = g_n(m).$$

Obviously $\bigcup A = \mathrm{rng}(F)$ and according to Proposition 6.28, $\bigcup A$ is a finite or countable class. □

6.6 Cuts on Natural Numbers

As before, ϑ denotes a fixed infinite natural number lying in the known land of the geometric horizon. Letters α, β, γ denote elements of the set $[\vartheta]$, letters $m, n \ldots$ elements of the semiset FN.

The letter M will be used to denote a headless cut on set $[\vartheta]$ such that $\mathrm{FN} \neq M$.[3]

Proposition 6.34. Let a set f be a function for which the following holds:

(i) $\mathrm{dom}(f) = [\gamma]$ where $\gamma \notin \mathrm{FN}$;

(ii) $\mathrm{rng}(f) \subseteq [\vartheta]$;

(iii) $f(n) \in \mathrm{FN}$ for every $n \in \mathrm{FN}$;

(iv) $\alpha < f(\alpha)$ for every $\alpha < \gamma$;

(v) if $\alpha < \beta < \gamma$, then $f(\alpha) < f(\beta)$.

Then there exists a cut M such that $M \subseteq [\gamma]$ and $f(\alpha) \in M$ for every $\alpha \in M$.

[3] This means that for every $\alpha, \beta < \vartheta$, $\alpha < \beta$, if $\beta \in M$, then also $\alpha \in M$; and $M = \{\alpha; \alpha < \gamma\}$ for no γ.

Proof. We form a sequence $\{\gamma_n\}_{n\in\mathrm{FN}}$ by induction, setting $\gamma_1 = \gamma$ and choosing γ_{n+1} to be the largest natural number such that $f(\gamma_{n+1}) < \gamma_n$. For every $n \in \mathrm{FN}$ obviously $\gamma_n \notin \mathrm{FN}$, and

$$M = \bigcap\{[\gamma_n]\}_{n\in\mathrm{FN}}$$

is a cut with the required properties. □

Proposition 6.35. Let γ be an infinite natural number $\gamma < \vartheta$. Then there exist cuts $M_1, M_2, M_3 \subseteq [\gamma]$ different from FN such that the following holds:

(i) $2\alpha \in M_1$ for every $\alpha \in M_1$;[4]

(ii) $\alpha^2 \in M_2$ for every $\alpha \in M_2$;[5]

(iii) $2^\alpha \in M_3$ for every $\alpha \in M_3$.

Proposition 6.36. Let M be a cut and assume that M is not a union of countably many sets. Then the class M is revealed.

Proof. Let $\{\alpha_n\}_{n\in\mathrm{FN}}$ be a sequence of numbers that belong to the class M. Obviously $\bigcup\{[\alpha_n]\}_{n\in\mathrm{FN}} \subseteq M$ and hence there exists $\gamma \in M$ such that $\{\alpha_n\}_{n\in\mathrm{FN}} \subseteq [\gamma]$. The sequence $\{\alpha_n\}_{n\in\mathrm{FN}}$ has an extension in the set $[\gamma]$, and therefore also in the class M. □

[4] So if $\alpha, \beta \in M_1$, then also $\alpha + \beta \in M_1$.
[5] So if $\alpha, \beta \in M_2$, then also $\alpha\beta \in M_2$.

Chapter 7

Two Important Kinds of Classes

7.1 Motivation – Primarily Evident Phenomena

In the natural real world, the most striking phenomena are those which are primarily evident. A **primarily evident phenomenon** is a phenomenon which we are able to evidence, that is we are able to see it and know that we are seeing it as soon as it can be evidenced, that is, as soon as it has appeared to us.

For example, if we look at this page of this book, we will probably agree that it is not red. In the sensorily perceptible world (and in fact in the natural real world in general), interpreted in the usual way as a community of objects, we interpret non-redness as a unary attendant phenomenon.[1] The phenomenon we are discussing is before our eyes on this page. But to be able to evidence it, we must first know redness from somewhere: in other words, we must have evidenced redness at an earlier date. Non-redness is thus not a primarily evident phenomenon. By contrast, redness is a primarily evident phenomenon because even if we had not come across this phenomenon before, we would notice the red colour as soon as we first saw it. Naturally, not only properties but also some relations are primarily evident; for example the binary relation "one solid is heavier than another one" and so on.

A child who is learning to recognise colours first encounters redness on a few equally-red things on which this phenomenon is sharply distinguished from other colours and other forms of redness. This enables us to interpret the collection of these things as a set. However, it soon turns out that this phenomenon is not as sharply defined as it might have seemed when first encountered, and the set of markedly red things is joined by another set of things which are also red, but in a slightly different way. This can occur a number of times. Still, in order to teach a child how to recognize redness, we do not have to expose him or her to all the different forms of this phenomenon existing in the sensorily perceptible world. It would be impossible and, surprisingly, it is not necessary. A child masters this phenomenon as soon as she or he grasps it in its

[1] That means that we do not interpret it as an independent individual – an object, but as a phenomenon accompanying other objects, appearing on them, characteristic for some of them: in other words as a property.

indistinctness. Only then does s/he begin to understand when somebody talks about a hint of red on a thing and so on. The indistinctness of redness is not something that just came to be associated with this phenomenon in the sensorily perceptible world, something that does not really belong there. Indistinctness is an essential component of it. If we do not grasp the phenomenon of redness in its indistinctness then we do not grasp it at all. In that case we could at best grasp some of its isolated sharp forms, but their connections, especially their organic affiliation to a unique phenomenon would escape us. If there were in some imaginary world only a few equally-red things and nothing else there was red, redness would represent a sharp phenomenon to its inhabitants. But it would not mean that these inhabitants knew this phenomenon more than we do. On the contrary, our knowledge of this phenomenon would be more comprehensive and in this sense more complete. For us, their concept of redness would capture only a special and very narrow form of the phenomenon of redness extracted from the broad span of this phenomenon. As opposed to this imaginary world, the natural real world presents the phenomenon of redness in its indistinctness. A view has opened which allows us to see, on the basis of a few initial forms of this phenomenon, as far as its limits. And the intention to look in the direction of this view has arisen in some spontaneous way straight upon the first few conscious encounters with the phenomenon. This is also the case with many other primarily evident phenomena.

The phenomenon of redness has no sharply defined boundaries but it does have some boundaries. Its boundaries can be interpreted as the horizon on which this phenomenon transforms to other phenomena. There is still redness before this horizon but none beyond; beyond this horizon there is non-redness. In other words, redness goes to up the horizon and no farther; the horizon bounces the phenomenon back and it can only be grasped in its entirety because of the fact that it is thus deflected. Not the individual forms of this phenomenon but the horizon seen through the durable view that opened in the entanglement of phenomena into which we are thrown, holds this phenomenon together.

The horizon in question is positioned differently to what we have been used to so far. It is not true that we can only see what lies before the horizon while we can only speculate what lies beyond it. We are looking at the whole issue, so to speak, from a different standpoint. It is a view from above; we are looking from a proper distance at what is captured by looking from an internal standpoint – a view through the entanglement of phenomena into which we are thrown as far as the horizon limiting the span of our primarily evident phenomenon. This view from above enables us to see what lies before the horizon of our internal look, that is, redness, as well as what lies beyond it, that is, non-redness.

A primarily evident property (and phenomenon in general) is not necessarily identifiable just by its name.

A phenomenon in whose name there is a negative prefix does not necessarily have to be a complement of a primarily evident phenomenon; it may itself be a primarily evident phenomenon. Sometimes we, so to speak, politely, refer to primarily evident antipoles of primarily evident phenomena in this way. For example the words "distasteful meal" usually do not describe a meal which is just not tasteful, but a meal which makes us sick. Analogously the words "unpleasant circumstances" often describe nasty circumstances and we therefore lack a word to describe circumstances that are neither pleasant nor nasty. Also "impolite person" is not just a complement of "polite person" but it denotes a primarily evident phenomenon that should in fact be referred to by the name "rude person".

The nature of primarily evident phenomena is positive in the sense that they present us with something that arises markedly from the entanglement of phenomena into which we are thrown, captures our attention and awakens us to activity. These phenomena (unless they are sharply defined) point to the horizon and hence also to the openness of the world.

By contrast, the nature of complements (that is, of negations) of primarily evident phenomena is negative in the sense that complements deny a presence to phenomena that had been laid so to speak directly before our eyes. Still, it is positive in the sense that they then divert our looks from mundane glistening targets and direct them – as long as we assist this – not into nothingness, but towards the truth hidden in the fascinating depths of terra incognita.

In a world abounding with phenomena which have been born indistinct, semisets are mostly secondary and to a great extent also artificial bearers of the phenomenon of indistinctness. For they arise there only after we begin to interpret the world – albeit unconsciously – as a collection of objects, when we enrich it with sets and classes of these objects and when, while creating classes, not only we do not suppress, but on the contrary we advance the attendant phenomena defining these classes.

For example, greenness, interestingness, dirtiness, largeness and similar are primary bearers of the phenomenon of indistinctness. In contrast, the class of all green leaves on the trees in some autumnal deciduous wood, the class of all interesting books in a given municipal library, the class of all dirty cars in Prague or the class of all heavy stones in a quarry are merely secondary bearers of this phenomenon and this is mediated through its primary bearers.

Still, precisely the secondary bearers of the phenomenon of indistinctness seem to provide a convenient tool for exploration of the phenomenon of indistinctness. That is why we will make them the starting point of our enquiry into this phenomenon. After all, direct exploration of this phenomenon on its primary bearers has, so far, not been very successful.

Moreover, following Bolzano's advice modified for our purposes, wherever secondary bearers of this phenomenon do not obviously arise by themselves or where there are very few of them, we will lay under the primary bearers some convenient if artificial secondary bearers; this means classes describing the distribution of the studied phenomenon.

The way in which we acquire primarily evident phenomena suggests that we should lay a class which is a stable increasing sequence of sets under each such phenomenon. The sets from this sequence are what we lay under the sharp forms of the phenomenon under consideration during its gradual discovery. Its stability corresponds to the enduring view through, thanks to which we can see as far as the horizon limiting the span of the given phenomenon.

The following section is devoted to the concepts motivated by these considerations.

7.2 Mathematization: σ-classes and π-classes

Recall that the letters m, n denote elements of the semiset FN.

Lemma 7.1. Let F be a stable function such that $F(n)$ is a set for every n. Then obviously functions F_+, F_- defined on the semiset FN by

$$F_+(n) = \bigcup \operatorname{rng}(F|[n]), \qquad F_-(n) = \bigcap \operatorname{rng}(F|[n])$$

are stable,

$$F_+(n) \subseteq F_+(m), \qquad F_-(m) \subseteq F_-(n) \quad \text{for } n \leq m$$

and

$$\bigcup \operatorname{rng} F_+ = \bigcup \operatorname{rng} F, \qquad \bigcap \operatorname{rng} F_- = \bigcap \operatorname{rng} F.$$

We say that X is a **σ-class** if there exists a stable function F such that for each n, $F(n)$ is a set, and

$$X = \bigcup \operatorname{rng} F, \quad (\text{that is, } \; X = F(0) \cup F(1) \cup F(2) \cup \dots).$$

Using Lemma 7.1, we can moreover assume about the function F without loss of generality that $F(m) \subseteq F(n)$ for every $m \leq n$. In that case we say that F is a **generating function** (or also generating sequence) of the σ-class X.

Note. By Proposition 6.8 a σ-class X is a set if and only if there is some m such that $X = F(m)$, where F is a generating function of the σ-class X.

If a σ-class X is not a set then it is a semiset, because $\bigcup \operatorname{rng}(F) \subseteq \bigcup \operatorname{rng}(f)$ where f is an extension of the function F such that $f(\alpha)$ is a set for every $\alpha \in \operatorname{dom}(f)$.

Note. If $X \subseteq \text{FN}$ then X is a σ-class since

$$X = \bigcup \operatorname{rng}(F), \qquad \text{where} \quad F(n) = X \cap [n], \quad n \in \text{FN}.$$

We say that X is a π-class if there exists a stable function F such that for each n, $F(n)$ is a set, and

$$X = \bigcap \operatorname{rng} F, \quad (\text{that is, } \; X = F(0) \cap F(1) \cap F(2) \cap \dots).$$

Using Lemma 7.1, we can moreover assume about the function F without loss of generality that $F(m) \subseteq F(n)$ for every $n \leq m$. In that case we say that F is a **generating function** (or also generating sequence) of the π-class X.

Note. Let w be a set, $X \subseteq w$. Then X is a π-class if and only if $w \setminus X$ is a σ-class.

This holds since if F is a generating function of the π-class X then the function G defined on the semiset FN by $G(m) = w \setminus F(m)$ is obviously a generating function of a σ-class, and $w \setminus X = \bigcup \operatorname{rng} G$. Also vice versa.

Note. Since $\bigcap \operatorname{rng} F \subseteq F(0)$, where F is a generating function of a π-class X, every π-class is either a set or a semiset.

By Proposition 6.18, every π-class is revealed.

Hence it is straightforward to check that the following holds:

Proposition 7.2. Let F be a generating function of a π-class X. Then X is a set if and only if there is some n such that $X = F(n)$.

Proposition 7.3. (i) Let X, Y be σ-classes. Then also $X \cup Y$, $X \cap Y$ are σ-classes.

(ii) Let X, Y be π-classes. Then also $X \cup Y$, $X \cap Y$ are π-classes.

Proof. (i) Let F, G be generating functions of classes X, Y. Let us set $K(n) = F(n) \cup G(n)$, $L(n) = F(n) \cap G(n)$. Obviously K, L are stable functions. It is straightforward to check that $X \cup Y = \bigcup \operatorname{rng} K$. We will prove that $X \cap Y = \bigcup \operatorname{rng} L$. Let $z \in X \cap Y$. Then $z \in F(n)$, $z \in G(m)$ for some m, n. Let for example $n \leq m$. Then $z \in F(m)$ and $z \in G(m)$ so $z \in L(m)$, and hence $z \in \bigcup \operatorname{rng} L$. Now let $z \in \bigcup \operatorname{rng} L$. Then $z \in L(n) = F(n) \cap G(n)$ for some n, and so $z \in \operatorname{rng} F \cap \operatorname{rng} G = X \cap Y$.

(ii) is derived from (i) using the De Morgan Rules. □

Proposition 7.4. Let X, Y be disjoint σ-classes (or disjoint π-classes). Then there exist disjoint sets u, v such that $X \subseteq u$, $Y \subseteq v$.

Proof. If X, Y are σ-classes then see Proposition 6.9. Let X, Y be π-classes and let F, G be their generating functions. Assume that $F(n) \cap G(n) \neq \emptyset$ for every n. Then by Proposition 6.10 there exists $z \in \bigcap \operatorname{rng} F$, $z \in \bigcap \operatorname{rng} G$, and so $z \in X \cap Y$, which is a contradiction. Therefore there is some m such that $F(m) \cap G(m) = \emptyset$. It suffices to set $u = F(m)$, $v = G(m)$. □

Proposition 7.5. Let X, Y be σ-classes (or π-classes), w a set, $w \subseteq X \cup Y$. Then there exist disjoint sets u, v such that

$$u \subseteq X, \quad v \subseteq Y, \quad w = u \cup v\,.$$

Proof. Obviously $w \setminus X$, $w \setminus Y$ are disjoint π-classes (or σ-classes). By Proposition 7.4 there is a set d such that $w \setminus X \subseteq d$, $d \cap (w \setminus Y) = \emptyset$. Let $v = w \cap d$. Then $w \setminus X \subseteq v$, and so $w \setminus v \subseteq X$. Analogously, since $v \cap (w \setminus Y) = \emptyset$, we have $w \setminus Y \subseteq w \setminus v$ and hence $w \cap v = v \subseteq Y$. It now suffices to set $u = w \setminus v$. □

Proposition 7.6. Let $X \subseteq Y$, where X is a σ-class and Y is a π-class (or X is a π-class and Y is a σ-class). Then there exists a set v such that $X \subseteq v \subseteq Y$.

Proof. Let w be a set, $Y \subseteq w$. Since X, $w \setminus Y$ are disjoint σ-classes (or disjoint π-classes), by Proposition 7.4 there is a set u such that $X \subseteq u$, $u \cap (w \setminus Y) = \emptyset$. Let us set $v = u \cap w$. Then $X \subseteq v$, $v \cap (w \setminus Y) = \emptyset$, and so $v \subseteq Y$. □

Proposition 7.7. (i) Let X be both a σ-class and a π-class. Then X is a set.

(ii) Let X, Y be disjoint σ-classes and let w a set such that $X \cup Y = w$. Then X, Y are sets.

(iii) Let X, Y be disjoint π-classes and let w a set such that $X \cup Y = w$. Then X, Y are sets.

Proof. (i) By Proposition 7.6 there is a set u such that $X \subseteq u \subseteq X$. Hence $X = u$.

(ii), (iii) Classes X, Y are at the same time σ-classes and π-classes. Further see (i). □

Note. (a) Let a σ-class A be a relation (or a function, or a bijective function). Let F be its generating function. Then $F(n)$ is a relation (or a function, or a bijective function) for every n, since $F(n) \subseteq A$ for every n.

(b) Let a π-class A be relation. Then there exists a generating function F of this relation such that $F(n)$ is a relation for every n because if G is some generating function of the π-class A then we can define $F(n)$ as the set of all ordered pairs that belong to the set $G(n)$. This function F has all the desired properties.

Proposition 7.8. Let a π-class A be a function (or a bijective function respectively). Then the following holds:

(i) There is a generating function F of the π-class A such that $F(n)$ is a function (or a bijective function respectively) for every n.

(ii) There is a function f (or a bijective function f respectively) such that $A \subseteq f$.

Proof. (i) Let G be a generating function of the π-class A such that $G(n)$ is a relation for every n. Let g be an extension of the function G, $\mathrm{dom}(g) = [\gamma]$ where $\gamma \neq \mathrm{FN}$ and such that $g(\beta) \subseteq g(\alpha)$ for $\alpha \leq \beta \in [\gamma]$. Let σ be the smallest number such that $\sigma \in [\gamma]$ and $g(\beta)$ is a (bijective) function for every β, where $\sigma \leq \beta \leq \gamma$. For $\beta \in [\gamma] \setminus \mathrm{FN}$, $g(\beta) \subseteq A$ so $g(\beta)$ is a (bijective) function; hence $\sigma \in \mathrm{FN}$. Let $F(n) = G(\sigma)$ for $n < \sigma$, $F(n) = G(n)$ for $\sigma \leq n$. Obviously F is the sought function.

(ii) It suffices to set $f = F(0)$, where F is a function guaranteed by the proposition 7.8(i). □

Proposition 7.9. Let a σ-class A be a relation. Then $\mathrm{dom}(A)$ is a σ-class.

Proof. Let G be a generating function of the σ-class A. Let $F(n) = \mathrm{dom}\, G(n)$. Obviously F is a stable function. Therefore it suffices to prove that

$$\bigcup \mathrm{rng}\, F = \mathrm{dom}(A).$$

If $x \in \mathrm{dom}(A)$, then there exists y such that $\langle y, x\rangle \in A$. So there exists n such that $\langle y, x\rangle \in G(n)$. Hence $x \in F(n) \subseteq \bigcup \mathrm{rng}\, F$.

If $x \in \bigcup \mathrm{rng}\, F$ then there exists n such that $x \in F(n)$. So there exists y such that $\langle y, x\rangle \in G(n)$. Hence $\langle y, x\rangle \in A$, so $x \in \mathrm{dom}(A)$. □

Proposition 7.10. Let a π-class A be a relation. Then $\mathrm{dom}(A)$ is a π-class.

Proof. If $A = \emptyset$, then it holds trivially. Let $A \neq \emptyset$ and let G be a generating function of the π-class A such that $G(n)$ is a relation for every n. Let $F(n) = \mathrm{dom}\, G(n)$. Obviously F is a stable function. So it suffices to prove that

$$\bigcap \mathrm{rng}(F) = \mathrm{dom}(A).$$

If $x \in \mathrm{dom}(A)$ then there exists y such that $\langle y, x\rangle \in A$, that is, $\langle y, x\rangle \in G(n)$ and $x \in F(n)$ for every n. Hence $x \in \bigcap \mathrm{rng}(F)$.

If $x \in \bigcap \mathrm{rng}(F)$ then $x \in \mathrm{dom}(G(n))$ for every n. Let $K(n)$ denote the set of all ordered pairs $\langle y, x\rangle$ for which $\langle y, x\rangle \in G(n)$. Obviously K is a stable function and $K(n) \neq \emptyset$, $K(n) \subseteq G(n)$ for every n and $K(m) \subseteq K(n)$ for every $n \leq m$. By Proposition 6.11 there exists y such that $\langle y, x\rangle \in K(n) \subseteq G(n)$ for every n. Hence $\langle y, x\rangle \in A$ and so $x \in \mathrm{dom}(A)$. □

Proposition 7.11. Let a σ-class A (or a π-class A) be a relation. Let w be a set such that $A \subseteq w^2$. Let B be the class of all $x \in w$ such that $\langle y, x\rangle \in A$ for every $y \in w$. Then B is also a σ-class (or a π-class).

Proof. It is straightforward to check that $B = w \setminus \mathrm{dom}(w^2 \setminus A)$. Obviously $w^2 \setminus A$ is a π-class (or a σ-class) and by Proposition 7.10 $\mathrm{dom}(w^2 \setminus A)$ is also a π-class (or according to Proposition 7.9 $\mathrm{dom}(w^2 \setminus A)$ is also a σ-class), and so $w \setminus \mathrm{dom}(w^2 \setminus A)$ is a σ-class (or a π-class). □

Proposition 7.12. Let X_1, X_2 be σ-classes (or π-classes). Then also their Cartesian product $X_1 \times X_2$ is a σ-class (or a π-class).

Proof. Let F_1, F_2 be generating functions of classes X_1, X_2. Let $F(n) = F_1(n) \times F_2(n)$ for $n \in \mathrm{FN}$. Obviously F is a stable function and $X_1 \times X_2 = \bigcup \mathrm{rng}(F)$ (or $X_1 \times X_2 = \bigcap \mathrm{rng}(F)$). □

7.3 Applications

We apply the mathematical results acquired in the previous section (and later on, almost all the results gained in this book) in the real world in the same way in which the results of ancient geometry have been applied ever since its origins. That is by substituting the geometric horizon with the horizon limiting the human looks into the natural real world – while evidencing inevitable distortions this brings.

Our above considerations imply that it is the σ-classes which correspond to the primarily evident phenomena when mathematization is carried out using classes of abstract objects. This means that we place σ-classes under these phenomena and we make judgements about the layout of the phenomena from the layout of these classes; in particular, about the form of the indistinctness of their definition. In other words, the indistinctness of primarily evident phenomena as such can be studied on the abstract frames of the secondary bearers of these phenomena – σ-classes in this case – and the results can then be translated to various primarily evident phenomena. To illustrate this point, we shall now demonstrate a few simple, but illuminating examples of such translations of mathematical results obtained earlier. Proposition 7.7 has simple and yet unfamiliar consequences applying to the layouts of primarily evident phenomena, even in the sensorily perceptible world.

For example the properties of "being a human" and "being a monkey" are primarily evident and, let us say, mutually exclusive. Let w be some set of higher living creatures and let X be the class of people belonging to the set w and Y the class of monkeys belonging to the set w. As X, Y are non-intersecting σ-classes, $X \cup Y \subseteq w$, only the following two cases are possible. Either $X \cup Y \neq w$, and therefore the set w has an element which is neither a human nor a monkey.[2] Or $X \cup Y = w$, and then according to Proposition 7.7, X, Y are sets and therefore humans belonging to the set w are distinctly separate from the monkeys which also belong to this set. This happens in the case where w is a set whose elements are only currently living humans and monkeys. The third possibility $X \cap Y \neq \emptyset$ was rather hastily excluded at the very beginning. This would be the case if a creature could be interpreted as being simultaneously a human and a monkey.

Also in case of other primarily evident properties and phenomena we can argue in the same way. For example "cold" and "warm" are mutually exclusive primarily evident phenomena and by the above proposition, there must be situations when it is neither cold nor warm. Cold is not the negation of warm. Still, if we had no other choice but a sultry desert or an Arctic iceberg then warm would be the negation of cold, although only because these two phenomena are distinctly separated.

The case of properties "being a living person" and "being a dead person" is worthy of special attention. Both these phenomena are primarily evident; let us also initially assume that they are mutually exclusive. Consider the process of some person dying, and suppose we lay a set of moments in rapid succession under it such that the person is still alive at the first moment but dead at the last moment. If the two phenomena covered the whole sequence then by the above proposition a moment would necessarily exist, at which the person would

[2] This would presumably be the case if w was the set of all members of the sequence starting from the Charlie the Monkey whose every subsequent element is the son of the previous member and which ends with Mr. Charles Darwin.

still be alive but s/he would be dead at the subsequent one. However, this could only happen if the intervals between the individual moments were long enough. If they were very short, which corresponds to this situation better, both the classes of moments when the person is still alive or already dead respectively, are semisets. Hence there must be moments (and there are quite a few) when a dying person is neither alive nor dead. For a surgeon carrying out a delicate organ transplant operation, the case when the two σ-classes intersect and thus a moment can be identified when a dying person is simultaneously alive and dead would be still more disheartening.

The propositions from the previous section imply that the complement of a σ-class X in its superset w (that is, the class $w \setminus X$) is not a σ-class unless the class X is a set. The negations of primarily evident properties (and phenomena in general) thus have other layouts than the primarily evident phenomena. Above all, the form of indistinctness present in their definition is different, unless we deal with sharply defined properties (phenomena).

For that matter, just the primarily evident attendant phenomena of objects were originally regarded as properties. It is only the contemporary logic which accords an entirely equal status to the negations of primarily evident properties, and moreover also to contents of a whole range of propositions which can be formed on their basis using logical connectives or even quantifiers. If we equalise primarily evident properties with their negations then it follows from Proposition 7.7 that we have no choice but to interpret properties as sharp phenomena. Also the other way round however, interpreting properties as sharp phenomena has made it possible to give equal status to primarily evident properties and their negations and then to any other properties describable from them using logical connectives.

Since we capture the layouts of our explored phenomena by means of suitable classes which we lay under these phenomena, we shall do so also in the case of negations of primarily evident properties (or phenomena). However, we will only study their layouts in close proximity to the negated property (or phenomenon); that means on some set such that the σ-class capturing the layout of the negated property (or phenomenon) is a subclass of it.

It follows that when we carry out matematization using classes of abstract objects, thus restricted negations of primarily evident properties (and phenomena in general) correspond to π-classes.

Recall again that π-classes are diametrically different from σ-classes. By Proposition 7.7, every class that is simultaneously a π-class and a σ-class is a set. Hence we can deduce that a phenomenon which is primarily evident and such that its restricted negation is primarily evident, is sharply defined.

A special attention must be paid to those π-classes which we lay under the gaps between pairs of mutually exclusive – but related as antipoles of the same more general phenomenon – primarily evident phenomena. We will call them **residual π-classes**.

They indeed are π-classes, since if X, Y are the σ-classes underlying the antipodal primarily evident phenomena then $X \cap Y = \emptyset$ and by Proposition 7.3, $X \cup Y$ is a σ-class. If w is a set such that $X, Y \subseteq w$ then $w \setminus (X \cup Y)$ is a π-class.

If at least one of the antipodal phenomena is indistinct, for example if X is not a set, then $X \cup Y$ is not a set by Proposition 7.7 and $w \setminus (X \cup Y)$ is therefore also not a set. Hence $w \setminus (X \cup Y) \neq \emptyset$.

For example, if we proceed along the frequently mentioned sequence leading from Charlie the Monkey to Mr. Darwin, then at the start of it there arises a phenomenon which (let zoologists forgive us) we will call a monkey and somewhere close to the end there arises the phenomenon of a human. These phenomena did not change suddenly one into the other. Both are primarily evident and they do not intersect on this sequence. Therefore we deal with two disjoint σ-classes and the gap between them is filled by that obscure and quite extensive residual π-class. In this gap, other primarily evident phenomena may arise, particular creatures different both from humans and from monkeys.[3] The corresponding σ-classes are subclasses of the residual π-class and Proposition 7.7 applies to them.

This example allows us to make the following preliminary observations.

(a) Above all, the residual π-class along with all that happens inside it, captures our attention and invites further exploration.

(b) Even though the natural infinity alone does not invoke the various phenomena of the natural real world, it nevertheless participates in invoking them, initiating it.

   To wit, the appearance of phenomena on the given sequence depends on the position of this sequence in relation to the horizon. We can realise this if we – with some stretch of imagination – picture all its elements as lying before the horizon; that is, with any two elements clearly distinguishable. In that case we would probably refuse to talk about humans and monkeys since we would fail to see them in the sequence; these phenomena would not be there. It would merely be a sequence of isolated individuals.

In contrast to the previous example where the residual π-class is extensive in a conspicuous way because this class is, so to speak, stretched out, in the example of a dying person the residual π-class (that is, the class underlying the gap where the dying person is neither alive nor dead) is contracted almost to a single point. However, this does not mean that the class in the second example should necessarily have to be less extensive. Under a different, much more detailed look at the dying person, this class would undoubtedly stretch out and new primarily evident phenomena would appear in the gap between life and death, similarly to the previous example. And this stretched-out residual π-class would naturally also capture our attention and interest.

We can see the utility of stretching-out the residual π-classes which have been contracted into a single point (that is, the residual π-classes which have

[3] For that matter, there are such phenomena evidenced by anthropologists there.

fallen beyond the horizon leaving only a trace of their existence on it in the form of a single point) on the following example.

Ice melts or water freezes (under normal and somewhat idealized conditions) exactly at the temperature of 0 degrees centigrade on a mercury Celsius thermometer with, let us say, scale marked from -20 to $+20$ degrees. Below zero is ice and above it is water. Both of these phenomena are primarily evident and they meet at one point only. If they do not intersect then this point should be the trace of a phenomenon which can be underlain by a π-class; if they do intersect then it should be a σ-class. In other words, at zero temperature something should be covertly going on with the water, or ice, that can be interpreted as a gradual change even if it does not appear like that. We can confirm this if we replace the temperature scale with a heat scale, that is, with a scale on which the marks defining the segments denote equal increases or decreases of heat in the investigated substance. Physicists inform us that in that case the original range from -20 to 0 degrees will be replaced with 10 segments, the original 0 point will be stretched to almost 80 segments and the original range from 0 to $+20$ will be represented by 20 segments. This means that there is much more going on at the temperature of 0 degrees than in the remaining range from -20 to $+20$ degrees.

The case of evaporation is analogous though more difficult since it is affected by air pressure.

7.4 Distortion of Natural Phenomena

It would be relatively easy to defend the thesis that, with the exception of small natural numbers, only those phenomena are sharply defined that have been created by people affected by the European civilization (or in whose creation such people significantly participated).

In contrast, what we may call natural phenomena and especially those amongst them which are primarily evident, are almost always indistinct.

Such a natural phenomenon becomes sharp only when it is possible to make a list of all its occurrences. However, if we produce such a list then we usually distort the original phenomenon and, strictly speaking, this distortion yields a new phenomenon.

So, for example, if we make a list of all living monkeys then the class of all creatures included in the list is a set and the phenomenon "living monkey" is sharply defined by this set. But then it is not the original phenomenon of a living monkey, rather a distortion thereof. A number of new monkeys may have been born and others which were included in the list may have already died since the moment when the list was produced.

Moreover such a distortion of the phenomenon "a living monkey" also distorts the original primarily evident phenomenon of a monkey. From this moment, a monkey may only be such a creature, regardless whether living, not yet born or already dead which is at least very similar to one of the monkeys on this list.

On the other hand, sharp distortions of original natural phenomena are very operable. Instead of working directly with the phenomena, we can operate with the respective lists.

Clearly it is not necessary to discuss in detail, that precisely such distortion of original natural phenomena is what lies at the foundations of the world of European civilization and its highly efficient operability.

It is a world where a citizen is s/he who is included in a list of citizens and a house owner is s/he who is included in the appropriate list at the land registry.

But it is also a world whose sharp laws are a distortion of justice and where the monetary price of an object is a distortion of its value.

It is a world in which splendid indefinite shapes and colours appearing on a meadow in bloom do not merit the attention of the science about this world.

Chapter 8

Hierarchy of Descriptive Classes

8.1 Borel Classes

In this section we will outline mathematization of some phenomena which are brought about by simple manipulation of primarily evident phenomena. We shall introduce a hierarchy of classes reminiscent of the hierarchy of Borel sets of a topological space in Cantor's classical set theory. However, in our case the hierarchy is not bound to any given topology.

In this whole section we assume that every class is a subclass of some previously chosen hard set w. This means that classes which we are dealing with are either sets or semisets.

The letters m,n denote elements of the semiset FN.

If we say that $\{X_n\}_{n\in \mathrm{FN}}$ (or in an abbreviated form $\{X_n\}$) is a **sequence of classes** then we mean that there is a function G such that

$$\mathrm{dom}(G) = \mathrm{FN} \qquad \text{and} \qquad G(n) = X_n \quad \text{for every } n.$$

By the **Cantor function** we mean the well-known bijective function H such that

$$\mathrm{dom}(H) = \mathrm{FN} \qquad \text{and} \qquad \mathrm{rng}(H) = \mathrm{FN} \times \mathrm{FN}.$$

By a **Borelian symbol** we mean a finitely long word (that is, a finitely long sequence, possibly empty) made up of signs σ, π only.

Letters $\mathcal{B}, \mathcal{B}_1, \mathcal{B}_2, \ldots$ denote Borelian symbols.

$\mathcal{B}\mathcal{B}_1$ denotes the Borelian symbol obtained by writing the word $\mathcal{B}_1$ behind the word $\mathcal{B}$.

The **dual symbol** for a Borelian symbol $\mathcal{B}$ is the Borelian symbol $\tilde{\mathcal{B}}$ obtained from $\mathcal{B}$ by substituting every σ with π and every π with σ.

The property X is a **$\mathcal{B}$-class** is defined recursively on the length of the word $\mathcal{B}$ using the following rules:

(i) If $\mathcal{B}$ is the empty word then X is a $\mathcal{B}$-class if and only if X is a set.

(ii) X is a $\sigma\mathcal{B}$-class if and only if there is a sequence of $\mathcal{B}$-classes $\{X_n\}_{n\in\mathrm{FN}}$ such that $X = \bigcup\{X_n\}_{n\in\mathrm{FN}}$.

(iii) X is a $\pi\mathcal{B}$-class if and only if there is a sequence of $\mathcal{B}$-classes $\{X_n\}_{n\in\mathrm{FN}}$ such that $X = \bigcap\{X_n\}_{n\in\mathrm{FN}}$.

Clearly the definitions of properties X is a σ-class and X is a π-class as given in Chapter 7 are in accordance with the more general definition set out above.

We say that a **class X is Borel** if X is a $\mathcal{B}$-class for some Borelian symbol $\mathcal{B}$.

Proposition 8.1. Let X be a $\mathcal{B}$-class, u a set and $\mathcal{B}$ a Borelian symbol. Then the following holds:

(i) $X \cap u$, $X \times u$, $u \times X$ are $\mathcal{B}$-classes;

(ii) $u \setminus X$ is a $\tilde{\mathcal{B}}$-class.

Proof. By induction on the length of the word $\mathcal{B}$. □

Proposition 8.2. (i) Let X be a $\sigma\sigma\mathcal{B}$-class. Then X is a $\sigma\mathcal{B}$-class.

(ii) Let X be a $\pi\pi\mathcal{B}$- class. Then X is a $\pi\mathcal{B}$-class.

Proof. (i) $X = \bigcup\{X_n\}_{n\in\mathrm{FN}}$ where every X_n is a $\sigma\mathcal{B}$-class, so

$$X_n = \bigcup\{X_{n,m}\}_{m\in\mathrm{FN}}$$

where every $X_{n,m}$ is a $\mathcal{B}$-class. Hence

$$X = \bigcup\{X_{n,m}; \langle n,m\rangle \in \mathrm{FN}\times\mathrm{FN}\}.$$

Using the Cantor function, we can see that X is a $\sigma\mathcal{B}$-class.

(ii) $X = \bigcap\{X_n\}_{n\in\mathrm{FN}}$ where every X_n is a $\pi\mathcal{B}$-class, so

$$X_n = \bigcap\{X_{n,m}\}_{m\in\mathrm{FN}}$$

where every $X_{n,m}$ is a $\mathcal{B}$-class. Hence

$$X = \bigcap\{X_{n,m}; \langle n,m\rangle \in \mathrm{FN}\times\mathrm{FN}\}.$$

Using the Cantor function, we can see that X is a $\pi\mathcal{B}$-class. □

The above proposition enables us to restrict the domain of Borelian symbols to those symbols in which no sign occurs twice in a row.

Proposition 8.3. Let X, Y be $\mathcal{B}$-classes. Then also $X\cup Y$, $X\cap Y$ are $\mathcal{B}$-classes.

Proof. We shall prove this proposition by induction on the length of the word $\mathcal{B}$. The case of the empty word is trivial. Assume that the proposition holds for $\mathcal{B}$-classes. We will prove that it then also holds for $\sigma\mathcal{B}$-classes (and $\pi\mathcal{B}$-classes).

First let X, Y be $\sigma\mathcal{B}$-classes,

$$X = \bigcup\{X_n\}_{n\in \mathrm{FN}}, \quad Y = \bigcup\{Y_n\}_{n\in \mathrm{FN}}$$

where X_n, Y_n are $\mathcal{B}$-classes. By the inductive hypothesis, $X_n \cup Y_m$ and $X_n \cap Y_m$ are $\mathcal{B}$-classes for every n, m. Since

$$X \cup Y = \bigcup\{X_n \cup Y_m; n, m \in \mathrm{FN}\}$$

$$X \cap Y = \bigcup\{X_n \cap Y_m; n, m \in \mathrm{FN}\},$$

using the Cantor function we can easily see that $X \cup Y$ and $X \cap Y$ are $\sigma\mathcal{B}$-classes. The case when X, Y be $\pi\mathcal{B}$-classes is similar. □

The following proposition is an almost immediate consequence of the proposition proven above.

Proposition 8.4. (i) X is a $\sigma\mathcal{B}$-class if and only if $X = \bigcup\{X_n\}_{n\in \mathrm{FN}}$, where X_n are $\mathcal{B}$-classes and $X_n \subseteq X_m$ for every $n \leq m$.

(ii) X is a $\pi\mathcal{B}$-class if and only if $X = \bigcap\{X_n\}_{n\in \mathrm{FN}}$, where X_n are $\mathcal{B}$-classes and $X_m \subseteq X_n$ for every $n \leq m$.

Proposition 8.5. Let X, Y be $\mathcal{B}$-classes. Then $X \times Y$ is also a $\mathcal{B}$-class.

Proof. This holds trivially if $\mathcal{B}$ is the empty word. Assume that the proposition holds for $\mathcal{B}$-classes. We will prove that it then also holds for $\sigma\mathcal{B}$-classes and for $\pi\mathcal{B}$-classes.

Let $X = \bigcup\{X_n\}_{n\in \mathrm{FN}}$, $Y = \bigcup\{Y_n\}_{n\in \mathrm{FN}}$ (or $X = \bigcap\{X_n\}$, $Y = \bigcap\{Y_n\}$), where X_n, Y_n are $\mathcal{B}$-classes such that for $n \leq m$, $X_n \subseteq X_m$, $Y_n \subseteq Y_m$ (or $X_m \subseteq X_n$, $Y_m \subseteq Y_n$). Then

$$X \times Y = \bigcup\{X_n \times Y_n\}_{n\in \mathrm{FN}}$$

(or $X \times Y = \bigcap\{X_n \times Y_n\}$.) According to the inductive hypothesis, $X_n \times Y_n$ is a $\mathcal{B}$-class. It follows that $X \times Y$ is a $\sigma\mathcal{B}$-class (or $\pi\mathcal{B}$-class). □

Some simple rules for operations with Borelian symbols.

Let $\mathcal{B}_1 \to \mathcal{B}_2$ stand for the following statement:

If X is a $\mathcal{B}_1$-class then X is also a $\mathcal{B}_2$-class.

Since the definition of the σ- (or π-) operation does not require the classes whose union (or intersection) this operation creates to be different, it is immediately obvious that the following holds:

Rule 8.6. $\mathcal{B} \to \sigma\mathcal{B}, \quad \mathcal{B} \to \pi\mathcal{B}$.

Rule 8.7. $\mathcal{B} \to \mathcal{B}\sigma, \quad \mathcal{B} \to \mathcal{B}\pi$.

Rule 8.8. $\mathcal{B} \to \mathcal{B}_1\mathcal{B}\mathcal{B}_2$.[1]

If σ (or π) is the first sign in a non-empty word $\mathcal{B}$ then $\gamma\mathcal{B}$ denotes the word $\pi\mathcal{B}$ (or the word $\sigma\mathcal{B}$).

If σ (or π) is the last sign in a non-empty word $\mathcal{B}$, then $\mathcal{B}\delta$ denotes the word $\mathcal{B}\pi$ (or $\mathcal{B}\sigma$).

Rule 8.9. The words $\gamma\mathcal{B}$ and $\tilde{\mathcal{B}}\delta$ are identical.[2]

Rule 8.10. The words $\mathcal{B}\delta$ and $\gamma\tilde{\mathcal{B}}$ are identical.

Hence we have:

Proposition 8.11. (i) Let X be a $\mathcal{B}$-class. Then X is a $\gamma\mathcal{B}$, $\mathcal{B}\delta$, $\gamma\tilde{\mathcal{B}}$, $\tilde{\mathcal{B}}\delta$-class.

(ii) Let X be a $\mathcal{B}$-class and let Y be a $\tilde{\mathcal{B}}$-class. Then $X \cap Y$ and $X \cup Y$ are $\gamma\mathcal{B}$, $\mathcal{B}\delta$, $\gamma\tilde{\mathcal{B}}$, $\tilde{\mathcal{B}}\delta$-classes.[3]

(iii) Let X, Y be $\mathcal{B}$-classes. Then $X \setminus Y$ is a $\gamma\mathcal{B}$, $\mathcal{B}\delta$, $\gamma\tilde{\mathcal{B}}$, $\tilde{\mathcal{B}}\delta$-class.

Proposition 8.12. (i) Let X be a π-class and let Y be a σ-class. Then $X \cup Y$, $X \cap Y$, $X \times Y$, $X \setminus Y$, $Y \setminus X$ are $\sigma\pi$-classes and also $\pi\sigma$-classes.

(ii)) Let X be a $\pi\sigma$-class and let Y be a $\sigma\pi$-class. Then $X \cup Y$, $X \cap Y$, $X \times Y$, $X \setminus Y$, $Y \setminus X$ are $\sigma\pi\sigma$-classes and at the same time $\pi\sigma\pi$-classes.

Proof. This is a special case of Proposition 8.11. □

8.2 Analytic Classes

The letters j, k, l, m, n denote elements of the semiset FN.

Proposition 8.13. (i) Let X be a π-class. Then also $\mathrm{dom}(X)$ is a π-class.

(ii) Let X be a σ-class. Then also $\mathrm{dom}(X)$ is a σ-class.

(iii) Let X be a $\sigma\pi$- class. Then also $\mathrm{dom}(X)$ is a $\sigma\pi$-class.

Proof. For (i) see Proposition 7.10. (ii) and(iii) are consequences of the identity

$$\mathrm{dom} \bigcup \{X_n\}_{n \in \mathrm{FN}} = \bigcup \{\mathrm{dom}(X_n)\}_{n \in \mathrm{FN}}$$

(and in the case (iii) also of the assertion (i)). □

We say that a **class** X is **analytic** (or also **Suslin** and briefly just X is an **A**-class) if $X = \mathrm{dom}(Y)$, where Y is a $\pi\sigma$ class.

[1] This follows directly from Rules 8.6 and 8.7.

[2] They have the same number of signs and the same first sign.

[3] See (i) and Proposition 8.3.

Proposition 8.14. Let $\{X_n\}_{n\in\mathrm{FN}}$ be a sequence of **A**-classes. Then

$$\bigcap\{X_n\}_{n\in\mathrm{FN}} \qquad \text{is an } \mathbf{A}\text{-class.}$$

Proof. (Adam Ráž) We seek a $\pi\sigma$ class Y such that $\bigcap\{X_n\} = \mathrm{dom}(Y)$. Let $X_n = \mathrm{dom}(Y_n)$ where Y_n is a $\pi\sigma$-class, so $Y_n = \bigcap_{k\in\mathrm{FN}} \bigcup_{l\in\mathrm{FN}} Y_n^{k,l}$ where $Y_n^{k,l}$ are sets. Let w be a set containing all domains and ranges of $Y_n^{k,l}$ as subsets and let $\gamma \notin \mathrm{FN}$ be a natural number lying in the known land. For any $n, k, l \in \mathrm{FN}$ we introduce the set

$$Z_n^{k,l} = \{\langle \{a_\beta\}_{\beta\in[\gamma]}, x\rangle \in w^{[\gamma]} \times w; \langle a_n, x\rangle \in Y_n^{k,l}\}.$$

We define $Y = \bigcap_{n,k\in\mathrm{FN}} \bigcup_{l\in\mathrm{FN}} Z_n^{k,l}$, which is clearly a $\pi\sigma$-class. We shall prove that Y is as required, that is, $\bigcap\{X_n\} = \mathrm{dom}(Y)$.

First let $x \in \bigcap\{X_n\}$. For each n, $x \in X_n = \mathrm{dom}(Y_n)$ and hence there exists $b_n \in w$ such that $\langle b_n, x\rangle \in Y_n$. Clearly for each n, k there exists l such that $\langle b_n, x\rangle \in Y_n^{k,l}$. Using the Silent Version of the Axiom of Choice we define a sequence $\{b_n\}$, which is stable. Let $\{b_\beta\}_{\beta\in[\gamma]} \in w^{[\gamma]}$ be some extension of it guaranteed to exist by the Axiom of Prolongation. Now we can easily check that for any n, k there exists l such that $\langle \{b_\beta\}_{\beta\in[\gamma]}, x\rangle \in Z_n^{k,l}$. Hence $\langle \{b_\beta\}_{\beta\in[\gamma]}, x\rangle \in Y$. We have proved that $x \in \mathrm{dom}(Y)$.

Conversely, let $x \in \mathrm{dom}(Y)$. Then there is c such that $\langle c, x\rangle \in Y$. For each $n, k \in \mathrm{FN}$ there exists $l \in \mathrm{FN}$ such that $\langle c, x\rangle \in Z_n^{k,l}$ and hence $c \in w^{[\gamma]}$ and it is of the form $\{c_\beta\}_{\beta\in[\gamma]}$, where $\langle c_n, x\rangle \in Y_n^{k,l}$. For each $n \in \mathrm{FN}$ we have $\langle c_n, x\rangle \in Y_n$ and thus $x \in \mathrm{dom}(Y_n) = X_n$. We have shown that $x \in \bigcap\{X_n\}$.
□

Proposition 8.15. Let $\{Y_n\}_{n\in\mathrm{FN}}$ be a sequence of sets such that for each $n \neq m$, $Y_m \cap Y_n = \emptyset$. Let $\{X_n\}_{n\in\mathrm{FN}}$ be a sequence of $\pi\sigma$-classes such that for each n, $X_n \subseteq Y_n$. Then

$$\bigcup\{X_n\}_{n\in\mathrm{FN}} \qquad \text{is a } \pi\sigma\text{-class.}$$

Proof. Let $X_n = \bigcap\{X_n^j\}_{j\in\mathrm{FN}}$, where X_n^j are σ-classes. As $X_n \subseteq Y_n$, we can suppose without loss of generality that for each n, j we have $X_n^j \subseteq Y_n$ and for $k \le j$ we have $X_n^j \subseteq X_n^k$. Let $Z_j = \bigcup\{X_n^j\}_{n\in\mathrm{FN}}$. Clearly every Z_j is a σ-class. It is sufficient to prove that

$$\bigcup\{X_n\}_{n\in\mathrm{FN}} = \bigcap\{Z_j\}_{j\in\mathrm{FN}}.$$

First let $x \in X_n$ for an n. Then $x \in X_n^j$ for every j, that is, $x \in Z_j$ for every j. Hence $x \in \bigcap\{Z_j\}_{j\in\mathrm{FN}}$.

Now let $x \in Z_j$ for every j. Then for every j there exists n such that $x \in X_n^j$. If $x \in X_m^i \cap X_n^j$, then $x \in Y_m \cap Y_n$ so $m = n$. This means that there exists n such that for every j we have $x \in X_n^j$, that is, $x \in \bigcap\{X_n^j\}_{j\in\mathrm{FN}} = X_n$. Hence $x \in \bigcup\{X_n\}_{n\in\mathrm{FN}}$. □

Note. If X, Y are **A**-classes and $X, Y \subseteq w$, where w is a set, then we can easily verify that also $X \times w$, $w \times Y$ are **A**-clasees, and in consequence to Proposition 8.14 also $X \times Y$ is an **A**-class.

Proposition 8.16. Let X be an **A**-class. Then also $\mathrm{dom}(X)$ is an **A**-class.

Proof. $X = \mathrm{dom}(Z)$ where Z is a $\pi\sigma$-class; $\mathrm{dom}(X) = \mathrm{dom}(\mathrm{dom}(Z))$. Hence we can assume without loss of generality that every element of the class Z is of the form $\langle z, \langle y, x\rangle\rangle$ and

$$\mathrm{dom}(X) = \{x; (\exists z)(\exists y)\langle z, \langle y, x\rangle\rangle \in Z\}.$$

Let Z' denote the class of all $\langle\langle z, y\rangle, x\rangle$ where $\langle z, \langle y, x\rangle\rangle \in Z$. We can easily show that Z' is a $\pi\sigma$-class. Moreover, $\mathrm{dom}(X) = \mathrm{dom}(Z')$. □

Proposition 8.17. Let $\{Y_n\}_{n\in\mathrm{FN}}$ be a sequence of **A**-classes. Then $\bigcup\{Y_n\}_{n\in\mathrm{FN}}$ is an **A**-class.

Proof. Y_n is $\mathrm{dom}(X_n)$ where X_n is a $\pi\sigma$-class and thus also $\{n\} \times X_n$ is a $\pi\sigma$-class. Let

$$X = \bigcup\{\mathrm{dom}(X_n)\}_{n\in\mathrm{FN}} = \mathrm{dom}\bigcup\{X_n\}_{n\in\mathrm{FN}}.$$

For every n, $\{n\} \times X_n \subseteq \{n\} \times w$, where $\{n\} \times w$ are pairwise disjoint sets. By Proposition 8.15 also $\bigcup\{\{n\} \times X_n\}_{n\in\mathrm{FN}}$ is a $\pi\sigma$-class. Hence

$$\mathrm{dom}\bigcup\{\{n\} \times X_n\}_{n\in\mathrm{FN}} = \bigcup\{X_n\}_{n\in\mathrm{FN}} \qquad \text{is an } \mathbf{A}\text{-class.}$$

By Proposition 8.16 also class X is an **A**-class. □

Propositions 8.14 and 8.17 almost directly imply the following fundamental proposition about analytic classes.

Proposition 8.18. Every Borel class is analytic.

Chapter 9

Topology

9.1 Motivation – Medial Look at Sets

If we are looking at a moderately large heap of sand from the distance of, let us say, ten centimetres, we can see and distinguish single grains belonging to this heap. But we are not able to see the whole heap since the sand takes up and exceeds our field of view.

If we look at the same heap of sand from a distance of ten metres, we can no longer see as far as the single grains. The heap, along with the whole set of grains belonging to it, has fallen beyond the horizon. Instead, we see something else, namely some connected body of a certain shape, a kind of continuum which we interpret as having been formed by the coalescence or merger of the single grains. If we look in the direction of this heap of sand from the distance of ten kilometres then we cannot see it at all. Not even a trace of it is left. If we did not know about it beforehand, it would not occur to us that there was something like that somewhere far beyond the horizon.

By a **medial look** at a set we mean a look at it under which, rather than seeing its individual elements, we see only traces left by them on the horizon, and instead of the whole set, we see only its trace; this trace possesses a certain **topological shape**.

Hence in the above example, a look from ten, but also twenty or thirty metres constitutes a medial look at the heap of sand.

Under a medial look, elements of the observed set no longer lie before the horizon. Still, the topological shape (usually a continuum) that we see is a trace left by this set on the horizon. This topological shape enables us to infer that the observed set (and therefore also its individual elements) still exists, albeit somewhere beyond the horizon; in other words, that this set has merely fallen beyond the horizon. Moreover, it indicates the direction in which we should look for the set. It is however also the limit of our look in the direction of the observed set; we cannot surpass this limit until we have moved the horizon farther back using a sharper look. The sighted topological shape on the horizon may thus be interpreted as a phenomenon lying on the horizon which challenges us to interpret it as a manifestation of a phenomenon lying beyond the horizon.

Under a medial look at some set we do not see its individual elements; they have fallen beyond the horizon and therefore it makes no sense to speak of their direct discernibility as we can do, for example, in the case of the sheets of paper lying on our desk. Some of the sheets, those which we have written on, are easily discernible, while the blank sheets of paper are less easily discernible or even not at all. But we can discern them indirectly, that is, without directly seeing them, namely by means of the trace that they have left on the horizon. So, for example, when we observe a table standing in front of us, we cannot see the molecules the table is made of and we cannot discern them directly. But two molecules, one of which is in the middle of the table and the other at its edge, are indirectly discernible, because it is easy to discern the middle of the table from its edge. (We can cut the table so that each of the two molecules is in a different part.) By contrast, molecules that blend for us into one point on the observed table are not even indirectly discernible in this way.

In fact, every look is accompanied by certain **indiscernibility**, namely by the indiscernibility that connects two objects which have fallen beyond the horizon just when they have left the same trace on the horizon. We can even interpret this indiscernibility as equivalence. More precisely, we can do this at least in some cases, like those molecules or atoms and similar.

Scholion

Another important phenomenon, second only to horizon, that has also been denied its place as a subject of study of modern European science is the phenomenon of indiscernibility.

This is well justified from the point of view of the pre-established approach characteristic for this science. Discernibility and therefore also indiscernibility are not sharp phenomena. Moreover, indiscernibility is regarded as imperfection, since what is indiscernible at one point very often becomes discernible if our powers of resolution are refined, that is, if the horizon limiting discernibility is moved farther back. The aim of modern science is not to explore indiscernibility, but to eliminate it.

However, even though we can overcome this or that form of the phenomenon of indiscernibility just as we can move the horizon farther back in some cases, we are not likely to overcome indiscernibility as such. Even if we allow that the world is merely a structure on a set of atomic objects, if we did manage to overcome indiscernibility entirely – that is, if we were able to discern every two such objects – the world, or at least some observed part of it, would appear as a set of isolated objects, and consequently the phenomenon of continuity as we know it would disappear. More precisely, if we excluded indiscernibility and continued to interpret continuity in terms of sets and to do so sharply (that is, if we did not admit the existence of indistinctly defined classes, namely the existence of semisets), we would have no other option than to interpret the world in the same way as the classical set-theoretical mathematics has done.

If we were consistent, then we should even demand discernibility of any two geometric points in space. But we alone cannot claim this and we would have to look for somebody who is able to do that and who would moreover be able, in turn, to re-join these points in such a way that they have some layout etc.

We will not embark on these speculations and accept that the phenomenon of indiscernibility is here to stay and that we are not able to overcome it. Our above considerations regarding overcoming it suggest that we should be able to use the phenomenon of indiscernibility to interpret the phenomenon of continuity, and that the phenomenon of indiscernibility forms the grounds for the cohesion of the world interpreted as a community of objects. We shall demonstrate that such an interpretation is indeed possible.

When carrying out mathematisation of the phenomenon of indiscernibility – in accordance with what we have already said – we shall limit ourselves to those forms of it that we can come across in worlds interpreted as communities of objects; namely indiscernibility of objects belonging to some such world. At the heart of any such indiscernibility there is the corresponding indiscernibility of two objects, that is, the binary relation into which two objects enter if and only if they are indiscernible. Using this binary relation we can capture also the indiscernibility of a greater number of objects. This is why, in what follows, by indiscernibility we will mean just its heart, that is the corresponding binary relation between objects.

It is not necessary to explain at length that on the domain of its effect, that is, on the explored class of objects, every indiscernibility is a reflexive, symmetric and sometimes also transitive relation.

Moreover, every indiscernibility is a negation of the corresponding discernibility; in other words, the negation of a primarily evident relation, because discernibility is primarily evident.

In order to avoid unnecessary diffusion in our considerations about indiscernibility and in order be able to focus on what we find important, we will always suppose that the domain of effect of the explored indiscernibility is some given (usually quite large) hard set.

Some very important indiscernibilities may be interpreted as equivalences on a given set. It is appropriate to start our enquiry into indiscernibilities from them since the structures which classical set-theoretical mathematics has introduced on sets in an effort to grasp the topological shapes arise spontaneously as organic derivatives of these indiscernibilities.

9.2 Mathematization – Equivalence of Indiscernibility

In the whole of this chapter, w denotes a non-empty hard set such that $\mathrm{Card}(w) \in [\vartheta]$.[1]

The letters x, y, z denote elements of the set w; the letters u, v denote subsets of the set w. The letters X, Y, Z denote subclasses of the set w.

[1] ϑ is as in the beginning of Chapter 6. [Ed]

Our previous considerations justify the formulation of the following definition:

We say that a relation R is a **symmetry of indiscernibility** on the set w if the following holds:

(i) R is a reflexive and symmetric relation on the set w,

(ii) $\mathrm{dom}(R) = w$,

(iii) R is a π-class.

We say that the relation R is an **equivalence of indiscernibility** on the set w when, moreover, the following holds:

(iv) The relation R is transitive on the set w.

In this section, R denotes some given equivalence of indiscernibility on the set w.

By the **monad** of an object x (denoted $\mathrm{Mon}(x)$) we understand the class of all y such that yRx.

Hence $\mathrm{Mon}(x)$ is the class of all y that have left the same trace on the horizon as x has. Since R is an equivalence, $\mathrm{Mon}(x) = \mathrm{Mon}(y)$ if and only if yRx. If it is not true that yRx then $\mathrm{Mon}(x) \cap \mathrm{Mon}(y) = \emptyset$.

By the **figure** of a class X (denoted $\mathrm{Fig}(X)$) we understand the class of all y for which there exists $x \in X$ such that yRx.

Therefore classes X, Y have left the same trace on the horizon if and only if $\mathrm{Fig}(X) = \mathrm{Fig}(Y)$ (that is, when they have the same figure). Obviously for every x we have $\mathrm{Mon}(x) = \mathrm{Fig}(\{x\})$ and for every X, $\mathrm{Fig}(X)$ is the union of the class of all $\mathrm{Mon}(x)$, where $x \in X$.

We say that a **class** X is a **figure** if for every x, y such that $x \in X$ and yRx also $y \in X$ (in other words, if for all $x \in X$ we have $\mathrm{Mon}(x) \subseteq X$).

The sets $\emptyset$, w are clearly figures.

The following proposition is trivial.

Proposition 9.1. For every X, Y the following holds:

(i) $X \subseteq \mathrm{Fig}(X)$;

(ii) if $X \subseteq Y$ then $\mathrm{Fig}(X) \subseteq \mathrm{Fig}(Y)$;

(iii) $\mathrm{Fig}(X) \cap \mathrm{Fig}(Y) = \emptyset$ if and only if $X \cap \mathrm{Fig}(Y) = \emptyset$;

(iv) $\mathrm{Fig}(X)$ is a figure;

(v) X is a figure if and only if $X = \mathrm{Fig}(X)$;

(vi) if Y is a figure and $X \subseteq Y$ then $\mathrm{Fig}(X) \subseteq Y$;

(vii) if X, Y are figures then $X \setminus Y$ is also a figure;

(viii) if $\mathcal{M}$ is a cluster of figures then $\bigcup \mathcal{M}$ is a figure. If $\mathcal{M}$ is non-empty then $\bigcap \mathcal{M}$ is a figure.

Proposition 9.2. Let X be a π-class. Then also $\mathrm{Fig}(X)$ is a π-class.[2]

Proof. It is straightforward to check that $\mathrm{Fig}(X) = \mathrm{dom}(R \cap (X \times w))$, where all the operations used make π-classes from π-classes. See Propositions 7.12, 7.3 and 7.10. □

We say that **classes** X, Y are **separable** if there exists a set u such that

$$\mathrm{Fig}(X) \subseteq u, \qquad \mathrm{Fig}(Y) \cap u = \emptyset.$$

Obviously classes X, Y are separable if and only if $\mathrm{Fig}(X)$, $\mathrm{Fig}(Y)$ are separable.

Separability of classes X, Y thus means that the set w can be separated into two set – and therefore into two sharply defined parts – u and $w \setminus u$, in such a way, that the traces left by X and Y on the horizon each lies in a different part.

So for example, to put it loosely, two blots on a sheet of paper are separable if this paper can be cut up into two pieces in such a way that each of the blots is on one of the pieces (of course each on a different one).

Proposition 9.3. (i) Let X, Y be π-classes such that $\mathrm{Fig}(X) \cap \mathrm{Fig}(Y) = \emptyset$. Then the classes X, Y are separable.

(ii) Let X, Y be π-classes such that $\mathrm{Fig}(X) \cap \mathrm{Fig}(Y) = \emptyset$. Then there is a set u such that $\mathrm{Fig}(X) \subseteq u$, $\mathrm{Fig}(u) \cap \mathrm{Fig}(Y) = \emptyset$.

(iii) Let x, y be such that it is not true that xRy. Then classes $\mathrm{Mon}(x)$ and $\mathrm{Mon}(y)$ are separable.

(iv) Let x, u be such that $\mathrm{Mon}(x) \subseteq u$. Then there is a set v such that $\mathrm{Mon}(x) \subseteq v \subseteq \mathrm{Fig}(v) \subseteq u$.

Proof. (i) See Proposition 7.4.

(ii) Let v be a set such that $\mathrm{Fig}(X) \subseteq v$, $v \cap \mathrm{Fig}(Y) = \emptyset$. Since $\mathrm{Fig}(X) \cap (w \setminus v) = \emptyset$, also $\mathrm{Fig}(X) \cap \mathrm{Fig}(w \setminus v) = \emptyset$ and hence there exists a set u such that

$$\mathrm{Fig}(X) \subseteq u, \qquad \mathrm{Fig}(u) \cap \mathrm{Fig}(w \setminus v) = \emptyset \quad \text{and so} \quad \mathrm{Fig}(u) \cap (w \setminus v) = \emptyset.$$

It follows that $\mathrm{Fig}(u) \subseteq v$ and hence $\mathrm{Fig}(u) \cap \mathrm{Fig}(Y) = \emptyset$.

(iii) is a consequence of (i), since $\mathrm{Mon}(x) \cap \mathrm{Mon}(y) = \emptyset$ and monads are π-classes.

(iv) By(ii) there exists a set v such that $\mathrm{Mon}(x) \subseteq v$ and $\mathrm{Fig}(v) \cap \mathrm{Fig}(w \setminus u) = \emptyset$, so $\mathrm{Fig}(v) \cap (w \setminus u) = \emptyset$. Hence $\mathrm{Fig}(v) \subseteq u$. □

By the **closure of a class** X (denoted $\overline{X}$) we understand the class of all x such that the classes $\mathrm{Mon}(x)$ and X are not separable.

The following proposition is trivial.

[2] Hence $\mathrm{Mon}(x)$, $\mathrm{Fig}(u)$ are π-classes for every x and for every u.

Proposition 9.4. For every X, Y the following holds:

(i) $\overline{X}$ is a figure;

(ii) $\overline{X} = \overline{\mathrm{Fig}(X)}$;

(iii) If $\mathrm{Fig}(X) = \mathrm{Fig}(Y)$, then $\overline{X} = \overline{Y}$.

Proposition 9.5. For every X, Y the following holds:

(i) $X \subseteq \overline{X}$;

(ii) if $X \subseteq Y$ then $\overline{X} \subseteq \overline{Y}$;

(iii) $\overline{X \cup Y} = \overline{X} \cup \overline{Y}$;

(iv) $\overline{X} = \overline{\overline{X}}$.

Proof. (i), (ii), (iii) are easy to prove. (iv) Obviously it is sufficient to show that $\overline{\overline{X}} \subseteq \overline{X}$. Let $x \in \overline{\overline{X}}$, $\mathrm{Mon}(x) \subseteq u$. We will prove $u \cap X \neq \emptyset$. Let v be some set such that $\mathrm{Mon}(x) \subseteq v \subseteq \mathrm{Fig}(v) \subseteq u$ (see Proposition 9.3(iv)). As $x \in \overline{\overline{X}}$, there exists $y \in v \cap \overline{X}$. Hence $\mathrm{Mon}(y) \subseteq \mathrm{Fig}(v) \subseteq u$. As $y \in \overline{X}$, there exists $z \in X$ such that $z \in u$, so $z \in X \cap u$. □

We say that a class X is **closed**, if $X = \overline{X}$. We say that a class X is **open**, if $w \setminus X$ is closed. A class is **clopen** if it is both closed and open.

Obviously every closed and every open class is a figure.

Some of the following propositions have two forms, one for closed classes and the other for open classes. It is easy to see that both forms are equivalent so it is sufficient if we prove only one of them.

Proposition 9.6. (i) The sets $\emptyset$, w are clopen.

(ii) Let X, Y be closed (or open) classes. Then also $X \cup Y$ is a closed (or $X \cap Y$ is an open) class.

(iii) Let $\mathcal{M}$ be a non-empty cluster of closed (or open) classes. Then also $\bigcap \mathcal{M}$ is a closed (or $\bigcup \mathcal{M}$ is an open) class.

Proof. (i) holds trivially.

(ii) According to Proposition 9.5 $X \cup Y \subseteq \overline{X \cup Y} = \overline{X} \cup \overline{Y}$; hence $X \cup Y = \overline{X \cup Y}$.

(iii) Let us set $X = \bigcap \mathcal{M}$. Obviously X is a figure and if $x \in \overline{X}$, then because for every $Y \in \mathcal{M}$, $X \subseteq Y$ holds, and Y is a closed class, $x \in Y$. Hence $x \in \bigcap \mathcal{M} = X$. □

Proposition 9.7. Let a class X be a figure and a π-class (or a σ-class). Then X is closed (or open).

Proof. If $x \notin X$, then $\mathrm{Mon}(x) \cap X = \emptyset$ and according to Proposition 9.3(i) there is some u such that $\mathrm{Mon}(x) \subseteq u$, $u \cap X = \emptyset$. Hence $x \notin \overline{X}$. □

Proposition 9.8. $\overline{\{x\}} = \mathrm{Mon}(x)$ for every x and so $\mathrm{Mon}(x)$ is a closed class.

Proof. As $\overline{\{x\}}$ is a figure, $\mathrm{Mon}(x) \subseteq \overline{\{x\}}$. If $y \notin \mathrm{Mon}(x)$, then according to Proposition 9.3(iii) $y \notin \overline{\{x\}}$. □

A version of the following proposition in classical topology is called the **axiom of regularity**.

Proposition 9.9. (axiom of regularity) Let X be a closed class, $x \notin X$. Then there is an open class Z such that $x \in Z$ and $\overline{Z} \cap X = \emptyset$.

Proof. Let u be a set such that $\mathrm{Mon}(x) \subseteq u$, $u \cap X = \emptyset$. Let v be a set such that $\mathrm{Mon}(x) \subseteq v \subseteq \mathrm{Fig}(v) \subseteq u$ (see Proposition 9.3(iv)). Therefore $\mathrm{Fig}(v) \cap X = \emptyset$. Since $\mathrm{Mon}(x) \cap (w \setminus v) = \emptyset$, $\mathrm{Mon}(x) \cap \mathrm{Fig}(w \setminus v) = \emptyset$. Let $Z = w \setminus \mathrm{Fig}(w \setminus v)$. Then Z is an open class, $\mathrm{Mon}(x) \subseteq Z$. As $Z \cap \mathrm{Fig}(w \setminus v) = \emptyset$, $Z \cap (w \setminus v) = \emptyset$, so $\overline{Z} \subseteq \overline{v} = \mathrm{Fig}(v)$ (see Proposition 9.7). Hence $\overline{Z} \cap X = \emptyset$. □

9.3 Historical Intermezzo

The real world is usually placed into the real space. For a long time, this space has been interpreted as a shaped field of emptiness. And the space of the geometric world has been interpreted similarly. Bodies of real (or geometric) objects take various shapes and they fill up various places in space. Places in space have shapes; they supply them to bodies or to emptiness. The shape of a body is the shape of the place in space that is occupied at that moment by the given body.

Places are in space. Space is the most comprehensive place; all other places are parts of it. All the possible shapes thus lie in space. Space decides on the possible shapes for various bodies, namely by providing places to be filled by them. Thus, if we want to study all possible shapes, we have to study space.

Until the birth of non-Euclidean geometries, the real space had been identified with the classical Euclidean space. However, even later on it has been – and sometimes still is – interpreted as a geometric, even though not necessarily Euclidean, space. Such interpretation makes it possible for the space to be a subject of a perfect – sharp – science: mathematics. Even though mathematics alone cannot decide which of the different geometric spaces the real space is, it can study all these spaces, and hence also the real space which is one of them.

Space as such, separated from everything that fills it, is empty. However, it is very difficult to study the places in an empty space. The shape of a place can be seen, and hence handled, only after this place has been filled. In the pre-set-theoretical mathematics, the space was studied by filling various places with geometric objects and investigating the laws which governed their shapes.

One of the most demanding challenges that the classical set-theoretical mathematics had to face on the journey to achieve its goals was the development of an interpretation of shapes in space. Its method of meeting this challenge has proved to be extremely efficient. Space is not interpreted as a molded emptiness; on the contrary, it is completely filled up to the tiniest place, namely with points. Space is then identified with the set of all the points that have been thus placed in it, and individual places in space are interpreted as subsets of this set. So for example a given straight line (or the place for it) is identified with the set of all the points lying on it (or occupying this place). Also conversely however, every subset of the set of space points now becomes a geometric object. This substantially enlarges the domain of the geometric objects studied so far (even in Euclidean spaces). Objects belonging to this extended domain of objects came to be called topological objects, and their shapes came to be called topological shapes. Moreover, the need to distinguish between a topological object and the place occupied by it in space, has disappeared.

However, from the point of view of a strict and rigorous science which does not admit any vagueness, no set alone – even if it is a set of points in some space – has a topological shape. Loosely said, if we break a space into individual points, we also break its shape and the shape of every place in it. That is why, for example, the interpretation of a straight line as a set of points lying on it, had been regarded for many centuries as unacceptable; and let us note that this was rightly so. At this point we are not thinking of the fact that this interpretation has imposed actualisation of classically interpreted infinity (since modern mathematics had no understanding of natural infinity). Rather, continuum broken into individual points ceases to be a continuum. If it is to become a continuum again, these points must be appropriately glued together. Classical mathematics solved this problem by capturing the structure of this gluing; for example by singling out open and closed sets. In this conception of mathematics, a space (we mean a topological space) is a set with some topological structure (for example, a set of all open sets or an operation of closure of sets, and similar). This approach resulted in a discovery of a range of other, hitherto unknown topological spaces and hence also other topological shapes.

Despite the fact that the classical set-theoretical mathematics captured the structure of such re-gluing of points together in great detail, the phenomenon of continuum as such and of shapes in general was merely set-theoretically described by it. The classical set-theoretical mathematics has not provided any deeper interpretation of how these phenomena, in particular continua, arise for example in the sensorily perceptible world. On the other hand it must be admitted that within narrow margins, pursuing just the goals of set theory, it coped with the problem of capturing and describing the variety of possible shapes in an excellent and essentially exhausting way.

9.4 The Nature of Topological Shapes

The objects belonging to the set w, which have entered into relation R where R is an equivalence of indiscernibility on the set w, have fallen beyond the horizon (if they were lying before the horizon, we would be able to discern them). In a way, we have been considering something that we cannot see. Still, somehow we do see it. We cannot see the semisets which have arisen on the set w; or more precisely, we do not see them as classes. We see neither subsets of the set w nor the set w; more precisely, we do not see them as sets and not even as classes. And since we cannot see the objects belonging to the set w, we cannot see how they enter into the equivalence of indiscernibility R.

What we do see are traces left by these semisets, sets and objects on the horizon or, if you wish, we see their backward projections onto the horizon. In other words, we do not see directly them but only their traces and as long as these traces remain, they bear witness to the fact that the sets, semisets and objects are still there, somewhere close beyond the horizon.

The trace left by an element of the set w on the horizon is a point. Under a look, using physical or mental sight, this trace is a geometric point; what Euclid referred to when he wrote:

> A point is that which has no part.

We add:

> [...] and that which shows in the geometric world,

since it is this world that is described and investigated in Euclid's *Elements*.

Geometric points have a layout even though it is not further divisible; they have a shape even though it is not developed in any of the many forms of shape; they have a size even though it is zero.

One and the same geometric point showing on the horizon can be the trace of many elements of the set w and in this sense it determines the corresponding monad on this set; also, a monad in the equivalence of indiscernibility R on the set w determines a corresponding geometric point on the horizon.

The indiscernibility captured by the relation R forms more than the monads on the set w; also, and in fact primarily, it ties them together. The corresponding ties between the traces of the monads, that is, between the geometric points, evokes the resulting topological shape in which we see the set w on the horizon.

Figures in the equivalence R leave traces on the horizon consisting of the traces of those monads that are their subclasses, so from the corresponding geometric points. The variety of traces of figures, their shapes, their relative layout and interlinking enables us to describe and study the resulting topological shape in which the set w has appeared through them.

If we substitute the corresponding geometric points (or just abstract points) for monads then Proposition 9.5 introduces to the class of all these points a topology in a form closely related to that used for the study of topological shapes by the classical set-theoretical mathematics.

However, the classical set-theoretical topology only studies the traces of figures on the horizon and not what it is that forms them from beyond the horizon; by contrast, to us topological shapes are phenomena derived from equivalences of indiscernibility. In this there lies both a correspondence and a difference between our conception and the classical one. The propositions proved above allow those familiar with the basics of topology to see that the shape evoked by an equivalence of indiscernibility is a regular topological space from the point of view of classical topology. In consequence, more general topological spaces have been left outside the sphere of our interest. On the other hand, matters like the delicate but far from negligible differences between a figure which is a figure of a set and a figure which is not a figure of a set would be hard to capture in classical topology. Moreover, the direction of further study of topological shapes is in our case determined by various additional yet justified conditions imposed on the equivalences of indiscernibility, which make it quite different from the direction of the topological research as carried out in the classical set-theoretical topology. Still, both these conceptions of study of topological shapes, the classical conception and our conception, do have also much in common. Writing about that on which the two agree would however be just a waste of time and paper.

Topological shapes arising under medial looks on the horizon as traces left there by sets which have fallen beyond the horizon, and by their subclasses, depend both on the equivalences of indiscernibility associated with these looks and on the observed sets and their subclasses. It is however possible to separate and free them and even to grant them a higher modality of being. For we perceive these shapes while the sets and classes that have alerted us to them, can be only interpreted as presumed. They have served as targets for our looks; but one can aim even without targets and even such looks are aimed at something, or at least somewhere (sometimes even at something that is hidden behind another visible object).

The classical mathematics also sensed that geometric and topological shapes can be interpreted as phenomena showing on the horizon. However, beside lying as far as the classically interpreted infinity, this horizon was understood as a sharp and rigid boundary. Even so it clearly was a horizon, albeit a horizon in the limit and a sharpened one; for it was being surpassed in a way which is typical for a horizon. This happened within the infinitesimal calculus. However, it was regarded as something inappropriate or even mystical and something to be avoided.

9.5 Applications: Invisible Topological Shapes

Although we understand looks in an unusually broad sense, our previous investigation of topological shapes as phenomena showing on the horizon under medial looks at sets whose elements have fallen behind the horizon, was primarily guided by visual or tactile looks (physical or mental). Still, other looks into a variety of worlds, especially into worlds interpreted as communities of objects, are limited by horizons and there are the corresponding fundamental triads for them. Even there we often look medially at some sets whose elements have fallen beyond the horizon, and therefore there are corresponding equivalences of indiscernibility.

An equivalence of indiscernibility determines the topological shape of the trace left by the observed set on the horizon, and hence even under these other looks the sets (or their subclasses) have topological shapes. The horizon itself however does not need to be as striking a phenomenon as it is in the case of visual or tactile looks. These topological shapes are likely to be inaccessible by sight or touch (physical or mental). We may not even be able to evidence the points making up these shapes.

Thus for example we may be looking – that means considering while employing some criteria of discernibility – at a set which contains the semiset of all living people as a subclass. The criteria may be various criteria of intelligence, interests, character, health, physiognomy and so on. Under any such look the human community leaves a certain topological shape on the horizon; this shape has a dimension, homology groups, components etc. Even though these shapes are hard to grasp because they are neither visible nor tangible, they are there. Only if we were to set such sharp criteria that every person would be distinguishable from every other person would we obtain a discrete shape of human society (that is, a shape composed from isolated points only). But then we could add nothing much about this society. Everything we do say about it concerns phenomena appearing on it which are underlain by topological shapes that arise on the horizon under various medial looks at it. Only our discourse about it is not conducted using topological concepts; more precisely, not concepts of classical topology.

More advanced developments of classical topology have involved forming concepts aimed predominantly at capturing notable phenomena, which show on topological shapes arising on the horizon under visual looks, albeit looks often just mental and sharpened to the limit. For looks other than the visual or tactile ones the classical topological concepts and research directions may prove to be much less useful. When building a topology based on equivalences of indiscernibility we should thus avoid senseless adoption of concepts of classical topology; each such concept needs to be considered anew, and if necessary new concepts developed suitable for topology which corresponds to other sorts of looks.

Summarising, no matter in which direction we focus our attention towards some more extensive community of objects (and this could be almost anywhere), we will always find topological shapes, as long as we are looking for them. The tangle of phenomena into which we have been thrown thus becomes a tangle of topological shapes, even if often invisible ones.

Chapter 10

Synoptic Indiscernibility

10.1 Synoptic Symmetry of Indiscernibility

In this section, R denotes a given symmetry of indiscernibility on the set w. The letters m, n denote elements of the semiset FN.

Proposition 10.1. There exists a generating function F of the relation R such that $F(n)$ is a symmetry on the set $\mathrm{dom}(F(n))$ for every n.

Proof. Let G be a generating function of the relation R such that $G(n)$ is a relation on the set w for every n. Let

$$F(n) = G(n) \cup \{\langle y, x\rangle; \langle x, y\rangle \in G(n)\} \cup \{\langle x, x\rangle; x \in dom(G(n)) \cup rng(G(n)\}.$$

F is clearly a stable function, $F(n)$ is a symmetry on the set $\mathrm{dom}(F(n))$ for every n and if $n < m$ then $F(m) \subseteq F(n)$. Obviously

$$\bigcap \mathrm{rng}(F) = \bigcap \mathrm{rng}(G) = R,$$

so F is a generation function of the relation R. □

In this chapter, r denotes a chosen fixed function such that the following holds:

(i) r is a set;

(ii) $\mathrm{dom}(r) = [\delta]$, where $\delta \in [\vartheta] \setminus \mathrm{FN}$;

(iii) $r(\beta) \subseteq r(\alpha)$ for every $\alpha < \beta < \delta$;

(iv) $r(\alpha)$ is a symmetry on the set $\mathrm{dom}(r(\alpha))$ for every $\alpha < \delta$;

(v) $r(0) = w \times w$;

(vi) the relation R is the intersection of all $r(\alpha)$ where $\alpha \in \mathrm{FN}$.

Proposition 10.2. The symmetry R is an equivalence, if and only if for every n there exists m such that $r(m) \circ r(m) \subseteq r(n)$.

Proof. First let R be an equivalence. For a proof by contradiction assume that there is some n such that for every m there are x_m, y_m and z_m satisfying $\langle x_m, y_m \rangle$, $\langle y_m, z_m \rangle \in r(m)$, $\langle x_m, z_m \rangle \notin r(n)$. Using the Axiom of Prolongation we can see that there exist $\gamma \in [\delta] \setminus \mathrm{FN}$ and x, y, z such that

$$\langle x, y \rangle, \langle y, z \rangle \in r(\gamma) \subseteq R, \qquad \langle x, z \rangle \notin r(n)$$

and so $\langle x, z \rangle \notin R$, which is a contradiction.

Now assume that R is not an equivalence. Then there exist x, y, z such that

$$\langle x, y \rangle \in R, \quad \langle y, z \rangle \in R, \quad \langle x, z \rangle \notin R.$$

Let α be the smallest number such that $\langle x, z \rangle \notin r(\alpha)$. Obviously $\alpha \in \mathrm{FN}$ and at the same time $\langle x, y \rangle \in r(\beta)$, $\langle y, z \rangle \in r(\beta)$ for every $\beta \in \mathrm{FN}$. □

We say that a set u is an **$r(\alpha)$-net** (where $\alpha < \delta$), if

$$\langle x, y \rangle \notin r(\alpha) \qquad \text{for every} \quad x, y \in u \text{ and } x \neq y.$$

Obviously the following holds:

Proposition 10.3. Let $a < \delta \in [\vartheta] \setminus \mathrm{FN}$.

(i) Let u be an $r(\alpha)$-net, $v \subseteq u$. Then v is also an $r(\alpha)$-net.

(ii) Let $\alpha \leq \beta < \delta$, let u be an $r(\alpha)$-net. Then u is also an $r(\beta)$-net.

We say that a set u is a **maximal $r(\alpha)$-net** (where $\alpha < \delta$) if u is an $r(\alpha)$-net and for any v satisfying $u \subseteq v$ and $u \neq v$, v is not an $r(\alpha)$-net.

We can easily see that the following holds:

Proposition 10.4. (i) Let u be an $r(\alpha)$-net. Then u is maximal, if and only if for every y there exists a $x \in u$ such that $\langle x, y \rangle \in r(\alpha)$.

(ii) Let u be an $r(\alpha)$-net. Then there exists a maximal $r(\alpha)$-net v such that $u \subseteq v$.

Proof. (i) Assume that u is maximal. If y were such that for no $x \in u$, $\langle x, y \rangle \in r(\alpha)$ then $y \notin u$ and $u \cup \{y\}$ would be an $r(\alpha)$-net.

Now assume that u is not maximal. Let $v \neq u$, $u \subseteq v$ be an $r(\alpha)$-net, $y \in v \setminus u$. Then $\langle x, y \rangle \in r(\alpha)$ for no $x \in u$.

(ii) From all the $r(\alpha)$-nets that contain u as a subset we choose one of those with the largest number of elements for v. □

Proposition 10.5. There exists a function t such that $\mathrm{dom}(t) = [\delta]$, $\delta \in [\vartheta] \setminus \mathrm{FN}$, $t(\alpha)$ is a maximal $r(\alpha)$-net for every $\alpha < \delta$ and $t(\alpha) \subseteq t(\beta)$ for every $\alpha \leq \beta < \delta$.

Proof. Let ϑ be the natural number giving the number of elements of the set w. Let us number the elements of the set w using natural numbers smaller than ϑ. If u is an $r(\alpha)$-net then $k(u,\alpha)$ denotes the set $u \cup \{x\}$ where x is the first element of the set w in the chosen numbering such that $u \cup \{x\}$ is an $r(\alpha)$-net. If there is no such element x then we set $k(u,\alpha) = u$. Clearly $k(u,\alpha) = u$, if and only if u is a maximal $r(\alpha)$-net. Using inductive construction, we can now easily create a function l of two variables such that for every $r(\alpha)$-net u, $l(u,\alpha)$ is a maximal $r(\alpha)$-net, $u \subseteq l(u,\alpha)$. Using inductive construction, we can then create the function t by choosing a maximal $r(0)$-net for $t(0)$ and setting $t(\alpha+1) = l(t(\alpha), \alpha+1)$ for $\alpha < \alpha+1 < \delta$. □

We say that **a symmetry of indiscernibility** R on w is **synoptic** if for every set $u \subseteq w$ with $\mathrm{Card}(u) \notin \mathrm{FN}$ there exist $x, y \in u$, $x \neq y$ such that $\langle x, y\rangle \in R$. In other words, it is impossible to place any "large" set $u \subseteq w$, in such a way that each two different elements of it are discernible, in the field of vision of the look at w which evokes the synoptic symmetry of indiscernibility R. So in very broad terms, it is such a symmetry of indiscernibility where magnification is limited by the same horizon as reduction; in both cases we deal with the horizon whose distance is captured by the semiset FN.

In what follows we assume that the symmetry of indiscernibility R is synoptic on the set w.

We can easily see that the following holds:

Proposition 10.6. (i) Let u be an $r(n)$-net. Then $\mathrm{Card}(u) \in \mathrm{FN}$.

(ii) For every n there exists m such that $\mathrm{Card}(u) \leq m$ for every $r(n)$-net u.

(iii) For every $\beta \notin \mathrm{FN}$ there exists $\alpha \in [\delta] \setminus \mathrm{FN}$ and an $r(\alpha)$-net u such that $\mathrm{Card}(u) \leq \beta$.

Proof. (i) is obvious.

(ii) For a given n, we choose β as the largest natural number for which there exists an $r(n)$-net u such that $\mathrm{Card}(u) = \beta$. According to (i), $\beta \in \mathrm{FN}$.

(iii) For a given β, we choose α as the largest natural number such that $\alpha < \beta$ and for every $r(\alpha)$-net u is $\mathrm{Card}(u) \leq \beta$. Obviously $\alpha \notin \mathrm{FN}$. □

We say that a **set** u is a **maximal** $r(n)$**-net inside a class** X if $u \subseteq X$, u is an $r(n)$-net and for every $x \in X$ there is $y \in u$ such that $\langle x, y\rangle \in r(n)$.

We can easily verify that the following holds:

Proposition 10.7. A set u is a maximal $r(n)$-net inside a class X if and only if u is an $r(n)$-net, $u \subseteq X$ and there is no $r(n)$-net v such that $u \subseteq v \subseteq X$, $u \neq v$.

Proposition 10.8. Let $X \subseteq w$, $n \in \mathrm{FN}$ and let $u_0 \subseteq X$ be an $r(n)$-net. Then there exists a set u which is a maximal $r(n)$-net inside the class X such that $u_0 \subseteq u$.

Proof. By Proposition 10.6(ii), the numbers $\mathrm{Card}(v)$ where v is an $r(n)$-net and $u_0 \subseteq v \subseteq X$ form a finite subset of the class FN. Hence there exists an $r(n)$-net u such that $u_0 \subseteq u \subseteq X$ and for every $r(n)$-net v for which $u_0 \subseteq v \subseteq X$, $\mathrm{Card}(v) \leq \mathrm{Card}(u)$ holds. This set u is obviously a maximal $r(n)$-net inside the class X and $u_0 \subseteq u$. □

Proposition 10.9. Let $u \subseteq w$ be an infinite set, $n \in \mathrm{FN}$. Then the following holds:

(i) There exists $x \in u$ such that the set of all $y \in u$ for which $\langle y, x\rangle \in r(n)$ is infinite.

(ii) There exists an infinite set v such that $v^2 \subseteq r(n)$.

Proof. (i) Let u_0 be a maximal $r(n)$-net inside the set u. For a proof by contradiction, let us assume that for every $x \in u_0$ the set of all $y \in u$ for which $\langle y, x\rangle \in r(n)$ is finite. But then the set u is a union of a finite number of finite sets and it is therefore finite, which is a contradiction.

(ii) Let $x_0 \in u$ be such that the set $s_0(x_0)$ of all $y \in u$, $y \neq x_0$, $\langle y, x_0\rangle \in r(n)$ is infinite. By (i) there exists $x_1 \in s_0(x_0)$ such that the set $s_1(x_1)$ of all $y \in s_0(x_0)$, $y \neq x_1$ $\langle y, x_1\rangle \in r(n)$ is infinite. Repeating this procedure, clearly we will obtain a sequence $x_0, x_1, x_2, \ldots$ of distinct elements of the set u such that for every $i, j \in \mathrm{FN}$, $\langle x_i, x_j\rangle \in r(n)$ holds. Let $\{x_\alpha\}_{\alpha<\delta}$ be a hard extension of this sequence. Let γ be the largest natural number such that $\gamma < \delta$, $x_\alpha \in u$ for every $\alpha < \gamma$ and $\langle x_\alpha, x_\beta\rangle \in r(n)$ for every $\alpha, \beta < \gamma$. Obviously $\gamma \notin \mathrm{FN}$ and the set $v = \{x_\alpha; \alpha < \gamma\}$ is infinite, $v^2 \subseteq r(n)$, $v \subseteq u$. □

Proposition 10.10. Let $u \subseteq w$ be an infinite set. Then there exists an infinite set $v \subseteq u$ such that $v^2 \subseteq R$.

Proof. Using Proposition 10.9 we can form a sequence $\{u_n\}_{n\in\mathrm{FN}}$ of infinite sets such that $u_n^2 \subseteq r(n)$ and $u_{n+1} \subseteq u_n \subseteq u$ for every n. Let $\{u_\alpha\}_{\alpha<\delta}$ be a hard extension of this sequence such that $u_{\alpha+1} \subseteq u_\alpha$ for every $\alpha < \delta$. If $\gamma \in [\delta] \setminus \mathrm{FN}$ then $u_\gamma^2 \subseteq r(n)$ for every n and so $u_\gamma^2 \subseteq R$. Now, it is sufficient to prove that there exists a $\gamma \in [\delta] \setminus \mathrm{FN}$ such that the set u_γ is infinite.

Assume that no such γ exists. Let γ_n denote the smallest $\gamma < \delta$ such that $\mathrm{Card}(u_\gamma) \leq n$ and let $D = \bigcap\{[\gamma_n]\}_{n\in\mathrm{FN}}$. Obviously, D is a π-class and $\mathrm{FN} = D$ which means that FN is a set, contradiction. □

10.2 Geometric Equivalence of Indiscernibility

Although in this section we build on the previous one, our focus now moves to geometrical-topological matters. Now we additionally assume that the relation R is an equivalence on the set w. We will call it a geometric equivalence of in discernibility because under backward projections, its monads and figures evoke the familiar classical compact separable topology on the geometric horizon. We also make use of various results about countable classes.

A **geometric equivalence of indiscernibility** is defined to be a synoptic equivalence on the set w, which is also a π-class.

Recall that m, n are elements of the semiset FN. By Proposition 10.2 the following holds for the function $r|\mathrm{FN}$: For every n there exists m such that $r(m) \circ r(m) \subseteq r(n)$.

Clearly we can consecutively choose numbers from the class FN to obtain a countable class $X \subseteq \mathrm{FN}$ so that for every $n \in X$, the required m is the smallest number from the class X larger than n. In other words the function $r|\mathrm{FN}$ can be modified so that $r(n+1) \circ r(n+1) \subseteq r(n)$ for every n. This property can be transferred to the hard extension of this modified function $r|\mathrm{FN}$ so that in the following we can assume that function r also meets the following requirement:

If $\alpha + 1 < \delta$, *then* $r(\alpha+1) \circ r(\alpha+1) \subseteq r(\alpha)$, *where* $\mathrm{dom}(r) = [\delta]$.

We say that a class Z is a **set base** (of the topology on the set w determined by the geometric equivalence of indiscernibility R) if the following holds:

(a) Every element of the class Z is a subset of the set w.

(b) Let $\mathrm{Mon}(x) \subseteq u$ where $x \in w$, $u \subseteq w$. Then there exists $v \in Z$ such that $\mathrm{Mon}(x) \subseteq v \subseteq u$.

Proposition 10.11. There exists a set base which is at most countable.

Proof. For $x \in w$ let $s(n, x)$ denote the set of all $y \in w$ so that $\langle x, y\rangle \in r(n)$. Let $\{v_n\}_{n\in\mathrm{FN}}$ be a sequence such that for every n, v_n is a maximal $r(n)$-net. Since every set v_n is finite, the class

$$C = \bigcup \{v_n\}_{n\in\mathrm{FN}}$$

is at most countable. Let Z be the class of all $s(n, y)$ where $y \in C$. Obviously Z is an at most countable class. We will prove that Z is a set base.

Let $\mathrm{Mon}(x) \subseteq u$. Let α be the smallest natural number such that $\alpha < \delta$ and for every $y \notin u$, $\langle x, y\rangle \notin r(\alpha)$. Since $\mathrm{Mon}(x) \cap (w \setminus u) = \emptyset$, $\alpha \in \mathrm{FN}$. This means that there exists n such that $\mathrm{Mon}(x) \subseteq s(n, x) \subseteq u$. Let $y \in C$ be such that $\langle x, y\rangle \in r(n+2)$. Since $s(n+1, y) \in Z$, it is sufficient to prove that

$$\mathrm{Mon}(x) \subseteq s(n+1, y) \subseteq s(n, x) \subseteq u.$$

First let $z \in \mathrm{Mon}(x)$. Then $\langle z, x\rangle \in r(n+2)$ and so $\langle y, z\rangle \in r(n+1)$, $z \in s(n+1, y)$. Now let $z \in s(n+1, y)$. Since $\langle z, y\rangle \in r(n+1)$, $\langle x, y\rangle \in r(n+1)$, also $\langle z, x\rangle \in r(n)$ and hence $z \in s(n, x)$. □

In the remaining part of this section, Z denotes some fixed and at most countable set base.

Proposition 10.12. (i) Let X be an open figure and let D be the class of all sets $u \in Z$ such that $u \subseteq X$. Then $X = \bigcup D$.

(ii) Every open figure is a σ-class and every closed figure is a π-class.

Proof. (i) Obviously $\bigcup D \subseteq X$. Let $x \in X$. The class $w \setminus X$ is closed, $x \notin w \setminus X$, and so classes $\mathrm{Mon}(x)$, $w \setminus X$ are separable. This means that there exists a set u such that $\mathrm{Mon}(x) \subseteq u$ and $u \cap (w \setminus X) = \emptyset$, so $u \subseteq X$. Therefore there exists $v \in Z$ such that $\mathrm{Mon}(x) \subseteq v \subseteq u \subseteq X$; it follows that $v \in D$ and $x \in \bigcup D$.

(ii) is an immediate consequence of (i) since the class D is at most countable.
□

Proposition 10.13. Let X be a figure. Then the following properties are equivalent:

(i) There exists a set u such that $X = \mathrm{Fig}(u)$.

(ii) X is a π-class.

(iii) X is a revealed class.

(iv) X is a closed class.

Proof. (i)⇒(ii), (ii)⇒(iii) see Propositions 9.2 and 6.18.

(iii)⇒(i) Let $\{v_n\}_{n \in \mathrm{FN}}$ be a sequence of maximal $r(n)$-nets on the class X such that for each $n \leq m \in \mathrm{FN}$, $v_n \subseteq w_n$. Since each set v_n is finite and since the class X is revealed, there exists a set $u \subseteq X$ such that for each n, $v_n \subseteq u$. Let $x \in X$. Then there exists a sequence $\{y_n\}_{n \in \mathrm{FN}}$ of elements of the set u such that for each n, $\langle y_n, x \rangle \in r(n)$. By the Axiom of Prolongation, there exists $y \in u$ such that for each n, $\langle y, x \rangle \in r(n)$ and hence $\langle y, x \rangle \in R$, $x \in \mathrm{Fig}(u)$, $X \subseteq \mathrm{Fig}(u)$. Since $u \subseteq X$, we have $\mathrm{Fig}(u) \subseteq \mathrm{Fig}(X) = X$. Consequently, $X = \mathrm{Fig}(u)$.

Therefore properties (i), (ii) and (iii) are equivalent.

(ii)⇒(iv) see Proposition 9.7.

(iv)⇒(ii) see Proposition 10.12(ii). □

In classical topology, the following proposition is called the axiom of normality.

Proposition 10.14. Let X_1, X_2 be closed classes, $X_1 \cap X_2 = \emptyset$. Then there exists an open class Y such that $X_2 \subseteq Y$, $X_1 \cap \overline{Y} = \emptyset$.

Proof. Recall that every closed class is a figure. Since X_1, X_2 are disjoint π-classes, there exists a set $u \subseteq w$ such that $X_1 \subseteq u$, $X_2 \cap u = \emptyset$ (see Proposition 7.4). Therefore also $X_2 \cap \mathrm{Fig}(u) = \emptyset$. Let us set $Y = w \setminus \mathrm{Fig}(u)$. Obviously Y is an open class $X_2 \subseteq Y$. If $x \in X_1$, $\mathrm{Mon}(x) \subseteq \mathrm{Fig}(u)$, therefore $x \notin \overline{Y}$. Hence $X_1 \cap \overline{Y} = \emptyset$. □

Proposition 10.15. Let A be a collection of open classes such that $\bigcup A = w$. Then there exist finitely many classes $X_0, X_1, \ldots, X_n \in A$ such that $X_0 \cup X_1 \cup \cdots \cup X_n = w$.[1]

[1] The author did not include a proof of this proposition. Since by Proposition 10.12(i) each X from A is a union of sets from the countable class Z, the set w is a union of countably many sets from Z, each of which is a subset of some X from A. Using Proposition 6.8 we can see that w is the union of finitely many of these sets and the result follows. [Ed]

The above propositions imply: If we replace monads by geometric points then from the point of view of classical topology, the topological shapes that show on the horizon of a geometric equivalence of indiscernibility form a topological normal separable compact space.

Chapter 11

Further Non-traditional Motivations

11.1 Topological Misshapes

Not every set w under a medial look leaves its trace on the horizon in the form of a topological shape. It may sometimes happen that the indiscernibility of elements of the set w is not a transitive relation, that is, the corresponding relation of indiscernibility R is only a symmetry of indiscernibility on the set w. In the sensorily perceptible world, examples of this kind can be most easily encountered with sounds. For example, we may be unable to distinguish a tone x from a slightly higher tone y (or from a slightly louder one, or from a tone with somewhat different timbre), and the tone y from a tone z, while being able to distinguish between the tones x and z.

We can still define the **monad** of an element $x \in w$ just as in the case of equivalences of indiscernibility, namely as a class of all $y \in w$ such that yRx. Since R is a π-class and $\mathrm{Mon}(x) = \mathrm{dom}(R \cap (\{x\} \times w))$, the monad of x is also a π-class for every $x \in w$. If $y \in \mathrm{Mon}(x)$ then naturally also $x \in \mathrm{Mon}(y)$ but the classes $\mathrm{Mon}(x)$ and $\mathrm{Mon}(y)$ can now differ.

If we call the trace left by an object $x \in w$ on the horizon a **mispoint** then the trace of another object $y \in \mathrm{Mon}(x)$ may now be a different mispoint. All the same, these two mispoints do somehow partially intersect.

If $\mathrm{Mon}(x) \cap \mathrm{Mon}(y) \neq \emptyset$ then this class is also a π-class, but it does not need to be the monad of any one of its elements; so the intersection of the two corresponding mispoints is not a trace of any element from the set w and it cannot be captured on the horizon in any other way than via these two mispoints – and naturally also via many other mispoints which also intersect with them in various ways. The interlinking of these mispoints represents the trace of the whole set w on the horizon. This trace is something that we will call a **misshape**.

Since the structure of monads on the set w, determined by a symmetry of indiscernibility R, may be very varied, the family of misshapes is equally varied. Currently there is no structure resembling these misshapes being studied in classical mathematics. Unlike in the case of equivalences of indiscernibility, knowledge of classical mathematics is of no help in the case of symmetries of indiscernibility, and the whole field of misshapes is still virgin territory.

And still, when we perceive some music, for example, the set of tones leaves a trace on the horizon which is just such a topological misshape. We leave answers to the question of how this misshape relates to our experience to a distant future.

11.2 Imaginary Semisets

The overflow of the known land of the geometric horizon is a place that invites being filled in a meaningful way, predominantly by forming various uncountable imaginary semisets.

By an **imaginary semiset** we understand a semiset such that its existence has to be asserted before the collection of its elements. In other words, the collection is evoked by its existence and not the other way round.[1]

To wit, we can tie uncountable classes together using various – possibly even mutually incompatible – imaginary semisets and thus enter various branches of the new theory of sets and semisets.

As an example of such a useful axiom we state just the following one:

Axiom of two cardinalities. *Let X, Y be uncountable classes. Then there exists a bijective function F such that $dom(F) = X$ and $rng(F) = Y$.*

By virtue of this axiom there are – as regards cardinality – only two sorts of infinite classes:

(a) countable classes

(b) uncountable classes, that is, classes of cardinality continuum.

Also, if X and Y are infinite classes of the same sort then there exists a bijective function F such that $dom(F) = X$ and $rng(F) = Y$. Recall that any infinite set is an uncountable class.

In many cases the function F (the subclass of the class $X \times Y$ consisting of the ordered pairs $\langle F(x), x\rangle$) is an imaginary semiset. For example, always when X, Y are infinite sets with different numbers of elements.

Axiom of two cardinalities is not a minimalist axiom. It does not minimalise the domain of uncountable classes as might superficially appear to be the case. On the contrary, the axiom enriches the domain significantly, namely by those bijective functions and by classes formed from them by various operations

The following weaker version of the above axiom is in its way a more interesting and stimulating axiom.

Axiom of maximal cardinality. *Let u, v be infinite sets. Then there exists a bijective function F such that $dom(F) = u$ and $rng(F) = v$.*

---

[1] In classical Cantor set theory, such imaginary sets are the sets whose existence is only guaranteed by the Axiom of Choice. In *Cinq lettres sur la théorie des ensembles*, amongst other things, Lebesgue wrote to Borel:

> We cannot build mathematics properly unless we accept that the existence of something cannot be proved without [first] specifying it.

Since any semiset is a part of some set, in the new theory of sets and semisets augmented by this (imaginary) axiom there exists a maximal cardinality, which is the cardinality of any infinite set.

In this theory, if u is an infinite set and Z a semiset such that $u \subseteq Z$ then also the semiset Z has the maximal cardinality. This follows since upon choosing a set v such that $u \subseteq Z \subseteq v$ we can, with some care, imitate the familiar classical proof of Cantor-Bernstein theorem and prove the existence of a bijective function F such that $dom(F) = u$ and $rng(F) = Z$.

The case of semisets that have no infinite subset is interesting for its own sake. Examples of such semisets is provided by well ordered semisets of natural numbers defined as follows.

We say that a semiset Z of natural numbers is well ordered by the relation $\leq$ if there is no sequence $\{\alpha_n\}$ of its elements such that for each $n \in \mathrm{FN}$ we have $\alpha_{n+1} < \alpha_n$.

Clearly the semiset FN is well ordered and if Z is a well ordered semiset of natural numbers, $Z \subseteq [\gamma]$ then also the semiset $\{\gamma + \alpha;\ \alpha \in Z\}$, is well ordered and so on.

The existence of an uncountable well ordered semiset of natural numbers contradicts the axiom of two cardinalities. On the other hand, using the axiom of maximal cardinality such a set can be created; we will not demonstrate it here. However, it cannot be decided if its cardinality is, or is not, the maximal cardinality. Hence at this point the new theory of sets and semisets branches, predominantly on account of the imaginary semisets.

As is apparent, the path we have taken in this section was to a large extent guided by an effort to imitate (and to some extent, explain) the path taken by Cantor set theory. We will not travel along this path of Cantor in this book any further.

Chapter 12

Search for Real Numbers

In more or less obvious ways, ever since its origins and up to now mathematics have relied on real numbers.

Also in the new infinitary mathematics these numbers (along with the natural numbers) play a key role. The class FN of finite natural numbers captures the length of the journey to the geometric horizon and the class Real of real numbers captures the cohesion and continuity of this horizon.

Since we, in contrast to the classical infinitary mahematics, do not place this horizon as far as some rigid and simply interpreted infinity, we have surrounded the real numbers lying on it by a deep and wide emptiness which invites meaningful content. We will make use of this in our treatise on real numbers.

12.1 Liberation of the Domain of Real Numbers

As is generally known, by integers we mean numbers

$$\ldots - 3,\ -2,\ -1,\ 0,\ 1,\ 2,\ 3,\ldots.$$

By rational numbers we mean those numbers that can be expressed in the form of a fraction

$$\frac{p}{q},\ \text{where } p, q \text{ are integers},\ q \neq 0.$$

By the **number line** we mean a line on which two different points $B(0)$, $B(1)$ are marked. The half-line of it determined by point $B(0)$, on which the point $B(1)$ lies (or does not lie), we call the **positive (or negative) half-line** of the number line.

We represent rational numbers on the number line by assigning the point $B(0)$ to the number 0, and then assigning the point $B(x)$ to the positive (or negative) rational number x, which lies on the positive (or negative) half-line of the number line and for which the ratio of the length of the line segment $B(x)B(0)$ to the length of the line segment $B(1)B(0)$ is equal to the number x.

In this way we have introduced a correspondence between the domain of rational numbers and a certain subdomain of the domain of all points lying on the number line, but not all of this domain. On the positive half-line of the number line there lies, for example, the point $B(\sqrt{2})$ such that the ratio of the length of the line segment $B(\sqrt{2})B(0)$ to the length of the line segment $B(1)B(0)$ is $\sqrt{2}$.

This reasoning (beside other considerations) has led to the extension of the domain of rational numbers to the domain of **real numbers**, intended to ensure a complete correspondence with the domain of all points on the number line; the complete correspondence should moreover extend the original correspondence involving just the rational numbers. Accordingly, the real number corresponding to a point X lying on the positive half-line of the number line has been interpreted as the ratio of the length of the line segment $B(0)X$ to the length of the line segment $B(0)B(1)$.

In other words, positive real numbers have been introduced so that they could be interpreted as ratios of line-segment lengths. The real numbers that are not rational (that is, those that have been added to the rational numbers in order to make the correspondence complete) came to be called **irrational numbers**.

As long as the domain of real numbers was defined only as the domain of addressees in this so-called canonical correspondence with the domain of points lying on the number line, the manipulation of real numbers was more or less tacitly derived from, and justified by, geometric intuition.

A striking example of the subordination of real numbers to geometric intuition was the assertion that

> every continuous function f on a closed interval $[a, b]$ of real numbers, for which $f(b) < 0 < f(a)$ acquires at some point in this interval the value 0.

This geometric assertion was used by **Carl Friedrich Gauss** (1777–1855) who referred to it when in 1799 he gave the first proof of the fundamental theorem of algebra, according to which

> every polynomial $a_n x^n + \ldots + a_0$ with real coefficients can be decomposed into a product of linear and quadratic polynomials (also with real coefficients).

At the same time, he was aware that he reached this "purely analytical solution" by using geometric intuition. To free the proof of the fundamental theorem of algebra from the geometrically justified assertion, Gauss gave in 1816 two more proofs of the algebraic theorem in which he no longer used this geometric statement and in which the presence of geometric intuition was barely noticeable.

Gauss's effort to exclude geometric intuition from the proof of the fundamental theorem of algebra prompted Bernard Bolzano (1781–1848) to develop a "purely analytical" proof of the above assertion about continuous functions. He did so in a work published as early as 1817.[1] When mathematicians read this work at the end of the nineteenth century, the above assertion about continuous functions came to be called Bolzano's theorem.

However, in order to do what he intended, Bolzano first needed to liberate the domain of real numbers: to separate it from the number line. More precisely, to find or at least to suggest a definition of this domain, which would be based only on manipulation of numbers, would not rely on geometric intuition and, of course, such that the canonical correspondence between the thus-liberated domain of real numbers and the domain of points on the number axis was preserved; now just as a correspondence between two domains of equal standing. Bolzano's solution of this task was much more far-reaching than just the "purely analytical" proof of the assertion that received his name. To explain Bolzano's contribution to the interpretation of real numbers, we will use sequences, as is now commonly done. Bolzano, however, used partial sums of infinite series, in keeping with the preference for series of his time.

We say that a real number a (or a point A lying on the number line) is the limit of the sequence $a_1, a_2 \ldots$ of real numbers (or the sequences $A_1, A_2, \ldots$ of points lying on the number line) if:

> no matter how small a positive number ϵ is chosen, it is always possible to arrive in this sequence at such a member a_n (or A_n) that for every $m \geq n$ we have
>
> $$|a - a_m| < \epsilon \quad (\text{or } |A - A_m| < \epsilon).$$

Recall that if x is a real number, then $|x|$ denotes its absolute value. If X, Y are points lying on the number axis, then $|X - Y|$ denotes the real number giving the ratio of the length of the line segment XY to the length of the line segment $B(0)B(1)$.

Clearly, if the sequence of real numbers $a_1, a_2, \ldots$ has a limit, then it is unique.

We say that the sequence of real numbers $a_1, a_2, \ldots$ (or sequence of points $A_1, A_2, \ldots$ lying on the number line) is Bolzano if:

> no matter how small a positive rational number ϵ is chosen, it is always possible in this sequence to arrive at such a member a_n (or A_n) that for every $m \geq n$
>
> $$|a_m - a_n| < \epsilon \quad (\text{or } |A_m - A_n| < \epsilon).$$

If a is the limit of the sequence of real numbers $a_1, a_2, \ldots$, then this sequence is Bolzano. For if we choose a rational number $\epsilon > 0$ and n such that for every

[1] See Bernard Bolzano, "Rein analytischer Beweis des Lehrsatzes dass zwischen je zwey Werthen, die ein entgegengesetzetes Resultat gewähren, wenigstens eine reelle Wurzel der Gleichung liege," *Abhandlungen der k. Gesellschaft der Wissenschaften* (Prag: Gottlieb Haase, 1817).

$m \geq n$ we have $|a_m - a| \leq \frac{\epsilon}{2}$ then for each $m \geq n$

$$|a_m - a_n| \leq |a_n - a| + |a - a_m| \leq \frac{\epsilon}{2} + \frac{\epsilon}{2} = \epsilon.$$

The case of the sequence of points lying on the number line is similar.

The question arises as to whether, also conversely, every Bolzano sequence of real numbers has a limit. In pre-set-theoretical mathematics this amounted to the following question: if some Bolzano sequence of real numbers is given, is its limit actualisable?

Given the original definition of the real number domain, we need to search for an answer to this question in geometry. We therefore ask whether, whenever a Bolzano sequence of points lying on the number line is given, a point is actualisable, that is its limit.

Not even Zeus can give us a satisfactory answer to the question posed in this way. His answer would have to be negative, because in most cases he would not be able to actualise such a point. Zeus may be able to follow the points of a given Bolzano sequence as they settle on the number line, from certain points, on ever shorter line segments (whose lengths tend to 0), but grasping the point into which all these line segments eventually shrink is beyond his capabilities; that is, provided he has not known this point for some reason to start with. For he would have to go through the whole sequence of these shrinking line segments to reach it. But he cannot do that; he can merely travel in this sequence as far as he pleases. No matter how far he goes, he will not reach the end.

However, a more powerful executor than Zeus enters the game. One that does not even have to go through the given sequence step by step, but can survey it with one look as far as the horizon at infinity, where his eyes rest on the point to which the Bolzano sequence tends. Christian Europe does know such a powerful executor. It was Him whom Bolzano had to rely on (even if subconsciouslý in this case) when he dared to utter the following postulate

Bolzano's postulate. Every Bolzano sequence of real numbers has a limit.[2]

For only He guarantees the actualisability of the limit of Bolzano sequence $B(a_1), B(a_2), \ldots$ of points lying on the number line. To this limit, which we will denote B, there corresponds a real number a, that is, $B = B(a)$. The canonical correspondence thus not only allows us, but even forces us to interpret the number a as the limit of the sequence of real numbers $a_1, a_2, \ldots$.

We remark that in the above work Bolzano, using this postulate, also proved the least upper bound theorem and thus enabled "purely analytical" proofs of those assertions that relied on the geometric precept according to which the breaking of a line segment into two parts always happens at some of its points. He also used the occasion to draw attention to common mistakes made by mathematicians when employing this precept.

Consistent and complete liberation of the domain of real numbers was not addressed until the 1870s. Two different methods were chosen to achieve this

[2] This means that the limit of any Bolzano sequence is actualisable.

goal. The essential ideas of both were already contained in the more than fifty-year-old Bolzano's work discussed above, which however nobody in Germany, let alone in France, bothered to read (although it was written German) for it was printed in Prague.

The first independent construction of real numbers, that is, separated from geometry, was proposed by Georg Cantor (1845–1916) as early as 1872.[3] Cantor's method of extending the domain of rational numbers to the domain of real numbers is as follows:

To each Bolzano sequence of rational numbers we assign as its limit an abstract object that we call a real number. But if this sequence has a rational number as its limit then the assigned object is this rational number. In other words, in this case, we do not assign any other limit to that sequence. If no rational number is the limit of the sequence, then the real number that was assigned to it as its limit is called an irrational number.

Cantor, of course, did not use the term Bolzano sequence. Later, as he returned to his theory of real numbers, he called it a fundamental sequence.

Richard Dedekind (1831–1916) modelled real numbers in mathematics based on the classical Cantor's set theory as certain cuts on the set of real numbers.[4] Dedekind's theory of real numbers was – somewhat loosely stated – based on Bolzano's least upper bound theorem.

Shortly afterwards, David Hilbert (1862–1943) described the structure of real numbers and captured it axiomatically in the work *Über den Zahlbegriff*.[5]

We will add the following consideration to close this section. We are used to considering numbers as separate individuals, and even if they enter into different relationships, these relationships between them are somehow discrete, also in the commonly used meaning of the word. These relationships need neither any intermediaries or just a few intermediaries and there is no basis underlying them, such as the space in the case geometric points. In other words, the domain of real numbers does not in itself resemble a continuum, until we represent these numbers on the number line. By this we wish to indicate that at the moment when the domain of real numbers was interpreted independently without the use of geometric intuition, the canonical correspondence broke through the psychological barrier that had been unbreakable since Aristotle; namely, the barrier that prevented breaking the line into individual points and subsequently interpreting the line as a set of points lying on it.

---

[3] Georg Cantor, "Über die Ausdehnung eines Satzes aus der Theorie der trigonometrischen Reihen," *Mathematische Annalen* 5 (1872): 123–132.

[4] Richard Dedekind, *Was sind und was sollen die Zahlen?* (Braunsweig: Friedrich Vieweg und Sohn, 1888).

[5] David Hilbert, "Über den Zahlbegriff," *Jahresbericht der Deutschen Mathematiker-Vereinung* 8 (1900): 180–184.

12.2 Relation of Infinite Closeness on Rational Numbers in Known Land of Geometric Horizon

In this section N denotes some fixed cut on natural numbers such that $\mathrm{N} \neq \mathrm{FN}$ and for every $\alpha, \beta \in \mathrm{N}$ also $\alpha\beta \in \mathrm{N}$. Rac denotes the class of all rational numbers that can be expressed as $\pm\frac{\alpha}{\beta}$, where $\alpha, \beta \in \mathrm{N}$. Letters $x, y, z, \ldots$ denote rational numbers belonging to the class Rac and $|x|$ denotes the absolute value of the number x.

Recall that the letters m and n denote finite natural numbers (that is, elements of the class FN). $\leq$ denotes the customary ordering of the class Rac.

We say that a **rational number** x is **bounded** if there exists an n such that $|x| < n$.

BRac denotes the class of all bounded rational numbers.

Obviously, it holds that:

Proposition 12.1. (i) If $x, y \in \mathrm{BRac}$ then also $x + y, xy \in \mathrm{BRac}$.

(ii) If $x \neq 0$, then $\frac{1}{x} \notin \mathrm{BRac}$ if and only if $|x| < \frac{1}{n}$ for every $n \neq 0$.

We say that **rational numbers** x and y are **infinitely close** (notation $x \doteq y$) if either $x, y \notin \mathrm{BRac}$, or $x, y \in \mathrm{BRac}$ and $|x - y| < \frac{1}{n}$ for every $n \neq 0$.

It is straightforward to check that $\doteq$ is an *equivalence of indiscernibility* of the class Rac.

Proposition 12.2. Let $x, x_1, y, y_1 \in \mathrm{BRac}$, $x \doteq y$, $x_1 \doteq y_1$. Then the following holds:

(i) $x + x_1 \doteq y + y_1$;

(ii) $xx_1 \doteq yy_1$;

(iii) if $x, y \neq 0$ then $\frac{1}{x} \doteq \frac{1}{y}$.

Proof. (i) Let $n \neq 0$. Then $|x - y|, |x_1 - y_1| < \frac{1}{2n}$. Hence

$$|x + x_1 - y - y_1| \leq |x - y| + |x_1 - y_1| < \frac{1}{n}.$$

(ii) Let $n \neq 0$. Let $m \neq 0$ be such that $|x|, |y_1| < m$. Clearly $|x - y| + |x_1 - y_1| < \frac{1}{mn}$ and hence

$$|xx_1 - yy_1| = |xx_1 - xy_1 + xy_1 - yy_1| \leq |x||x_1 - y_1| + |y_1||x - y| < \frac{1}{n}.$$

(iii) If $\frac{1}{x} \notin \mathrm{BRac}$ then $x \doteq 0$, so also $y \doteq 0$; hence $\frac{1}{y} \notin \mathrm{BRac}$.

So let $\frac{1}{x}, \frac{1}{y} \in \mathrm{BRac}$. Let $n \neq 0$. Since we do not have $x, y \doteq 0$, $|x - y| < \frac{1}{n}|xy|$. Hence $\left|\frac{1}{x} - \frac{1}{y}\right| = \frac{|y-x|}{|xy|} < \frac{1}{n}$. □

We say that a **rational number** x is **finite** if there exist m and n such that $n \neq 0$ and $|x| = \frac{m}{n}$.

FRac denotes the class of all finite rational numbers.

Obviously, the class FRac is countable; FRac $\subseteq$ BRac.

Proposition 12.3. (i) Assume that $x < y$ and not $x \doteq y$. Then there exists $z \in$ FRac such that $x < z < y$.

(ii) Let $x, y \in$ FRac. Then $x \doteq y$ if and only if $x = y$.

Proof. (i) First let $y \notin$ BRac. Then $x \in$ BRac and there exists m such that $x < m < y$. The case $x \notin$ BRac is analogous.

So let $x, y \in$ BRac. Let $m \neq 0$ be such that $\frac{1}{m} < y - x$, so $x + \frac{1}{m} < y$. Let n be the smallest natural number such that $mx < n$, so $x < \frac{n}{m}$. Since $\frac{n}{m} - x \leq \frac{1}{m}$, also

$$\frac{n}{m} \leq x + \frac{1}{m} < y.$$

(ii) Assume that $x \neq y$. Then $0 \neq |x - y| \in$ FRac and so there exists $m \neq 0$ such that $\frac{1}{m} < |x - y|$. It follows that it is not true that $x \doteq y$. The converse implication holds trivially. □

Proposition 12.4. Let $u \subseteq$ Rac be an infinite set. Then there exist $x, y \in u$, $x \neq y$, such that $x \doteq y$.

Proof. If there exist $x, y \in u$, $x \neq y$, $x, y \notin$ BRac, then there is nothing to prove. This means that in the following we can assume that $u \subseteq$ BRac. Obviously, the class of all $\langle x, y \rangle$, where $x, y \in u$, $x \leq y$, is a set. Hence there exists a bijective function g such that

$$\mathrm{dom}(g) = [\gamma], \quad \text{where} \quad \mathrm{Card}(u) = \gamma \notin \mathrm{FN}, \quad \mathrm{rng}(g) = u$$

and for every $\alpha < \beta < \gamma$,

$$g(\alpha) < g(\beta)$$

(since every set is finite from the classical point of view).

Let β be the smallest natural number such that for every $\alpha + 1 < \gamma$,

$$\frac{1}{\beta} < g(\alpha + 1) - g(\alpha).$$

If $\beta \notin$ FN, then there exists an $\alpha + 1 < \gamma$ such that $\frac{1}{\beta} < g(\alpha + 1) - g(\alpha) < \frac{1}{\beta - 1}$, and so $g(\alpha + 1) \doteq g(\alpha)$.

For a proof by contradiction, assume that no two different numbers belonging to the set u are infinitely close. By the above considerations there exists $m \neq 0$ such that $\frac{1}{m} < g(\alpha + 1) - g(\alpha)$ for every $\alpha + 1 < \gamma$. It follows by induction on m that for every $\alpha < \gamma$,

$$g(0) + \frac{\alpha}{m} < g(\alpha)$$

and therefore also
$$\frac{\gamma-1}{m} < g(\gamma-1) - g(0).$$
Let n be the smallest natural number such that $g(\gamma-1) - g(0) < n$. Then $\gamma - 1 < nm$, so $\gamma \in \mathrm{FN}$, which is a contradiction. □

In the remaining part of this section, w denotes a set such that $\mathrm{FRac} \subseteq w \subseteq \mathrm{Rac}$.[6]

Proposition 12.5. For every x there exists $y \in w$ such that $x \doteq y$.

Proof. Let γ be the largest natural number such that $\frac{\alpha}{\gamma} \in w$ for every $|\alpha| \leq \gamma^2$. Obviously $\gamma \notin \mathrm{FN}$, $\gamma \in w$. If $x \notin \mathrm{BRac}$ then $x \doteq \gamma$ so assume $x \in \mathrm{BRac}$. Then there exists $|\alpha| < \gamma^2$ such that
$$\frac{\alpha}{\gamma} \leq x < \frac{\alpha+1}{\gamma}.$$
Hence $x \doteq \frac{\alpha}{\gamma} \in w$. □

Define $x \overset{\circ}{=} y$ if and only if
$$x \doteq y \text{ and } x, y \in w.$$

Obviously $\overset{\circ}{=}$ is an equivalence on the set w.

Proposition 12.6. The relation $\overset{\circ}{=}$ is a geometric equivalence of indiscernibility on the set w.

Proof. Let $\langle x, y\rangle \in r(n)$ if and only if $x, y \in w$ and either $n < |x|, |y|$, or $|x|, |y| \leq n$ and $|x - y| < \frac{1}{n}$. It is straightforward to check that for every $x, y \in w$, $x \overset{\circ}{=} y$ if and only if $\langle x, y\rangle \in r(n)$ for every n. Therefore the equivalence $\overset{\circ}{=}$ is a π-class. By Proposition 12.4, the equivalence $\overset{\circ}{=}$ is synoptic on the set w. □

Proposition 12.7. Let $X \neq \emptyset$ be a closed figure in the equivalence $\overset{\circ}{=}$ on the set w. Then there exists $z \in X$ such that for every $x \in X$ either $x < z$ or $x \overset{\circ}{=} z$.

Proof. By Proposition 10.13 there exists a non-empty set $u \subseteq X$ such that $\mathrm{Fig}(u) = X$. If z is the greatest element of set u then obviously z has the desired property. □

Real$^+$ denotes the class of all monads $X \neq \emptyset$ in the equivalence $\overset{\circ}{=}$ on the set w such that $X \subseteq \mathrm{BRac}$.

For $X, Y \subset \mathrm{Real}^+$ we define:

(i) $X < Y$ if and only if $x < y$ for every $x \in X$, $y \in Y$.

[6] We can show that such a set exists, for example, using the Axiom of Prolongation.

(ii) $X \leq Y$ if and only if either $X < Y$ or $X = Y$.

It is easy to show that the thus defined relation $\leq$ is a linear ordering of the cluster Real^+.

For $X, Y \in \mathrm{Real}^+$ we denote the class of all $x + y \in w$ (or $xy \in w$), where $x \in X$, $y \in Y$, by $X + Y$ (or $X \cdot Y$ respectively).

Obviously, $X + Y$, $XY \in \mathrm{Real}^+$.

If $0 \notin X$ then $\frac{1}{X}$ denotes the class of all $\frac{1}{x}$ where $x \in X$.

In such case $\frac{1}{X} \in \mathrm{Real}^+$.

Using the propositions stated and proved in this section, it is straightforward to check that the ordering and operations of addition, multiplication and division on the cluster Real^+ defined above meet all the requirements of the structure of real numbers. In particular, Proposition 12.7 implies the least upper bound theorem.

In the new theory of set and semisets real numbers are thus modelled very intuitively as monads in the geometric equivalence of indiscernibility on the class of rational numbers.

Under the corresponding geometric look at the class of rational numbers, real numbers appear as traces left by these monads on the horizon, that is, they appear as geometric points. As in the case of other geometric looks we can attach these points to abstract ur-objects and call them **real numbers**.

12.3 Real Numbers

Dedekind's cuts on rational numbers are not real numbers. No real number is a set or a class. Dedekind's cuts are models of real numbers in Cantor's set theory; faithful models, because their structure meets all the conditions required of the structure of real numbers (for example, as written down by David Hilbert in 1900).[7]

When dealing with real numbers, it is best to forget their Dedekind's models and work with the formal structure of real numbers. The importance of these models lies in the fact that through them, Cantor's set theory has taken responsibility for the theory of real numbers. To wit, any contradiction in the theory of real numbers theory would immediately become a contradiction Cantor's set theory.

For the same reasons as Dedekind's cuts, the class Real^+ is not the class of real numbers, but merely a class of their models. The case of Cantor's theory of real numbers is similar, although the real numbers appear in it more clearly. If the limit of some Bolzano sequence of rational numbers is not a rational number, then this sequence points to a gap in the ordering of rational numbers, so essentially to emptiness. To fill this gap with an irrational number is the task which Bolzano poses as a postulate (see page 204).

[7] David Hilbert, "Über den Zahlbegriff," *Jahresbericht der Deutschen Mathematiker-Vereinung* 8 (1900): 180–184.

In all three cases we thus deal with models of real numbers. But that means that there is something, namely the real numbers, whose models they are and whose formal structure was described by David Hilbert.

In their raw form as ratios of lengths of line segments, real numbers have been present already in ancient civilizations. **Thales of Miletus** (630–548) brought them to ancient Greece from his travels through Egypt in the following precept:

> The lengths u, v are in the same ratio as the lengths u', v' if in the right-angled triangle with legs of lengths u, v, the angle lying opposite the leg of length v is the same as the angle in the right-angled triangle with legs of lengths u', v' lying opposite the leg of length v'.

This is documented in the following quotations from the pre-Socratic thinkers, referring to Eudemus of Rhodes (4th century BC):[8]

> Thales was the first to go to Egypt and bring back to Greece this study; he discovered many propositions and disclosed the underlying principles of many others to his successors.
>
> Eudemus in his History of Geometry attributes this theorem to Thales. For he says that the method by which Thales showed how to find the distance of ships at sea necessarily involves this method.

Along with the discovery of the ancient geometric world, what we now call (positive) real numbers today was discovered. To wit, ratios of lengths of line segments lying in this geometric world. Natural and positive rational numbers are their organic components.

From Plato's strict point of view, positive real numbers are ideas, unique and timeless. The way to them was illuminated by Eudoxus[9] and executed by Bolzano.

In the new infinitary mathematics the geometric horizon limiting the ancient geometric world is no firm and impassable boundary of this world; it is a horizon in the real sense of the word. Our new interpretation of real numbers has to be ruled by this.

As we have said before, monads in the equivalence of infinite closeness on the class BRac (see Section 12.2) are not real numbers. The class Real^+ of all these monads is not the class of all real numbers. On the other hand, backward projections of the monads belonging to the class Real^+ to the geometric horizon

[8] Cf. fragments A11 and A20 from Proclus, here quoted from Ivor Thomas, *Greek mathematical works I* (Loeb Classical Library: William Heinemann Ltd., 1980).

[9] Vopěnka refers to Eudoxus' solution to the tetrahedron problem (to show that two tetrahedra (triangular pyramids) with congruent bases and equal heights have the same volume, cf. Euclid, *Elements, Book X,* trans. T. L. Heath, The Thirteen Books of Euclid's Elements (Dover Publications, 1956). In the Czech version of New Infinitary Mathematics Vopěnka devotes a section to this proof. [Ed.]

are real numbers. The class Real of all these backward projections is the class of all real numbers.

Recall that FRac denotes the class of all finite rational numbers. Clearly if $X \in \text{Real}^+$ then $X \neq \emptyset$ and by Proposition 12.3 there exists at most one number z such that $z \in \text{FRac} \cap X$.

If $X \in \text{Real}^+$ then $\text{Proj}(X)$ denotes the real number y that is the backward projection of precisely those numbers z belonging to the class X.

Real is thus the class of all

$$\text{Proj}(X) \qquad \text{where} \quad X \in \text{Real}^+.$$

If $z \in X \in \text{Real}^+$, then clearly $z \in \text{FRac}$ if and only if $z = \text{Proj}(X)$. If $X \in \text{Real}^+$, $X \cap \text{FRac} = \emptyset$ then

$$\text{Proj}(X) \in \text{Real} \setminus \text{FRac}.$$

In other words: in such case, $\text{Proj}(X)$ is an irrational number.

Hence the limits of Bolzano sequences of rational numbers are the backward projections of those of their members that fell through beyond the geometric horizon.

Now we make the point (c) in the definition of hard sets[10] of type 1 more precise, namely in the following way:

(iii) sets of real numbers.

Any sequence $\{a_n\}_{n \in \text{FN}}$ of real numbers is stable and by the Axiom of Prolongation it has a set extension (similarly as the sequence of finite natural numbers does).

12.4 Intermezzo About the Stars in the Sky

During the whole development of European natural sciences, the interpretation of the *horizon* represented by the starry firmament had been the one to change most. There were four most significant phases.

(a) In antiquity and in Europe until the 17th century, this horizon was the boundary of the real world of the time. The enigmatic system of stars spread on it, their constellations and the travel of planets amongst them could not be but a mysterious message, the deciphering of which would have been undoubtedly considered the greatest success of human ingenuity.

(b) The second phase of these changes happened when the planets were taken out of the horizon represented by the sphere of stars and each of them obtained its own sphere closer to the Earth (or the Sun).

[10] See Section 6.2, page 147.

(c) The third phase of changes was started by the telescope. Using it, its discoverers Galileo Galilei (1564–1642) and Johannes Kepler (1571–1630) found many other stars and details, invisible to the naked eye. Since this did not involve damaging the standing system of stars and planets, but just adding to it, it was in fact an expansion of the starry sky. This enabled making the interpretation of previously known phenomena on the firmament more precise and more complete, rather than changing it.

(d) Only after the death of the above two astronomers the fourth phase of interpretations of the starry sky started, which has continued ever since. It was initiated already by Giordano Bruno (1548–1600), for he abolished the firmament. More precisely, he took it out of the real world and declared it to be merely an illusory phenomenon. By its removal from this world, man started to conquer the distant universe.

However, naturally, the firmament has not disappeared, only its interpretation has radically changed. Since the stars distributed on it are backward projections of celestial bodies to this horizon, it continues to serve for orientation in the near universe and in fact also in our everyday life on Earth.

12.5 Interpretation of Real Numbers Corresponding to the First and Second phase in Interpreting Stars in the Sky

Positive real numbers entered the ancient geometric world via ratios of lengths of line segments (and, as Eudoxus demonstrated, also via ratios of sizes of areas of Euclidean planary shapes and via ratios of volumes of Euclidean bodies). In India, at the latest during the life of **Brahmagupta**(597–668), these numbers were augmented by negative numbers. This happened in Europe almost a thousand year later.

We shall approach real numbers simply as phenomena that show on the geometric horizon – in a similar way to how ancient astronomers approached stars in the sky. Thus we will not, for example, investigate all of what these numbers could be back projections etc.

Even though the class Real is a phenomenon of the ancient geometric world, most of its elements lie as far as the geometric horizon. From there some numbers have been successively picked and placed so to speak inside the ancient geometric world. First rational numbers of the form $\pm\frac{n}{m}$, where $m, n \in \mathrm{FN}$, $m \neq 0$; then the numbers obtained from them by taking square roots of nonnegative numbers. Later also **algebraic numbers**, that is, roots of polynomials with integer coefficients.

Furthermore, already in the Greek antiquity from amongst the numbers lying on the geometric horizon, the number π, denoting the ratio of the area of a circle and the area of the square of its radius, had been placed inside the ancient geometric world.

In 1873 **Charles Hermite**(1822–1901) proved that Euler number e is not algebraic but **transcendental.**[11] The same was proved about the number π in 1882 by **Carl Louis Ferdinand von Lindemann** (1852–1939).[12] Only this proved the impossibility of squaring the circle.

In spite of many ingenious proofs of trancendence of various clearly defined numbers there will always remain on the geometric horizon uncountably many transcendental numbers that can be reached only as the limits of Bolzano sequences, as was proved in 1873 by **Georg Cantor.**[13]

We will demonstrate that the new theory of sets and semisets makes it possible to interpret and investigate real numbers also in the way that corresponds to the third phase of the interpretation of stars in the sky.

The remaining chapters of this book contain various suitable ways of such investigation.

[11] Published in Charles Hermite, *Cours d'Analyse de l'École Polytechnique. Première Partie* (Paris: Gauthier-Villars, 1874).

[12] See Ferdinad Lindemann, "Über die Zahl π," *Mathematische Annalen* 20 (1882): 213–225.

[13] The work was published in the following year: Georg Cantor, Über eine Eigenschaft des Inbegriffes aller reelen algebraischen Zahlen, *Journal für die Reine und Angewandte Mathematik* 77 (1874): 258–262.

Chapter 13

Classical Geometric World

Modern mathematics does not consider the intuitive ancient geometric world to be the subject of study in geometry; more than that, it refuses to acknowledge it altogether. A geometric point is always incomparably smaller than a bacterium and it is impossible to place in it a single atom, let alone a bacterium. Notions such as horizon or standpoint are subjective and have no place in modern European science, which prides itself on its objectivity. It is the classical geometric world what constitutes the subject of study of modern geometry.

In the classical geometric world, points, line segments, circles and other geometric objects have fallen into an absolutely infinite depth beyond the horizon limiting the potential of our eye's perception. In other words, into a depth so breathtaking that not even a super-human can see that far. The route to a geometric point in the classical geometric world – by this we understand a ceaseless reduction – is absolutely infinitely long, just like a ceaseless prolongation of a line segment by a given length or an effort to reach the completely exact boundaries of a geometrical object and so on.

So despite the fact that we cannot see into the classical geometric world – nobody has the ability to imagine a circle and its tangent in it, not even a single point, because we are not able to see so deep – we still have a perfect science about it: geometry. This is so, because knowledge about the intuitive ancient geometric world is taken (without us actually being aware of it) to be knowledge of the classical geometric world. That is why, for example, the axioms of Euclidean geometry, etc. hold in the classical geometric world. In this sense – and only in this sense – both these worlds are identical. They are internally indistinguishable, every proposition that holds in one of them also holds in the other.

To wit, the classical geometric world is the geometric world of the God-man. By a God-man we understand a being which is God, for he rules absolute infinity, but also human, for he moves in a human body and perceives the world from the same standpoint as a human. In other words the God-man perceives everything as a human or a super-human, but unlike them he also sees as far as the absolute infinity. Consequently, the absolute infinity is for him the natural infinity and he interprets it similarly as we interpret the ancient geometric infinity. His interpretation of the absolute infinity became our classical interpretation and absolute infinity interpreted in his way became our classical infinity.

The classical geometric world is therefore the intuitive ancient world expanded as far as the classical infinity. But because the God-man is also God, the horizon on which his geometric objects appear lies at the very end, or, the very bottom of the real world.

The existence of the classical geometric world is thus clearly more problematic than the existence of the intuitive ancient geometric world. Classical geometric world is not what mathematicians (and others) believe it to be. Its natural numbers cannot go as far as the absolute infinity. For in that case it would not be limited by a horizon, but by an impenetrable, sharp boundary lying in the absolute infinity. This would make it possible (at least with a reference to the God of medieval and baroque rational theology) to actualise the domain of all natural numbers in the way that Bernard Bolzano and later Georg Cantor have done it. The classical geometric world thus would not be a mere illusion but transcendental reality; as it is in fact still taken to be today.

Naturally this does not mean that we should not even speak of the classical geometric world. On the contrary, if we realize what this world is, we lose nothing of its previous and current usefulness. To wit, the classical geometric world is an expansion of the ancient geometric world not as far as some traditionally interpreted absolute infinity, but merely somewhere into the overlap of the known land of the geometric horizon. This, and in fact only this, can be used to explain the internal indiscernibility of the ancient and classical geometric worlds.

At the time of the birth of infinitesimal calculus, the ancient geometric world was already being replaced with the classical geometric world. This duality of the worlds, although pushed into the subconscious, yielded a remarkable and, in its way, safe intuition for working with infinitely small quantities: an intuition, which **Archimedes** probably found quite difficult to establish and which none of his potential ancient and medieval followers were able to establish.

Unlike the former interpretation of the classical geometric world, our new interpretation makes it possible – in a manner of speaking – to manipulate this world: draw it nearer or push it further away and also rotate it a little, but always only within the overlap of the known land of the geometrical horizon. We will shortly make use of this possibility.

Unless necessary, we will not introduce any new terminology for our new interpretation of the classical geometric world. We will just be aware of what this classical geometric world is and what it is not.

Part III

Infinitesimal Calculus Reaffirmed

Introduction

The discovery of infinitesimal calculus by Newton and Leibniz has always been regarded as one of the greatest achievements of human intellect. And rightly so, since it equipped mathematics with a tool of unprecedented efficiency making it possible to transcend the horizon of existing geometrical intuition, and it provided modern European natural sciences (and later all sciences) with an indispensable tool without which they would never have achieved many of their triumphant results.

Higher mathematics (the part of mathematics founded on infinitesimal calculus) had been based on calculations using infinitely small quantities. Quantities comprehensible to the prevailing geometrical intuition of the time could be gained primarily with the help of quotients of two and sums of infinitely many of these quantities and – accordingly –infinitesimal calculus has traditionally been divided into differential and integral calculus.

The mysterious nature of the infinitely small quantities added a mystical flavour to infinitesimal calculus and as a consequence, its very discovery assumed, in the eyes of many, proportions almost undeserved; however, at the same time it aroused doubts in the minds of more than one strictly rational mathematician. A suspicion that these concepts and arguments using them, to say the very least, lack clarity, was boosted by awkward errors that anybody carelessly calculating with infinitely small quantities could easily have been making. And yet, calculations guided by certain unconscious new geometric intuition concerning infinitely small quantities undoubtedly did bring astonishing and unquestionable results.

Mathematicians, whose science had always been perceived as the most exact of the sciences, naturally could not accept that the most productive part of mathematics should be based on vague concepts. After several failed attempts at bringing the enigmatic notion of infinitely small quantities from the unconscious to light, so as to clarify and sharpen the sphere of its applicability and to satisfy the demands of mathematical exactness, they eventually gave up the concept of an infinitely small quantity altogether. An opportunity to do so was provided by d'Alembert who in his effort to eliminate actual infinity from mathematics introduced the concept of limit in its different variations. This concept became the basis for mathematical analysis (which, from then on, has no longer analysed infinitely small quantities) and replaced infinitely small quantities in various arguments; this has continued to be the case even when actual infinity became firmly established in mathematics.

The transformation of infinitesimal calculus into $\epsilon\delta$-analysis (this rather mocking term arising from the techniques typical for handling limits) not only met the mathematicians' demands for exactness and clarity, but also opened up new wide fields of activity. Replacing arguments involving infinitely small quantities by arguments involving limits was not entirely mechanical and it often required creative approaches. Moreover, the thus developed procedures and methods were fit for generalization and for use in the exploration of much more comprehensive areas than the domain of real numbers and functions. In conjunction with the entry of set theory into mathematics, this also gave rise to disciplines such as topology or functional analysis.

The successful mastering of all that infinitesimal calculus had achieved by the mid-nineteenth century, the elimination of various mistakes and misconceptions that had occurred while handling infinitely small quantities, the clarification of some misunderstandings and also the unprecedented productiveness of this direction in which mathematicians had set off in an effort to eliminate infinitely small quantities all combined to produce a general belief that $\epsilon\delta$-analysis indeed was the sought-after answer to the questions concerning foundations of infinitesimal calculus. Although the original notion of infinitesimal calculus could not be expunged from the history of mathematics (and what is more, for example physicists have never really yielded to the dictates of $\epsilon\delta$-analysis), mathematicians perceived the original infinitesimal calculus to be at best a kind of historical slip or at worst absolute nonsense.

It was not through some vain enmity towards mathematicians that physicists preferred the original conception of infinitesimal calculus, but because the $\epsilon\delta$-analysis suppressed that which gave infinitesimal calculus its power. Namely the intuition gained from the peculiar and yet most useful insight regarding infinitely small quantities, and hence also the transparency of calculations with these quantities based on this intuition. A much larger number of quantifiers was needed in $\epsilon\delta$-analysis in order to capture the originally intuitive concepts, and simple calculations were frequently replaced by propositions with far from transparent proofs. Some mathematicians in fact secretly worked with infinitely small quantities and only then laboriously translated their results into $\epsilon\delta$-analysis. This insincerity on the part of mathematicians may be illustrated, for example, by the fact that many undergraduate courses in mathematical analysis involved solving differential equations, but it was tacitly assumed that students would learn how to form these equations in courses on physics. In the twentieth century some key and illustrative concepts developed by physicists with the help of infinitely small quantities could only be detached from these quantities by means of intricate modeling using functional analysis. The well-known Dirac's function is a telling example of this.

A change came half way through the twentieth century, in connection with the discovery of non-standard models of set theory. The sequence of the original – that is, standard – natural numbers has been extended in these models by the so-called non-standard numbers which are indistinguishable within the model from the original ones. There is no doubt that inverse values of non-standard natural numbers can be interpreted as infinitely small numbers. This potential was seized upon by Abraham Robinson, who laid foundations for the *non-standard analysis*. He and his followers thus successfully vindicated the techniques used in calculations with infinitely small quantities and also justified some other useful principles related to them; even those that would have been only reluctantly accepted by the discoverers of infinitesimal calculus.

Mathematicians specializing in $\epsilon\delta$-analysis conceded this rehabilitation of infinitely small quantities and calculations with them (in fact one could not but concede it) but they raised various objections to non-standard analysis, the following two of them substantial.

First of all, the objection that rehabilitation of the techniques used in calculations with infinitely small quantities does not necessarily mean a rehabilitation of the original intuition needed when handling these quantities. Non-standard analysis did not capture infinitely small quantities in their original form, it only modeled them. As a consequence, the original intuition was substituted with a different intuition, albeit a congeneric one, which can only be acquired by a person who is an expert in non-standard models of set theory. However, this presupposes extensive knowledge of modern set theory and mathematical logic. This is why non-standard analysis cannot commonly be a subject of undergraduate courses at universities. For physicists (or for anybody who still uses the original infinitesimal calculus), the discovery of non-standard analysis has been highly satisfactory, but regrettably only of little help, as this satisfaction is purely formal.

The other serious objection raised against non-standard analysis was that it only translates results obtained via $\epsilon\delta$-analysis into the language of infinitely small quantities. However, it brings no new results. Until it demonstrates something that the $\epsilon\delta$-analysis cannot do, it remains a mere interesting curiosity.

This second objection is hypocritical. For $\epsilon\delta$-analysis (at least in its initial stages) did not do anything but translating concepts and results of infinitesimal calculus into the language of limits. Hence it should be up to $\epsilon\delta$-analysis to demonstrate something that non-standard analysis cannot do. However, there is no such thing since the language of non-standard analysis is a (conservative) extension of the language of $\epsilon\delta$-analysis and everything that is expressed and proved in $\epsilon\delta$-analysis is automatically expressed and proved in non-standard analysis.

Even if non-standard analysis did not bring any result that cannot be obtained by $\epsilon\delta$-analysis, it would not be a sufficient reason to reject it. After all, complex numbers are conservative extension of real numbers (meaning that nothing can be proved about real numbers using complex numbers that cannot be proved using real numbers alone) but they who try to prove for example the assertion about factorizing a polynomial with real coefficients into quadratic and linear polynomials with real coefficients (that is, an assertion concerning exclusively real numbers) come to truly appreciate the usefulness of complex numbers.

Moreover the proponents of $\epsilon\delta$-analysis implicitly require of non-standard analysis that the result which is to be beyond the possibilities of $\epsilon\delta$-analysis should be formulated in the language of $\epsilon\delta$-analysis. Non-standard analysis is thus a priori degraded to a mere tool for gaining results of $\epsilon\delta$-analysis. However, this is as if we recognized just those concepts and results in the theory of complex numbers that concern real numbers only, giving merely ancillary status to concepts like complex conjugate numbers, trigonometric representations of complex numbers etc., and had no special value for the result $e^{2\pi i} = 1$.

On the contrary, $\epsilon\delta$-analysis should consider if some of its methods are not reminiscent of the labour that would be required if we forbade ourselves to use complex numbers when arguing about real numbers. Then again, some may object that the comparison with complex and real numbers is not suitable for our case or even that the situation in $\epsilon\delta$-analysis is by such a comparison intentionally misrepresented.

Arguing along these lines, proponents of either side appear to miss a much more important fact: namely that the birth of non-standard analysis made it possible to state clearly the question whether $\epsilon\delta$-analysis has indeed achieved its original aims. That means, if it really succeeded in eliminating infinitely small quantities without inflicting any notable damage on mathematics. This question can be stated for example as follows: is it possible to express any property of real functions such that in its definition (only) real and infinitely small numbers are quantified by an equivalent meaningful definition in which only real numbers are quantified (that is, when quantification of infinitely small quantities is no longer allowed)?

In the Appendix to this part there are instructions on how to translate some concepts and propositions containing no more than two alternating quantifications of infinitely small numbers into $\epsilon\delta$-analysis in a fairly straightforward manner. However, the obvious way to generalize this rule for the case of a larger number of alternating quantifications of infinitely small numbers does not work, as has been proved by Karel Čuda.

This then defines the range of concepts and propositions of infinitesimal calculus that were,so to speak, child's play for $\epsilon\delta$-analysis. It includes limits, continuity, first derivatives and derivatives of a higher order etc. However, if one is to cope with the more complex concepts of infinitesimal calculus without the use of infinitely small quantities, one must involve quite complex set structures, consequently capturing intuitively clear concepts through concepts that are much more demanding and much less clear. This also explains why the translation of infinitesimal calculus into $\epsilon\delta$-analysis was smooth to start with but from a certain point it started to require methods of functional analysis, topology, advanced modern differential geometry etc.

Hence it appears that $\epsilon\delta$-analysis can achieve the given goal only with the substantial help of set theory, utilizing its extensive supply of various set structures which conveniently include or complement the domain of real numbers and functions. It is interesting to note that it was mainly due to the part played by set theory in the problems of mathematical analysis that mathematicians were reluctantly forced to acknowledge set theory as a proper mathematical discipline; in fact, they could hardly do anything else. However, if we do allow $\epsilon\delta$-analysis to make use of structures offered by set theory to achieve its goal, then it is exactly the non-standard models of set theory which represent the most convenient set structures provided by set theory for this purpose. In other words, the very creation of these models and the related constitution of non-standard analysis may be judged to be the final acquisition of its goal by $\epsilon\delta$-analysis.

On the other hand, if we do not consider it admissible that $\epsilon\delta$-analysis should be using set theory to achieve its goal then we have to conclude that $\epsilon\delta$-analysis has not reached this goal. Not only that, its sometimes almost fanatical opposition to infinitely small quantities barred mathematics from pursuing one of promising lines of its development. Namely a direct investigation of concepts with definitions involving a larger number of alternating quantifications of infinitely small quantities, and establishing corresponding rules of calculations. Consequently, the conviction of mathematicians about $\epsilon\delta$-analysis being an appropriate substitute for infinitesimal calculus, which has lasted for many decades, ranks amongst the greatest errors in the intellectual development of European mankind.

Hence it is the first of the above mentioned objections against non-standard analysis that carries much more weight. We cannot reject it unless we adopt the position of pure formalists recognizing only the syntactic side of mathematical theories and ignoring the question of what these theories concern. Such a decision can be made by anybody for their own self but it cannot be forced upon other mathematicians let alone physicists or other users of infinitesimal calculus. Notwithstanding all the effort invested by some mathematicians into simplifying the initial assumptions of non-standard analysis to make it as accessible as possible, this objection still stands. Non-standard analysis as such is unable to deal with it; it cannot overcome the limits inherent in its conception.

This objection can only be refuted if the original infinitesimal calculus is rehabilitated in its undistorted form; that means reestablishing not only the techniques for calculations with infinitely small quantities but also the original intuition needed for dealing with these quantities. To this end it is necessary to guide the enigmatic geometric insight regarding infinitely small quantities that attended the birth of infinitesimal calculus from unconsciousness to consciousness.

In other words, we need to say clearly what are the infinitely small quantities; not those studied by non-standard analysis but those handled fearlessly but safely by Newton, Leibniz and a number of later mathematicians and physicists.

To reestablish infinitesimal calculus in this sense is what this part is concerned with.

Chapter 14

Expansion of Ancient Geometric World

This text on infinitesimal calculus contains only basic results from the differential calculus of real functions of one variable, corresponding approximately to an introductory course on the topic. Consequently all concepts and results considered here can be translated in a straightforward manner into $\epsilon\delta$-analysis, as can be checked in the Appendix. Since it is an introductory text, the number of initial principles in it is limited to the bare minimum, even at the cost of forfeiting still simpler demonstrations of various results which could be achieved by accepting some further, equally intuitive principles, or through a more detailed analysis of the accepted principles.

In order to avoid frequent references to the previous parts, we will now briefly summarize out of them what is sufficient for the purposes of this third part.

14.1 Ancient and Classical Geometric Worlds

Geometric world contains not just geometric objects but all that shows on them. Along with line segments, it contains also their lengths, and this enables us to include in this world also real numbers and various relations on them, in particular real functions, sequences of numbers or functions and similar.

Recall that the intuitive ancient geometric world is the Euclidean geometric world bounded by a horizon which we have called the geometric horizon. We denote the class of natural numbers that lie in it by FN.

In modern times an inconspicuous (but enormous) expansion of the ancient geometric world to absolute infinity occurred. The letter N denotes the class of all natural numbers lying in this geometric world. (As there is no *set of all natural numbers* N is not a set of all natural numbers either. In this third part of new infinitary mathematics this fact can be ignored).

Hence we have two Euclidean geometric worlds: the ancient and the classical geometric worlds. To be able to express ourselves more concisely, we shall denote the former by the letter $\mathcal{A}$ (ancient) and the latter by $\mathcal{C}$ (classical). As long as we consider only one of them it does not matter which of the two it is. Results achieved for one of them apply equally for the other. However, we can also study both of them simultaneously, as they relate to each other. That is, in their relationship of expansion of $\mathcal{A}$ to $\mathcal{C}$ since the classical geometric world arose through an expansion of the intuitive ancient geometric world.

We can capture this inner indistinguishability of the two geometric worlds as follows.

14.2 Principles of Expansion

The first principle of expansion. If ψ is a property of objects belonging to the world $\mathcal{A}$ described exclusively by the means of the world $\mathcal{A}$ (that is, without taking into account the existence of the world $\mathcal{C}$) then ψ is also a property of objects belonging exclusively to the world $\mathcal{C}$ (that is, without taking into account the existence of the world $\mathcal{A}$). Furthermore, in the world $\mathcal{A}$ there exists an object with the property ψ, if and only if there exists an object with the property ψ in the world $\mathcal{C}$.

This directly implies that:

All objects of the world $\mathcal{A}$ have the property ψ if and only if all objects of the world $\mathcal{C}$ have the property ψ.

Once we include numbers in the geometric worlds then it naturally must be the case that numbers as 1,2,3,..., belong both to $\mathcal{A}$ and $\mathcal{C}$. Similarly, for example, their square roots do, since the real number which when squared yields for instance the number 3 must certainly lie in both these worlds. This applies to all real numbers that we have some grasp of, and thus in fact it applies to all numbers that lie in the world $\mathcal{A}$.

The identification of points of geometric worlds with ordered n-tuples of real numbers then allows us to interpret not only real numbers, but all objects belonging to $\mathcal{A}$ in this way. This leads us to adopt the following principle which captures the apparent identity of the worlds $\mathcal{A}$ and $\mathcal{C}$, and which will moreover enable us to express ourselves in a much more concise way.

The second principle of expansion. The objects belonging to the world $\mathcal{A}$ also belong to the world $\mathcal{C}$ (they entered $\mathcal{C}$ during the expansion of $\mathcal{A}$) and they have the same properties and are set in the same interrelations (formulated exclusively by the means of the world $\mathcal{A}$, i. e. also exclusively by the means of the world $\mathcal{C}$) in the world $\mathcal{C}$ as in the world $\mathcal{A}$.

For the object belonging to the world $\mathcal{A}$ we use the same notation also in the world $\mathcal{C}$. This implies the following useful rule.[1]

[1] The second principle of expansion corresponds to the transfer principle of nonstandard analysis, and the usual convention would mean adding a star to indicate the object in the world $\mathcal{C}$. Vopěnka is not entirely consistent in "using the same notation" in both worlds, for example FN becomes N in $\mathcal{C}$. In the next Part, he introduces the notation $\mathrm{Ex}(X)$, see page 331. In the Czech original, four-volumed version, this Part III was the last, fourth book. For the English version, Vopěnka left out some material from the Czech third book and turned it

The rule of defined objects. Let ψ be a property of objects described exclusively by the means of the world $\mathcal{A}$ (and therefore also exclusively by means of the world $\mathcal{C}$). Let there only be one object A in the world $\mathcal{C}$ that has the property ψ. Then the object A belongs to the world $\mathcal{A}$ and it is the only object of the world $\mathcal{A}$ which has the property ψ.

Proof. This is so, because by the first principle of expansion there is precisely one object B in the world $\mathcal{A}$ which has the property ψ in this world. By the second principle of expansion B also has the property ψ in the world $\mathcal{C}$ and therefore $B = A$. □

If $\{a_1, a_2, \ldots, a_n\}$ is a sequence of objects belonging to the world $\mathcal{A}$ which is finite in this world then it means that this whole sequence can be surveyed from its first member to the last, even if it required an almost superhuman effort.

By the second principle of expansion this sequence $\{a_1, a_2, \ldots, a_n\}$ as well as any of its members $a_1, a_2, \ldots, a_n$ also belongs to the world $\mathcal{C}$. Furthermore, no new member joins it in the world $\mathcal{C}$. For if there existed a member x of this sequence $a_1, a_2, \ldots, a_n$ in the world $\mathcal{C}$ such that $x \neq a_1$, $x \neq a_2$, $\ldots$, $x \neq a_n$ then according to the first principle of expansion such a member of it would exist in the world $\mathcal{A}$ too, which is a contradiction.

With infinite sequences the situation is different, the sequence FN representing the key case. Under expansion of $\mathcal{A}$ to $\mathcal{C}$ this sequence FN extends to the sequence N.

Hence FN is clearly a cut on the class N. In consequence, under the expansion of $\mathcal{A}$ to $\mathcal{C}$ any infinite sequence belonging to $\mathcal{A}$ extends in a similar way.

14.3 Infinitely Large Natural Numbers

Those natural numbers in the world $\mathcal{C}$ that belong to the world $\mathcal{A}$, that is, the elements of the class FN, are called **finitely large natural numbers**.

Those natural numbers in the world $\mathcal{C}$ that were added to the class N only after the expansion of $\mathcal{A}$ to $\mathcal{C}$ are called **infinitely large natural numbers**. We shall denote the class of all such numbers IN. Hence we have

$$\mathrm{IN} = \mathrm{N} \setminus \mathrm{FN}\,.$$

Furthermore, $n \in \mathrm{FN}$ or $n \in \mathrm{IN}$ denotes n being a finitely large or infinitely large natural number respectively.

The classes FN and IN cannot be isolated out of the class (sequence) N by any internal means of the world $\mathcal{C}$. In fact, these classes do not belong to

into Part IV; both Czech third and fourth books had much of what is now this present Chapter 14 in common, with the third book using "$x \in Ex(\mathrm{Real})$" etc. in place of the (retained) "x is a real numbers belonging to $\mathcal{C}$". [Ed] the world $\mathcal{C}$ (understood on its own). They got in from outside. They can be isolated out of N only when we have the possibility to compare *simultaneously* both the geometric worlds $\mathcal{A}$ and $\mathcal{C}$.in their relationship of expansion of $\mathcal{A}$ to $\mathcal{C}$.

The following proposition is obvious.

Proposition 14.1.

(i) If $n \in \mathrm{FN}$ and $m < n$ then $m \in \mathrm{FN}$.

(ii) If $m \in \mathrm{IN}$ and $m < n$ then $n \in \mathrm{IN}$.

In the world $\mathcal{A}$ the class N consists only of those natural numbers that belong to FN. In other words: in the world $\mathcal{A}$ all natural numbers are finitely large. Consequently, the following propositions hold in the world $\mathcal{C}$.

Proposition 14.2. If $m, n \in \mathrm{FN}$, then also

$$m + n \in \mathrm{FN} \quad nm \in \mathrm{FN}, \quad m^n \in \mathrm{FN}$$

and so on.

Proposition 14.3. If $0 \neq m \in \mathrm{FN}$, $n \in \mathrm{IN}$ then there exist $k, l \in \mathrm{N}$ such that $k < \frac{n}{m} \leq k + 1$, $l^m < n \leq (l + 1)^m$ and $k, l, n - m \in \mathrm{IN}$.

Note. Even though in the world $\mathcal{C}$ the class IN is a non-empty part of the class (sequence) of all natural numbers, it has no least element.

14.4 Infinitely Large and Small Real Numbers

Real denotes the class of all real numbers lying in the world $\mathcal{A}$. With the expansion of the world $\mathcal{A}$ to the world $\mathcal{C}$ this class expands to the class of all real numbers of the world $\mathcal{C}$.

We say that a real number x belonging to $\mathcal{C}$ is **infinitely large** if $n < |x|$ for every $n \in \mathrm{FN}$.

Proposition 14.4. A real number x belonging to $\mathcal{C}$ is infinitely large if and only if there exists $n \in \mathrm{IN}$ such that $n < |x|$.

Proof. Let x be infinitely large. Let n be the largest natural number such that $n < |x|$. Then $n \in \mathrm{IN}$ since if $n \in \mathrm{FN}$ then $|x| \leq n + 1 \in \mathrm{FN}$, which is a contradiction. The other implication is trivial. □

We say that a real number x belonging to $\mathcal{C}$ is **finitely large** if it is not infinitely large.

The following proposition is obvious.

Proposition 14.5. (i) A real number x belonging to $\mathcal{C}$ is finitely large if and only if $|x| < n$ for every $n \in \mathrm{IN}$.

(ii) A real number x belonging to $\mathcal{C}$ is finitely large if and only if there exists $n \in \mathrm{FN}$ such that $|x| < n$.

We say that a real number x belonging to $\mathcal{C}$ is **infinitely small** (or **infinitesimal**) if either $x = 0$, or $x \neq 0$ and $\frac{1}{x}$ is infinitely large.

The following two propositions are also obvious.

Proposition 14.6. (i) A real number x belonging to $\mathcal{C}$ is infinitely small if and only if $|x| < \frac{1}{n}$ for every $0 \neq n \in \mathrm{FN}$.

(ii) A real number x belonging to $\mathcal{C}$ is infinitely small if and only if there exists an $n \in \mathrm{IN}$ such that $|x| < \frac{1}{n}$.

Proposition 14.7. (i) Every real number belonging to $\mathcal{A}$ is finitely large.

(ii) No real number belonging to $\mathcal{A}$ other than 0 is infinitely small.

Proposition 14.8. If x, y are real numbers belonging to $\mathcal{C}$ such that $x \cdot y$ is an infinitely small number then either x or y is an infinitely small number.

Proof. If there were $m, n \in \mathrm{FN}$ such that $\frac{1}{m} < |x|$ and $\frac{1}{n} < |y|$ then $\frac{1}{mn} < |x \cdot y|$ and by Proposition 14.2, $x \cdot y$ would not be an infinitely small number. □

Proposition 14.9. The product of infinitely small and finitely large real number is an infinitely small real number.

Proof. Let x, y be real numbers belonging to $\mathcal{C}$, $m \in \mathrm{FN}$, $n \in \mathrm{IN}$ such that $|y| < m$, $|x| < \frac{1}{n}$. Then $|x \cdot y| < \frac{m}{n}$. Let $k \in \mathrm{N}$ be such that $k < \frac{n}{m} \leq k+1$. By Proposition 14.3, $k \in \mathrm{IN}$ and $|x \cdot y| < \frac{m}{n} < \frac{1}{k}$. □

Proposition 14.10. The sum of finitely many infinitely small real numbers is an infinitely small real number.

Proof. Let $x_1, x_2, \ldots, x_n$ be infinitely small real numbers belonging to $\mathcal{C}$, $n \in FN$. Let us set $x = \max\{|x_1|, |x_2|, \ldots, |x_n|\}$. Obviously x is an infinitely small real number, $0 \leq x$. At the same time,

$$|x_1 + x_2 + \cdots + x_n| \leq |x_1| + |x_2| + \cdots + |x_n| \leq nx,$$

holds and nx is an infinitely small number by Proposition 14.9. □

Proposition 14.11. Let x be a real number belonging to $\mathcal{C}$, $m \in \mathrm{FN}$. Let the number x^m be infinitely small. Then also the number x is infinitely small.

Proof. Assume that $n \in \mathrm{FN}$ is such that $\frac{1}{n} < |x|$. Then $\frac{1}{n^m} < |x^m|$. But by Proposition 14.2, $n^m \in \mathrm{FN}$ and so x^m is not an infinitely small number, which is a contradiction. □

14.5 Infinite Closeness

We say that real numbers x, y belonging to $\mathcal{C}$ are **infinitely close**,[2] denoted $x \doteq y$, if the number $x - y$ is infinitely small.

Proposition 14.12. Let x, y, z be real numbers belonging to $\mathcal{C}$. Then clearly:

(i) $x \doteq x$.

(ii) If $x \doteq y$ then $y \doteq x$.

(iii) If $x \doteq y$ and $y \doteq z$ then $x \doteq z$.

(iv) Assume $x \doteq y$. If x is infinitely large, also y is infinitely large; if x is finitely large, also y is finitely large; if x is infinitely small, also y is infinitely small.

Proposition 14.13. Let x, y, u, v be real numbers belonging to $\mathcal{C}$ such that $x \doteq y$ a $u \doteq v$.

(i) $x + u \doteq y + v$, $x - u \doteq y - v$.

(ii) If x, u are finitely large, then $x \cdot u \doteq y \cdot v$.

(iii) If x is not infinitely small, then $\frac{1}{x} \doteq \frac{1}{y}$.

(iv) If x, u are finitely large and if x is not infinitely small then $\frac{u}{x} \doteq \frac{v}{y}$.

Proof. Case (i) is a trivial consequence of Proposition 14.10. Case (iv) is a consequence of cases (ii) and (iii). To prove (ii) it suffices to realize that

$$y \cdot v = (x + (y - x)) \cdot (u + (v - u)) = x \cdot u + x \cdot (v - u) + u \cdot (y - x) + (y - x) \cdot (v - u),$$

where the last three summands are infinitely small numbers by Proposition 14.9. To prove (iii) it suffices to realize that

$$\frac{1}{x} - \frac{1}{y} = \frac{y - x}{x \cdot y},$$

where the number in the denominator is not infinitely small. If this quotient were not infinitely small, then the infinitely small number $y - x$ would be the product of numbers $x \cdot y$ and $\frac{y-x}{x \cdot y}$, neither of which is infinitely small, which would contradict Proposition 14.8. □

Proposition 14.14. If x, y are numbers belonging to $\mathcal{A}$, then

$$x \doteq y \quad \text{if and only if} \quad x = y\,.$$

In other words no two different real numbers belonging to $\mathcal{A}$ are infinitely close.

[2] This relation of infinite closeness of real numbers in $\mathcal{C}$ differs from the relation introduced in Section 12.2 for rational numbers. [Ed]

Proof. The number $x - y$ belongs to $\mathcal{A}$ and according to Proposition 14.7 this number is infinitely small if and only if $x - y = 0$. □

Proposition 14.15. Let x, y be finitely large real numbers belonging to $\mathcal{C}$ that are not infinitely close, $x < y$. Then there exist real numbers r, s belonging to $\mathcal{A}$ such that

$$x < r < s < y\,.$$

Also $x < r < \frac{r+s}{2} < s < y$ and so on.

Proof. First let $0 \le x < y$. Let $m, n \in \mathrm{FN}$ be such that

$$\frac{1}{n} < \frac{1}{3}(y - x), \qquad y < m\,.$$

Let k be the smallest natural number such that $x < \frac{k}{n}$. Then $x < \frac{k}{n} < \frac{k+1}{n} < y < m$, so $k < mn$, in consequence of which $k \in \mathrm{FN}$. But this means that the numbers $\frac{k}{n}$, $\frac{k+1}{n}$ belong to $\mathcal{A}$.

If $x < y \le 0$, then $0 \le -y < -x$, which converts this case to the previous case.

Let $x < 0 < y$. Then either it is not true that $x \doteq 0$, or it is not true that $y \doteq 0$. According to the previous part of this proof there exist r, s belonging to $\mathcal{A}$ such that if it is not true that $x \doteq 0$, then $x < r < s < 0$ and if it is not true that $y \doteq 0$, then $0 < r < s < y$. □

14.6 Principles of Backward Projection

If we try to see as far as a geometric point X lying in the world $\mathcal{C}$ but not in the world $\mathcal{A}$ (that is, somewhere beyond the geometric horizon) we find ourselves unable to see the point. Our gaze rests on a point Y lying in the world $\mathcal{A}$ which is concealing the point X. (In other words, *the point X from $\mathcal{C}$ is reflected back to the point Y*). After the expansion of the Euclidean world $\mathcal{A}$ to the world $\mathcal{C}$, during which the point Y falls through into the world $\mathcal{C}$, the points X, Y are in this world infinitely close.

If we transfer this consideration into the language of real numbers, we get the following principle:

The first principle of backward projection. Under the backward projection of the world $\mathcal{C}$ onto the world $\mathcal{A}$, every finitely large real number x belonging to $\mathcal{C}$ projects onto an infinitely close real number y belonging to $\mathcal{A}$.

Since by Proposition 14.14 no two distinct real numbers belonging to $\mathcal{A}$ are infinitely close, the following assertion holds:

Proposition 14.16. If x is a finitely large real number belonging to $\mathcal{C}$, then there is a unique real number belonging to $\mathcal{A}$ (denoted by $\mathrm{Proj}(x)$) so that

$$x \doteq \mathrm{Proj}(x).$$

Proposition 14.13 directly yields the following:

Proposition 14.17. Let x and y be finitely large real numbers belonging to $\mathcal{C}$.

(i) $\mathrm{Proj}(x+y) = \mathrm{Proj}(x) + \mathrm{Proj}(y)$,

(ii) $\mathrm{Proj}(x-y) = \mathrm{Proj}(x) - \mathrm{Proj}(y)$,

(iii) $\mathrm{Proj}(x \cdot y) = \mathrm{Proj}(x) \cdot \mathrm{Proj}(y)$,

(iv) If $\mathrm{Proj}(y) \neq 0$ then

$$\mathrm{Proj}\left(\frac{x}{y}\right) = \frac{\mathrm{Proj}(x)}{\mathrm{Proj}(y)}.$$

Proposition 14.18. Let x and y be finitely large real numbers belonging to $\mathcal{C}$ and $x < y$. Then

$$\mathrm{Proj}(x) \leq \mathrm{Proj}(y),$$

where equality holds if and only if $x \doteq y$.

Proof. By Proposition 14.17,

$$\mathrm{Proj}(y-x) = \mathrm{Proj}(y) - \mathrm{Proj}(x).$$

Hence

$$x \doteq y \quad \text{if and only} \quad \mathrm{Proj}(y-x) = 0,$$

which holds if and only if $\mathrm{Proj}(x) = Proj(y)$.

Let $\mathrm{Proj}(y-x) \neq 0$. Assuming $\mathrm{Proj}(y-x) < 0$ gives

$$(y-x) - \mathrm{Proj}(y-x) > |\mathrm{Proj}(y-x)|.$$

As $\mathrm{Proj}(y-x)$ belongs to $\mathcal{A}$, this means that the numbers $(y-x)$ and $\mathrm{Proj}(y-x)$ are not infinitely close, contradiction. Therefore $\mathrm{Proj}(y-x) > 0$ and hence $\mathrm{Proj}(y) > \mathrm{Proj}(x)$. □

If we try to see as far as an infinitely large real number belonging to $\mathcal{C}$, our gaze rests on some sort of a vanishing point that has appeared behind all the real numbers belonging to $\mathcal{A}$. This leads us to accept the following principle.

The second principle of backward projection. In the backward projection of $\mathcal{C}$ onto $\mathcal{A}$, every positive (or negative) infinitely large real number belonging to $\mathcal{C}$ projects onto a number which has additionally appeared in the world $\mathcal{A}$, denoted by ∞ (or $-\infty$ respectively).

The world $\mathcal{A}$ is therefore extended by the numbers ∞ and $-\infty$. These numbers originally did not belong to this world: they appeared in it after the expansion of $\mathcal{A}$ to $\mathcal{C}$. They are thus *improper real numbers* belonging to $\mathcal{A}$. (And by the second principle of expansion they entered also the world $\mathcal{C}$.)

The operation of backward projection can now be extended to *all real numbers belonging to* $\mathcal{C}$, namely in the following way:

Let x be an infinitely large real number belonging to $\mathcal{C}$. Then we define

$$\mathrm{Proj}(x) = \begin{cases} \infty & \text{if } x > 0, \\ -\infty & \text{if } x < 0. \end{cases}$$

In accordance with the thus-extended backward projection, we can extend the relation of infinite closeness in the world $\mathcal{C}$ to the numbers ∞ and $-\infty$, namely by setting

$$\begin{array}{rcll} \infty & \doteq & x & \text{if and only if } x \text{ is an infinitely large number and } x > 0, \\ -\infty & \doteq & x & \text{if and only if } x \text{ is an infinitely large number and } x < 0, \\ \infty & \doteq & -\infty & \text{does not hold.} \end{array}$$

Furthermore we define the relation $\doteq$ to remain reflexive and symmetric. However, transitivity of infinite closeness is broken by the improper numbers ∞ and $-\infty$ since even if $x \doteq \infty$ and $\infty \doteq y$, $x \doteq y$ does not need to be the case. Similarly for $-\infty$.

14.7 Arithmetic with Improper Numbers ∞, $-\infty$

The rules of arithmetic of infinite closeness of real numbers belonging to the world $\mathcal{C}$ as stated in Proposition 14.13 can, by virtue of Proposition 14.17, be also interpreted as follows: the basic arithmetic operations on (proper) real numbers belonging to $\mathcal{A}$ are backward projections of the corresponding operations on finitely large real numbers belonging to the world $\mathcal{C}$. These backward projections of basic arithmetic operations in $\mathcal{C}$ to $\mathcal{A}$ are merely crude reflections of the corresponding finer operations in the world $\mathcal{C}$.

With this interpretation in mind we can extend the basic arithmetic operations on real numbers in the world $\mathcal{A}$ to improper numbers in an obvious manner. Some arithmetic expressions containing improper numbers or the number 0 may be indeterminate. This means that they cannot be evaluated, as they do not provide sufficient information about the real numbers belonging to $\mathcal{C}$ that were projected onto the numbers ∞, $-\infty$ (or 0).

First of all, for every proper real number x belonging to $\mathcal{A}$ we define

$$-\infty < x < \infty.$$

In the rules listed below, A represents a real number belonging to $\mathcal{A}$, possibly even an improper one, that is ∞ or $-\infty$.

Rule 14.19.

(i) If $-\infty < A$ then $A + \infty = \infty + A = \infty$.

(ii) If $A < \infty$ then $A - \infty = -\infty + A = -\infty$.

(iii) $\infty - \infty$ and $-\infty + \infty$ are indeterminate expressions (since the difference between two positive infinitely large real numbers may be any number).

Rule 14.20. Let $A > 0$. Then

(i) $A \cdot \infty = \infty \cdot A = \infty$.

(ii) $(-A) \cdot \infty = A \cdot (-\infty) = -\infty$.

(iii) $0 \cdot \infty$, $\infty \cdot 0$ are indeterminate expressions.

Rule 14.21. Let $0 < A < \infty$. Then

(i) $\frac{\infty}{A} = \frac{-\infty}{-A} = \infty$.

(ii) $\frac{-\infty}{A} = \frac{\infty}{-A} = -\infty$.

Rule 14.22.

(i) If $A \neq 0$ then $\frac{A}{0} = \pm\infty$.

(ii) $\frac{0}{0}$ is an indeterminate expression.

Rule 14.23. If $-\infty < A < \infty$ then $\frac{A}{\infty} = \frac{A}{-\infty} = 0$.

Rule 14.24.

(i) $\frac{\infty}{\infty}$ and $\frac{-\infty}{-\infty}$ are indeterminate expressions but their value is non-negative.

(ii) $\frac{-\infty}{\infty}$ and $\frac{\infty}{-\infty}$ are indeterminate expressions but their value is non-positive.

The above definitions of basic operations for the improper numbers in $\mathcal{A}$ enable us to generalize Proposition 14.17 in the following way:

Proposition 14.25. Let x and y be real numbers belonging to $\mathcal{C}$. Then

(i) $\operatorname{Proj}(x+y) = \operatorname{Proj}(x) + \operatorname{Proj}(y)$,

(ii) $\operatorname{Proj}(x-y) = \operatorname{Proj}(x) - \operatorname{Proj}(y)$,

(iii) $\operatorname{Proj}(x \cdot y) = \operatorname{Proj}(x) \cdot \operatorname{Proj}(y)$,

(iv)

$$\operatorname{Proj}\left(\frac{x}{y}\right) = \frac{\operatorname{Proj}(x)}{\operatorname{Proj}(y)}.$$

Naturally, as long as the expressions on the right are not indeterminate.

14.8 Further Fixed Notation for this Part

Letters a, b, c, d (or p, q, r, u, v) denote proper real numbers belonging to the world $\mathcal{A}$. Letters A, B, C, D denote both proper and improper real numbers belonging to the world $\mathcal{A}$.

By a **closed** (or **left-closed, right-closed, open**) **interval** I with end points A, B, where $A < B$, we understand the class

$$[A, B]$$

(or $[A, B)$, $(A, B]$, (A, B)) of all real numbers x belonging to the world $\mathcal{A}$ such that

$$A \leq x \leq B$$

(or $A \leq x < B$, $A < x \leq B$, $A < x < B$ respectively).

A point from the interval I which is not its end point is called an **inner** point of I.

When after the expansion of the world $\mathcal{A}$ to the world $\mathcal{C}$ the interval I enters the world $\mathcal{C}$ then it contains all the real numbers that are related to the end points of I as above.

Letters f, g, h denote real functions defined on an interval I that belong to the world $\mathcal{A}$. By the second principle of expansion they also enter the world $\mathcal{C}$.

Letters $\alpha, \beta, \gamma, \delta$ denote infinitely small non-zero real numbers. These numbers belong to the world $\mathcal{C}$ but not to the world $\mathcal{A}$.

Letters m, n, k denote natural numbers belonging to the world $\mathcal{A}$, that is, elements of the class FN. After the expansion of the world $\mathcal{A}$ to the world $\mathcal{C}$ they also denote elements of the class N.

Chapter 15

Sequences of Numbers

15.1 Binomial Numbers

We define

$$\begin{aligned} 0! &= 1, \\ k! &= 1 \cdot 2 \cdot 4 \cdot \ldots (k-1) \cdot k, \\ \binom{r}{0} &= 1, \\ \binom{r}{k} &= \frac{r \cdot (r-1) \ldots (r-k+1)}{k!} \qquad \text{for } \; k \neq 0,. \end{aligned}$$

Proposition 15.1. If $n < k$ then

$$\binom{n}{k} = 0, \qquad \text{in particular} \quad \binom{0}{k} = 0\,.$$

Proposition 15.2.

$$\binom{r+1}{k+1} = \binom{r}{k} + \binom{r}{k+1}.$$

Proof. From the definition it follows that

$$\binom{r}{k} + \binom{r}{k+1} = \binom{r}{k} \cdot \left(1 + \frac{r-k}{k+1}\right) = \binom{r}{k} \cdot \frac{r+1}{k+1} = \binom{r+1}{k+1}$$

□

Proposition 15.3. (Binomial theorem)

$$(a+b)^n = \sum_{k=0}^{n} \binom{n}{k} a^k \cdot b^{n-k}.$$

Proof. The proof is by induction on n. For $n = 0$ we have

$$1 = (a+b)^0 = \binom{0}{0} a^0 \cdot b^0 = 1,$$

and hence the assertion holds. Now assume that it holds for some n. Then

$$\begin{aligned}
(a+b)^{n+1} &= (a+b) \cdot (a+b)^n = \\
&= (a+b) \cdot \sum_{k=0}^{n} \binom{n}{k} a^k \cdot b^{n-k} = \\
&= \sum_{k=0}^{n} \binom{n}{k} a^{k+1} \cdot b^{n-k} + \sum_{k=0}^{n} \binom{n}{k} a^k \cdot b^{n-k+1} = \\
&= \sum_{k=1}^{n+1} \binom{n}{k-1} a^k \cdot b^{n-k+1} + \sum_{k=0}^{n} \binom{n}{k} a^k \cdot b^{n-k+1} = \\
&= \sum_{k=1}^{n} \left[\binom{n}{k-1} + \binom{n}{k} \right] a^k \cdot b^{n-k+1} + \binom{n}{n} a^{n+1} + \binom{n}{0} b^{n+1} = \\
&= \sum_{k=1}^{n} \binom{n+1}{k} a^k \cdot b^{n-k+1} + a^{n+1} + b^{n+1} = \\
&= \sum_{k=0}^{n+1} \binom{n+1}{k} a^k \cdot b^{n+1-k}.
\end{aligned}$$

□

Hence in particular

Proposition 15.4.

(i) $2^n = (1+1)^n = \displaystyle\sum_{k=0}^{n} \binom{n}{k}$,

(ii) $0 = (1-1)^n = \displaystyle\sum_{k=0}^{n} (-1)^k \binom{n}{k}$.

Proposition 15.5.

$$\sum_{k=0}^{n} (-1)^k \binom{r}{k} = (-1)^n \binom{r-1}{n}.$$

Proof. We prove this proposition by induction on n. For $n = 0$ it holds trivially. So assume that it holds for a number n. Then

$$\begin{aligned}\sum_{k=0}^{n+1}(-1)^k\binom{r}{k} &= \sum_{k=0}^{n}(-1)^k\binom{r}{k} + (-1)^{n+1}\binom{r}{n+1} = \\ &= (-1)^n\binom{r-1}{n} + (-1)^{n+1}\binom{r}{n+1} = \\ &= (-1)^{n+1}\left[\binom{r}{n+1} - \binom{r-1}{n}\right] = \\ &= (-1)^{n+1}\binom{r-1}{n+1}.\end{aligned}$$

□

15.2 Limits of Sequences

The following numerical sequences belong to $\mathcal{A}$. After the expansion of the world $\mathcal{A}$ to the world $\mathcal{C}$ they enter $\mathcal{C}$ where they are extended to classical infinity.

$$\begin{aligned}&\{a_n\} \quad a_1, a_2, a_3, \dots,\\ &\{b_n\} \quad b_1, b_2, b_3, \dots,\\ &\{c_n\} \quad c_1, c_2, c_3, \dots .\end{aligned}$$

We say that a number A is the **limit of the sequence** $\{a_n\}$, denoted

$$A = \lim_{n\to\infty} a_n,$$

if for each $n \in \mathrm{IN}$ we have $A \doteq a_n$.

Proposition 15.6. Every sequence has at most one limit.

Proof. By proposition 14.14 no two different real numbers from $\mathcal{A}$ are infinitely close, which clearly holds also including the improper numbers. □

We say that a sequence $\{a_n\}$ is **convergent** if it has a limit and this limit is a proper number A. We also say that the sequence **converges to** A.

If $\{a_n\}$ is not convergent then we say that it is **divergent**. If its limit is ∞ (or $-\infty$) then we say that the sequence **diverges** to ∞ (or to $-\infty$ respectively).

A sequence that has no limit (proper or improper) is called **oscillatory**.

The following proposition clearly does not need proving.

Proposition 15.7.

(i) If $\lim_{n\to\infty} a_n = A$ then $\lim_{n\to\infty} |a_n| = |A|$.

(ii) $\lim_{n\to\infty} a_n = 0$ if and only if $\lim_{n\to\infty} |a_n| = 0$.

Proposition 15.8. Let $\{a_n\}$, $\{b_n\}$ a $\{c_n\}$ be sequences of real numbers.

(i) Let $\lim_{n\to\infty} a_n = 0$. Assume that there exists k such that for each $n > k$, $|b_n| \leq |a_n|$. Then $\lim_{n\to\infty} b_n = 0$.

(ii) Let $\lim_{n\to\infty} a_n = A$, $\lim_{n\to\infty} b_n = B$. Assume that there exists k such that for each $n > k$, $a_n \leq b_n$. Then $A \leq B$.

(iii) Let $\lim_{n\to\infty} a_n = \lim_{n\to\infty} b_n = A$. Assume that there exists k such that for each $n > k$, $a_n \leq c_n \leq b_n$. Then $\lim_{n\to\infty} c_n = A$.

Proof. If k is the least natural number guaranteed to exist in the above assertions, then $k \in \mathrm{FN}$ by the rule of defined objects. Hence for each $n \in \mathrm{IN}$ we have:

(i) $|b_n| \leq |a_n| \doteq 0$;

(ii) $a_n \leq b_n$ so by Proposition 14.18 $\mathrm{Proj}(a_n) \leq \mathrm{Proj}(b_n)$, which is clearly also true for improper numbers.

(iii) If A is proper then $a_n \leq c_n \leq b_n$ and so $a_n \doteq c_n \doteq b_n$. If A is improper, for example $A = \infty$ then $a_n \doteq \infty$, $a_n \leq c_n$ and so $c_n \doteq \infty$. □

Proposition 15.9. Let $\lim_{n\to\infty} a_n = A$, $\lim_{n\to\infty} b_n = B$, $A < B$. Then there exists k such that for each $n > k$ je $a_n < b_n$.

Proof. If no such k existed then there would exist some $n \in \mathrm{IN}$ such that $b_n \leq a_n$. Hence by Proposition 14.18 $B \leq A$, which is clearly also true for improper numbers. □

The following Proposition is an immediate consequence of Proposition 14.25.

Proposition 15.10. Let $\lim_{n\to\infty} a_n = A$, $\lim_{n\to\infty} b_n = B$.

(i) Assume that $A + B$ is not an indeterminate expression. Then

$$\lim_{n\to\infty} (a_n + b_n) = A + B\,.$$

(ii) Assume that $A - B$ is not an indeterminate expression. Then

$$\lim_{n\to\infty} (a_n - b_n) = A - B\,.$$

(iii) Assume that $A \cdot B$ is not an indeterminate expression. Then

$$\lim_{n\to\infty} (a_n \cdot b_n) = A \cdot B\,.$$

(iv) Assume that $\frac{A}{B}$ is not an indeterminate expression and that for each n, $b_n \neq 0$. Then

$$\lim_{n\to\infty} \left(\frac{a_n}{b_n}\right) = \frac{A}{B}.$$

We say that the sequence $\{a_n\}$ satisfies **Bolzano's condition**, if for each $n, m \in \mathrm{IN}$

$$a_n \doteq a_m.$$

If the sequence $\{a_n\}$ is convergent then it clearly satisfies Bolzano's condition. Moreover, the following converse assertion holds.

Proposition 15.11. Assume that the sequence $\{a_n\}$ satisfies Bolzano's condition. Then this sequence is convergent.

Proof. The sequence $\{a_n\}$ clearly has a limit, let

$$A = \lim_{n\to\infty} a_n.$$

It suffices to show that A is a proper, that is, finitely large, real number. Let k be the smallest natural number such that for each $n > k$

$$a_k - 1 < a_n < a_k + 1.$$

If $m \in \mathrm{IN}$ and $m < n$ then $a_m \doteq a_n$ and hence

$$a_m - 1 < a_n < a_m + 1.$$

It follows that $k \in \mathrm{FN}$. This means that a_k belongs to $\mathcal{A}$ and the numbers $a_k - 1$ a $a_k + 1$ are finitely large. Obviously

$$a_k - 1 \leq A \leq a_k + 1,$$

and hence A is a finitely large real number. □

We say that the sequence $\{a_n\}$ is **non-decreasing** (or **non-increasing**), if for each $n < m$

$$a_n \leq a_m \qquad (\text{or} \quad a_m \leq a_n \text{ respectively}).$$

Proposition 15.12. Assume that $\{a_n\}$ is a non-decreasing (or non-increasing) sequence. Then this sequence has a limit.

Proof. Let $m < n$ be natural numbers such that $m, n \in \mathrm{IN}$ and $\mathrm{Proj}(a_m) < \mathrm{Proj}(a_n)$. Let b be such that

$$\mathrm{Proj}(a_m) < b < \mathrm{Proj}(a_n).$$

Let k be the least natural number such that $b < a_k$. By the rule of defined objects $k \in \mathrm{FN}$. However, that is a contradiction since $k < m$ and $a_m < a_k$. The case of a non-increasing sequence is similar.

□

We say that a sequence $\{a_n\}$ is **bounded from above** (or **from below** respectively), if there exists a proper real number b such that for any n

$$a_n < b \qquad \text{(or} \quad b < a_n \text{ respectively).}$$

We say that a sequence $\{a_n\}$ is **bounded** if it is bounded both from above and from below.

Proposition 15.13. Let $\{a_n\}$ be a non-decreasing sequence bounded from above (or a non-increasing sequence bounded from below respectively). Then this sequence is convergent.

Proof. By Proposition 15.12 the sequence $\{a_n\}$ has a limit. Let

$$A = \lim_{n\to\infty} a_n .$$

Let b be a real number such that for each $n \in \mathrm{FN}$, $a_1 \leq a_n < b$. Then also for each $n \in \mathrm{IN}$, $a_1 \leq a_n < b$ holds and by Proposition 14.18

$$a_1 \leq \mathrm{Proj}(a_n) = A \leq b,$$

so A is proper.

The case of a non-increasing sequence is similar. □

By a **geometric sequence (with quotient q)** we understand a sequence of real numbers

$$\{a \cdot q^n\} .$$

Proposition 15.14. For the geometric sequence $\{q^n\}$ we have:

(i) If $1 < q$ then $\lim_{n\to\infty} q^n = \infty$.

(ii) If $1 = q$ then $\lim_{n\to\infty} q^n = 1$.

(iii) If $|q| < 1$ then $\lim_{n\to\infty} q^n = 0$.

(iv) If $q \leq -1$ then the sequence $\{q^n\}$ oscillates. Furthermore,
if $q = -1$ then the sequence is $-1, 1, -1, 1, -1, \ldots$
if $q < -1$ and $n \in \mathrm{IN}$ then $q^n \doteq \infty$ for n even and $q^n \doteq -\infty$ for n odd.

Proof. (i) Let $q = 1 + p$. Then

$$q^n = (1+p)^n = 1 + np + \cdots > 1 + np,$$

because the left-out summands are positive. For $n \in \mathrm{IN}$

$$1 + np \doteq \infty \qquad \text{and so} \qquad q^n \doteq \infty.$$

The case (ii) is trivial and the case(iv) follows from (i), (ii).

It remains to prove (iii). If $q = 0$, the sequence is $0, 0, 0, \ldots$ If $q \neq 0$ then $\frac{1}{|q|} > 1$. By (i)

$$\lim_{n\to\infty} \frac{1}{|q^n|} = \infty, \quad \text{so} \quad \lim_{n\to\infty} |q^n| = 0,$$

and hence $\lim\limits_{n\to\infty} q^n = 0$. □

Proposition 15.15. Let $\{a_n\}$ be a sequence of positive numbers.

(i) Assume that there exists a natural number k such that for any $n \geq k$, $\frac{a_{n+1}}{a_n} \leq q < 1$. Then $\lim\limits_{n\to\infty} a_n = 0$.

(ii) Assume that there exists $\lim\limits_{n\to\infty} \frac{a_{n+1}}{a_n} < 1$. Then $\lim\limits_{n\to\infty} a_n = 0$.

(iii) Assume that there exists a natural number k such that for any $n \geq k$, $\frac{a_{n+1}}{a_n} > 1$. Then $\lim\limits_{n\to\infty} a_n > 0$ exists.

(iv) Assume that there exists $\lim\limits_{n\to\infty} \frac{a_{n+1}}{a_n} > q > 1$, where q belongs to $\mathcal{A}$. Then $\lim\limits_{n\to\infty} a_n = \infty$.

Proof. (i) By the rule of defined objects the least natural number k with the given property belongs to the class FN. If $k < n$ then

$$a_n = a_k \cdot \frac{a_{k+1}}{a_k} \cdot \frac{a_{k+2}}{a_{k+1}} \cdot \cdots \cdot \frac{a_n}{a_{n-1}} \leq a_k \cdot q^{n-k}.$$

If $n \in \mathrm{IN}$ then $n - k \in \mathbf{IN}$ and hence $q^{n-k} \doteq 0$. Hence $a_n \doteq 0$.

(ii) Let q belong to $\mathcal{A}$ and satisfy

$$\lim_{n\to\infty} \frac{a_{n+1}}{a_n} < q < 1.$$

For each $n \in \mathrm{IN}$ we have $\frac{a_{n+1}}{a_n} < q$. If k is the smallest natural number such that for each $n \geq k$, $\frac{a_{n+1}}{a_n} < q$ then $k \in \mathrm{FN}$. Hence by (i), $\lim\limits_{n\to\infty} a_n = 0$.

(iii) As in (ii) we may assume that $k \in \mathrm{FN}$. Then the sequence $a_k, a_{k+1}, a_{k+2}, \ldots$ is non-decreasing and by Proposition 15.12 it has a limit which obviously satisfies

$$0 < a_k \leq \lim_{n\to\infty} a_n.$$

(iv) Clearly there exists k such that for $n > k$

$$a_k \cdot q^{n-k} < a_n.$$

Hence the sequence $\{a_n\}$ is non-decreasing and by Proposition 15.12 it has a limit. Since $\{a_n\}$ is not bounded from above, the limit is improper. □

Proposition 15.16. For each d,

$$\lim_{n \to \infty} \frac{d^n}{n!} = 0.$$

Proof. The case $|d| \leq 1$ is trivial. Let $|d| > 1$. Let $a_n = \frac{|d|^n}{n!}$. Then

$$\frac{a_{n+1}}{a_n} = \frac{|d|}{n+1}, \qquad \text{hence} \quad \lim_{n \to \infty} \frac{a_{n+1}}{a_n} = 0.$$

By Proposition 15.15, $\lim\limits_{n \to \infty} a_n = 0$. □

Proposition 15.17. Let $k \in \mathrm{FN}$. Let $\{a_n\}$ be a sequence such that for each $n \in \mathrm{IN}$

$$\text{either} \quad 1 \leq a_n \leq a_n^n \leq n^k, \quad \text{or} \quad \frac{1}{n^k} \leq a_n^n \leq a_n \leq 1.$$

Then $\lim\limits_{n \to \infty} a_n = 1$.

Proof. Let $n \in \mathrm{IN}$. First let $1 \leq a_n$. Then $a_n = 1 + p$, where $0 \leq p$, and hence

$$a_n^n = (1+p)^n =$$

$$= 1 + np + \cdots + \binom{n}{k+1} p^{k+1} + \cdots + p^n \geq \binom{n}{k+1} p^{k+1},$$

since the left-out summands are non-negative. Hence $n^k \geq \binom{n}{k+1} p^{k+1}$ so

$$\frac{n^k}{n(n-1)\cdots(n-k)} \geq \frac{p^{k+1}}{(k+1)!}.$$

Moreover

$$\frac{1}{n}\left(\frac{n}{n-k}\right)^k > \frac{n^k}{n(n-1)\cdots(n-k)} \geq \frac{p^{k+1}}{(k+1)!}.$$

Furthermore

$$\frac{1}{n} \doteq 0, \qquad \left(\frac{n}{n-k}\right)^k = \frac{1}{\left(1 - \frac{k}{n}\right)^k} \doteq 1,$$

since $\dfrac{k}{n} \doteq 0$. Hence

$$\frac{p^{k+1}}{(k+1)!} \doteq 0, \quad \text{so} \quad p^{k+1} \doteq 0, \quad \text{so} \quad p \doteq 0.$$

It follows that $a_n \doteq 1$.

If $\frac{1}{n^k} \le a_n^n \le a_n \le 1$ then

$$1 \le \frac{1}{a_n} \le \frac{1}{a_n^n} \le n^k,$$

and hence by the above part of the proof, $\dfrac{1}{a_n} \doteq 1$. Hence $a_n \doteq 1$. □

15.3 Euler's Number

Proposition 15.18. $\{(1+\frac{1}{n})^{n+1}\}$ in a non-increasing sequence of positive numbers.

Proof. Every member of this sequence is obviously greater than 1. The following inequalities are equivalent.

$$\left(1+\frac{1}{n}\right)^{n+1} > \left(1+\frac{1}{n+1}\right)^{n+2},$$

$$\left(\frac{n+1}{n}\right)^{n+1} > \left(\frac{n+2}{n+1}\right)^{n+2},$$

$$\left(\frac{n+1}{n}\right)^{n+2} > \frac{n+1}{n}\cdot\left(\frac{n+2}{n+1}\right)^{n+2},$$

$$\left(\frac{(n+1)^2}{n(n+2)}\right)^{n+2} > 1+\frac{1}{n},$$

$$\left(\frac{n^2+2n+1}{n^2+2n}\right)^{n+2} > 1+\frac{1}{n},$$

$$\left(1+\frac{1}{n(n+2)}\right)^{n+2} > 1+\frac{1}{n}.$$

The last inequality does hold since on the right there are only the first two summands from the binomial expansion of the left hand side and the omitted summands are positive. Hence all the previous inequalities also hold. □

By Proposition 15.13 the sequence $\left\{\left(1+\frac{1}{n}\right)^{n+1}\right\}$ is convergent. Its limit, the **Euler number**, is denoted e.

Proposition 15.19.

$$e = \lim_{n\to\infty}\left(1+\frac{1}{n}\right)^n.$$

Proof.

$$\lim_{n\to\infty}\left(1+\frac{1}{n}\right)^n = \lim_{n\to\infty}\frac{\left(1+\frac{1}{n}\right)^{n+1}}{1+\frac{1}{n}} = \frac{\lim\limits_{n\to\infty}\left(1+\frac{1}{n}\right)^{n+1}}{\lim\limits_{n\to\infty}\left(1+\frac{1}{n}\right)} = \frac{e}{1}.$$

□

Since for each $n > 2$, $\left(1 + \frac{1}{n}\right)^n$ is larger than the first two summands of the binomial expansion of this expression, that is

$$1 + n \cdot \frac{1}{n} = 2,$$

we have $2 < e$.

Since for each n, $e \leq \left(1 + \frac{1}{n}\right)^{n+1}$, we have

$$e \leq \left(1 + \frac{1}{3}\right)^4 = \left(\frac{5}{4}\right)^5 < 3.$$

Hence

$$2 < e < 3.$$

Chapter 16

Continuity and Derivatives of Real Functions

16.1 Continuity of a Function at a Point

We say that a function f is **continuous** (or **right-continuous**, or **left-continuous**) **at a point** c, if for every α (or $\alpha > 0$, or $\alpha < 0$ respectively) the following holds:

$$f(c) \doteq f(c+\alpha).$$

The following assertions are obvious.

Proposition 16.1.

(i) If $f(c+\alpha) = f(c)$ for every α, then f is continuous at the point c.

(ii) If $f(c+\alpha) = c+\alpha$ for every α, then f is continuous at the point c.

(iii) If f is continuous at a point c, $f(c) \neq 0$, then $f(c+\alpha) \neq 0$ for every α.

The next proposition follows directly from Proposition 14.17.

Proposition 16.2. Let functions f, g be continuous at a point c. Then the following holds:

(i) The function $f+g$ is continuous at the point c.

(ii) The function $-f$ is continuous at the point c.

(iii) The function $f \cdot g$ is continuous at the point c.

(iv) If $f(x) \neq 0$ for every $x \in I$, then the functions $\frac{1}{f}$, $\frac{g}{f}$ are continuous at the point c.

Here $f+g$ is the function h defined by

$$h(x) = f(x) + g(x)$$

for every $x \in I$ and analogously in the other cases.

An analogous assertion naturally holds also for right-continuous (or left-continuous) functions, where $c \in I$ may be the left (or right) endpoint of interval I.

The next assertion follows directly from the previous propositions.

Proposition 16.3. (Continuity of polynomial and rational functions)

(i) If f is given by a polynomial, then f is continuous at every point.

(ii) If f is a rational function then f is continuous at every point at which the denominator is not equal to 0.

By a **composite function** $g(f)$ we understand a function h given by

$$h(x) = g(f(x))$$

for every x, for which the expression on the right hand side is defined.

Proposition 16.4. (Continuity of a composite function) Let function f be continuous at a point c. Let function g be defined on an interval whose interior point is $d = f(c)$. Let g be continuous at the point d. Then the composite function $g(f)$ is continuous at the point c.

Proof. Since f is continuous at c, that is $f(c+\alpha) \doteq f(c)$ for every α, we have

$$\begin{aligned} g(f(c+\alpha)) = g(f(c) + f(c+\alpha) - f(c)) = \\ = g(d + (f(c+\alpha) - f(c))) \doteq g(d) = g(f(c)) \end{aligned}$$

since $f(c+\alpha) - f(c) \doteq 0$. □

16.2 Derivative of a Function at a Point

We say that a number $f'(c)$ (possibly improper) belonging to $\mathcal{A}$ is the **derivative of a function f at a point** c if for every α the following holds:

$$f'(c) \doteq \frac{f(c+\alpha) - f(c)}{\alpha}.$$

Right-sided and **left-sided derivatives of a function** f at a point c are defined in the obvious way.

Proposition 16.5. Let a function f have a proper derivative at a point c. Then f is continuous at the point c.

Proof. For every α,

$$\frac{f(c+\alpha) - f(c)}{\alpha} = f'(c) + \beta,$$

where $\beta \doteq 0$. Hence $f(c+\alpha) - f(c) = \alpha \cdot f'(c) + \alpha \cdot \beta \doteq 0$. □

The following proposition is easy to prove.

Proposition 16.6. (i) If $f(c+\alpha) = f(c)$ for every α, then $f'(c) = 0$.

(ii) If $f(c+\alpha) = c+\alpha$ for every α, then $f'(c) = 1$.

(iii) Let f have a derivative at a point c and let b be a proper real number. Then the function $b \cdot f$ also has a derivative at the point c and

$$(b \cdot f)'(c) = b \cdot f'(c)\,,$$

the expressions $0 \cdot \infty$, $0 \cdot (-\infty)$ having in this case the value 0. So in particular

$$(-f)'(c) = -f'(c).$$

(iv) Let f, g have derivatives at a point c and assume that the expression $f'(c) + g'(c)$ is not indeterminate. Then also the function $f+g$ has a derivative at the point c and

$$(f+g)'(c) = f'(c) + g'(c)\,.$$

Proposition 16.7. (Product rule) Let f, g have derivatives at a point c. Assume that f is continuous at the point c and that the expression

$$f(c) \cdot g'(c) + g(c) \cdot f'(c)$$

is not indeterminate. Then also the function $f \cdot g$ has a derivative at the point c and

$$(f \cdot g)'(c) = f(c) \cdot g'(c) + g(c) \cdot f'(c).$$

Proof. For every α we have

$$\frac{(f \cdot g)(c+\alpha) - (f \cdot g)(c)}{\alpha} = \frac{f(c+\alpha) \cdot g(c+\alpha) - f(c) \cdot g(c)}{\alpha} =$$
$$= \frac{f(c+\alpha) \cdot g(c+\alpha) - f(c+\alpha) \cdot g(c) + f(c+\alpha) \cdot g(c) - f(c) \cdot g(c)}{\alpha} =$$
$$= \frac{f(c+\alpha) \cdot (g(c+\alpha) - g(c))}{\alpha} + \frac{g(c) \cdot (f(c+\alpha) - f(c))}{\alpha} \doteq$$
$$\doteq f(c) \cdot g'(c) + g(c) \cdot f'(c).$$

□

Proposition 16.8. (Derivative of power function) Let f_n be the function defined by $f_n(x) = x^n$ for every x. Then

$$f_n'(c) = n \cdot c^{n-1}.$$

Proof. By induction on n. For $n = 1$ see Proposition 16.6(ii). Assume that the proposition holds for n. Then also $f_{n+1}(x) = f_n(x) \cdot x$ for every x. By Proposition 16.7

$$f_{n+1}'(c) = f_n(c) + f_n'(c) \cdot c = c^n + n \cdot c^{n-1} \cdot c = (n+1) \cdot c^n.$$

□

Proposition 16.9. (Quotient rule) Let functions f, g have derivatives at a point c. Let $f(x) \neq 0$ for every $x \in I$. Let f be continuous at the point c. Then the following holds:

(i) The function $\frac{1}{f}$ has derivative at the point c and

$$\left(\frac{1}{f}\right)'(c) = -\frac{f'(c)}{f^2(c)}.$$

(ii) Assume that the expression $f(c) \cdot g'(c) - f'(c) \cdot g(c)$ is not indeterminate. Then the function $\frac{g}{f}$ has derivative at the point c and

$$\left(\frac{g}{f}\right)'(c) = \frac{f(c) \cdot g'(c) - f'(c) \cdot g(c)}{f^2(c)}.$$

Proof.(i) For every α it holds that

$$\frac{1}{\alpha}\left(\frac{1}{f(c+\alpha)} - \frac{1}{f(c)}\right) = \frac{1}{\alpha}\frac{f(c) - f(c+\alpha)}{f(c+\alpha) \cdot f(c)} \doteq -\frac{f'(c)}{f^2(c)}.$$

(ii) follows from Proposition 16.7 and case (i). □

Proposition 16.10. (Derivative of composite function) Let f have a derivative at a point c and be continuous at this point. Let g be defined on an interval whose interior point is $d = f(c)$. Let g have derivative at the point d. Assume that the expression $g'(d) \cdot f'(c)$ is not indeterminate.

Then the composite function $g(f)$ has a derivative at the point c and

$$(g(f))'(c) = g'(d) \cdot f'(c).$$

Proof. If α is such that $f(c+\alpha) - f(c) \neq 0$, then

$$\frac{g(f(c+\alpha)) - g(f(c))}{\alpha} = \frac{g(f(c+\alpha)) - g(f(c))}{f(c+\alpha) - f(c)} \cdot \frac{f(c+\alpha) - f(c)}{\alpha} \doteq g'(d) \cdot f'(c).$$

If at the same time there exists a β such that $f(c+\beta)-f(c) = 0$, then necessarily $f'(c) = 0$, and so the number $g'(d)$ is proper. Hence again

$$\frac{g(f(c+\beta)) - g(f(c))}{\beta} = 0 = g'(d) \cdot f'(c).$$

Finally assume that $f(c+\alpha) - f(c) = 0$ for every α. Then $f'(c) = 0$, the number $g'(d)$ is proper and $g(f(c+\alpha)) = g(f(c))$ for every α, so according to Proposition 16.6(i), $(g(f))'(c) = 0 = g'(d) \cdot f'(c)$. □

16.3 Functions Continuous on a Closed Interval

In this section f, g denote real functions defined on the closed interval $I = [a, b]$ that are continuous at every point of this interval, right-continuous at the point a, left-continuous at the point b.

We will prove the following theorems, familiar from the classical $\epsilon\delta$-analysis, by the means of the reaffirmed infinitesimal calculus.

Proposition 16.11. (Weierstrass extreme value theorem) There exists $c \in [a, b]$ such that for every $x \in [a, b]$

$$f(x) \leq f(c) \quad (\text{or } f(x) \geq f(c) \text{ respectively}).$$

Proof. Let $m \in \mathrm{IN}$ and let $\delta = \frac{b-a}{m}$. Clearly $\delta \doteq 0$. Letters k, l denote natural numbers such that $0 \leq k, l \leq m$.

Let k be such that for each l

$$f(a + l\delta) \leq f(a + k\delta)\,.$$

Let $c = Proj(a + k\delta)$. Hence $c \doteq a + k\delta$. Let $x \in [a, b]$. Then there is l such that $x \doteq a + l\delta$. Hence

$$f(x) \doteq f(a + l\delta) \leq f(a + k\delta) \doteq f(c)\,.$$

The case of minimum is analogical. □

Proposition 16.12. (Bolzano's intermediate value theorem) Let $f(a) \leq u \leq f(b)$ (or $f(b) \leq u \leq f(a)$ respectively). Then there exists $v \in [a, b]$ such that

$$f(v) = u\,.$$

Proof. Let $m \in \mathrm{IN}$ and let $\delta = \frac{b-a}{m}$. Clearly $\delta \doteq 0$. Letters k, l denote natural numbers such that $0 \leq k, l \leq m$.

Let k be the largest natural number such that for each $l \leq k$

$$f(a + l\delta) \leq u\,.$$

Then also

$$f(a + k\delta) \leq u < f(a + (k + 1)\delta)\,.$$

Then for $v = Proj(a + k\delta)$ we have $f(v) = u$. □

Proposition 16.13. (Rolle's theorem) Let a function f have a derivative (possibly improper) at every point x such that $a < x < b$. Let $f(a) = f(b) = 0$. Then there exists c such that $a < c < b$ and

$$f'(c) = 0.$$

Proof. If $f(x) = 0$ for every $x \in [a, b]$ then Rolle's theorem holds trivially. So let there for example exist $y \in (a, b)$ such that $f(y) > 0$. Let $c \in [a, b]$, guaranteed by Weierstrass theorem, be such that $f(x) \leq f(c)$ for every $x \in [a, b]$. Obviously $f(c) \neq 0$, so $a < c < b$. The number $\frac{1}{\alpha}(f(c+\alpha) - f(c))$ is non-positive for $\alpha > 0$ and non-negative for $\alpha < 0$. Since f has derivative at the point c,

$$\frac{f(c+\alpha) - f(c)}{\alpha} \doteq f'(c) = 0.$$

The case of existence of $y \in [a, b]$ such that $f(y) < 0$ is analogous. □

Proposition 16.14. (Lagrange's mean value theorem) Let a function f have a derivative (possibly even improper) at every point x such that $a < x < b$. Then there exists c such that $a < c < b$ and

$$f'(c) = \frac{f(b) - f(a)}{b - a}.$$

Proof. For $x \in [a, b]$ let

$$h(x) = f(x) - f(a) - (x - a)\frac{f(b) - f(a)}{b - a}.$$

It is straightforward to check that the function h satisfies the requirements of Rolle's theorem on the interval $[a, b]$. So there exists $c \in (a, b)$ such that $h'(c) = 0$. But $h'(c) = f'(c) - \frac{f(b)-f(a)}{b-a}$. □

Corollary 16.15. Let $f'(x) = 0$ for every $x \in (a, b)$. Then the function f is constant on the interval $[a, b]$.

Proof. Let $a < x < y < b$. Then by Lagrange's theorem there exists $c \in (x, y)$ such that $\frac{f(y)-f(x)}{y-x} = f'(c) = 0$, so $f(x) = f(y)$. As f is right-continuous or left-continuous at the points a, b respectively, also $f(a) = f(x) = f(b)$. □

Proposition 16.16. (Cauchy's mean value theorem) Let f, g have derivatives at every point $x \in (a, b)$, proper for the function g, possibly improper for the function f and such that $g'(x) \neq 0$. Then there exists $c \in (a, b)$ such that

$$\frac{f'(c)}{g'(c)} = \frac{f(b) - f(a)}{g(b) - g(a)}.$$

Proof. By Lagrange's theorem there exists $d \in (a, b)$ such that

$$g'(d) = \frac{g(b) - g(a)}{b - a}.$$

Hence $g(b) - g(a) \neq 0$. For $x \in [a, b]$ let

$$h(x) = (f(x) - f(a)) \cdot (g(b) - g(a)) - (g(x) - g(a)) \cdot (f(b) - f(a)).$$

So
$$h'(x) = f'(x) \cdot (g(b) - g(a)) - g'(x) \cdot (f(b) - f(a))$$
for $x \in (a, b)$. It is straightforward to check that on the interval $[a, b]$ the function h satisfies the requirements of Rolle's theorem. So there exists $c \in (a, b)$ such that $h'(c) = 0$ and hence
$$f'(c)(g(b) - g(a)) = g'(c)(f(b) - f(a)).$$
Since $g'(c) \neq 0$, $g(b) - g(a) \neq 0$,
$$\frac{f'(c)}{g'(c)} = \frac{f(b) - f(a)}{g(b) - g(a)}.$$
□

16.4 Increasing and Decreasing Functions

We say that a function f is **increasing** (or **decreasing**) on an interval I, if
$$f(x) < f(y) \qquad (\text{or} \quad f(x) > f(y))$$
for every $x, y \in I$, $x < y$.

We say that f is **increasing** (or decreasing) **at a point** $x \in I$ if
$$f(x - \alpha) < f(x) < f(x + \alpha) \qquad (\text{or} \quad f(x - \alpha) > f(x) > (x + \alpha) \text{ respectively})$$
for every $\alpha > 0$ as long as $x - \alpha \in I$, $x + \alpha \in I$.

Obviously if f is increasing (or decreasing) on an interval I, then f is increasing (or decreasing) at every point of this interval.

Proposition 16.17. Let a function f be increasing (or decreasing) at every point of an interval I. Then f is increasing (or decreasing) on the interval I.

Proof. Let f be increasing at every point of interval I. Let $a, b \in I$ be such that
$$a < b \text{ and } f(a) \geq f(b)\,.$$

Since $f(b - \alpha) < f(b)$ for every $\alpha > 0$, there exists $\overline{b}$ such that
$$a < \overline{b} < b \text{ and } f(a) > f(\overline{b})\,.$$

By the first principle of expansion such a point $\overline{b}$ exists in $\mathcal{A}$. This means that in the following we can assume that b is an interior point of the interval I and $f(a) > f(b)$.

Let $\overline{I}$ be the class of all x such that $a < x$ and
$$\text{if } a < y \leq x \text{ then } f(a) < f(y).$$

Obviously $\overline{I}$ is an interval and its left endpoint is the number a. Since $b \notin \overline{I}$, the inequality $c \leq b$ holds for right endpoint c of the interval $\overline{I}$. Since f is increasing at the point c,

$$f(c) < f(c+\alpha) \text{ for every } \alpha > 0.$$

If $a = c$ then $c + \alpha \in \overline{I}$, which is a contradiction. If $a < c$ then

$$f(a) < f(c-\alpha) < f(c) < f(c+\alpha),$$

so again $c + \alpha \in \overline{I}$, which is again a contradiction. The case of a decreasing function is analogous. □

16.5 Continuous Bijective Functions

Let $\overline{I} = \{f(x); x \in I\}$. This means that $\overline{I}$ is a set of all numbers $f(x)$ where x is a point in I.

We say that a function f is **continuous on an interval** I if it is continuous at every point of I (right or left continuous respectively at the right or left endpoints, if they belong to I).

Constant functions, the identity function and polynomial functions are clearly continuous on I.

Proposition 16.18. Let a function f be continuous on I. Then $\overline{I}$ is either a single element set, or an interval.

Proof. Obviously $\overline{I}$ is a single element set if and only f is constant on I. Let $u, v \in \overline{I}$; $u < w < v$. Then there exist $x, y \in I$ such that $u = f(x)$, $v = f(y)$. By Bolzano's intermediate value theorem there exists $z \in (x, y)$ or $z \in (y, x)$ and hence $z \in I$ such that $w = f(z)$. Consequently $w \in \overline{I}$. □

We say that a function f is **bijective** on I if for every $x, y \in I$, $x \neq y$ we have $f(x) \neq f(y)$.

Obviously if a function f is increasing (or decreasing) on I, then f is bijective on I.

Proposition 16.19. Let $\overline{I}$ be an interval, f increasing (or decreasing) on I. Then f is continuous on I.

Proof. Let $c \in I$ and let $\alpha > 0$. Assume that $f(c) \doteq f(c+\alpha)$ does not hold. Since f is increasing, there exists $n \in \mathrm{FN}$ such that

$$f(c) + \frac{1}{n} < f(c+\alpha)$$

and since $\overline{I}$ is an interval, there exists a unique $d \in I$ such that

$$f(d) = f(c) + \frac{1}{n}.$$

Since $f(c) < f(d) < f(c+\alpha)$, we have $c < d < c+\alpha$, so $c = d$, which is a contradiction. Similarly if $f(c) \doteq f(c-\alpha)$ does not hold. The case of a decreasing function is analogous. □

Proposition 16.20. Let f be continuous and bijective on I. Let $u < v < w$. Then either $f(u) < f(v) < f(w)$ or $f(u) > f(v) > f(w)$.

Proof. Let for example $f(u) < f(w)$. If $f(w) < f(v)$ then by Bolzano's intermediate value theorem there exists c such that $u < c < v$, $f(c) = f(w)$, which is a contradiction. If $f(v) < f(u)$ then there exists c such that $v < c < w$ and $f(c) = f(u)$, which is again a contradiction. The case $f(u) > f(w)$ is analogous. □

Proposition 16.21. Let f be continuous and bijective on I. Then f is increasing on I or f is decreasing on I.

Proof. Let $u < v$. Let for example $f(u) < f(v)$. Let $x < y$. If $x = u$ or $x = v$ or $y = u$ or $y = v$ then by Proposition 16.20 $f(x) < f(y)$. So let numbers u, v, x, y be different. Then one of the following cases occurs:

$$x < y < u < v, \qquad x < u < y < v, \qquad x < u < v < y,$$
$$u < x < y < v, \qquad u < x < v < y, \qquad u < v < x < y.$$

However by Proposition 16.20, in any of these cases $f(x) < f(y)$. The case $f(u) > f(v)$ is analogous. So the function f is increasing (or decreasing) on I. □

16.6 Inverse Functions and Their Derivatives

In this section, let f denote a bijective function defined on an interval I and $\overline{I} = \{f(x); x \in I\}$.

The function f^{-1} defined on $\overline{I}$ by

$$f^{-1}(x) = y, \qquad \text{where} \quad x = f(y),$$

is called the **inverse function to the function** f on I.

The following assertions are obvious.

Proposition 16.22. (i) The function f^{-1} is bijective and maps $\overline{I}$ onto I.

(ii) The function f is an inverse function to the function f^{-1} on $\overline{I}$.

(ii) If f is increasing on I, then f^{-1} is increasing on $\overline{I}$.

(ii) If f is decreasing on I, then f^{-1} is decreasing on $\overline{I}$.

From the propositions in the previous section it follows that:

Proposition 16.23. If f is continuous on I, then f^{-1} is continuous on $\overline{I}$.

We say that a function f is **weakly differentiable** on I, if f is continuous on I and f has a derivative (possible improper) at every point $x \in I$; at the relevant end points of the interval I it has right or left derivative respectively. (We remark that this definition applies equally to functions that are not necessarily bijective.)

Proposition 16.24. Let f be bijective and weakly differentiable on the interval I. Then f^{-1} is weakly differentiable on $\overline{I}$ and for every $x \in \overline{I}$

$$(f^{-1})'(x) = \frac{1}{f'(y)}, \qquad \text{where} \quad x = f(y)\,.$$

If $f'(y) = 0$ then

$$(f^{-1})'(x) = \begin{cases} \infty & \text{if } f \text{ (and so also } f^{-1}\text{) is increasing,} \\ -\infty & \text{if } f \text{ (and so also } f^{-1}\text{) is decreasing.} \end{cases}$$

Proof. Let $x \in \overline{I}$, $x = f(y)$ and let α be such that $x + \alpha \in \overline{I}$. (This condition is stated here for the case that x is an end point of the interval $\overline{I}$).

Since f is bijective and continuous on I, there exists β such that $(x + \alpha) = f(y + \beta)$. Hence $\alpha = f(y + \beta) - f(y)$. It follows that

$$\frac{f^{-1}(x+\alpha) - f^{-1}(x)}{\alpha} = \frac{y + \beta - y}{\alpha} = \frac{\beta}{\alpha} = \frac{\beta}{f(y+\beta) - f(y)} \doteq \frac{1}{f'(y)}.$$

Moreover if numbers β and $\alpha = f(y + \beta) - f(y)$ have the same sign then

$$\frac{\beta}{\alpha} > 0 \;\text{ so }\; \frac{1}{f'(y)} \geq 0\,.$$

If they have a different sign then $\frac{1}{f'(y)} \leq 0$. □

16.7 Higher-Order Derivatives, Extrema and Points of Inflection

In this section, let I denote an open interval.

Let a function f have a proper derivative at every $x \in I$. The function whose value at x is the number $f'(x)$ is called the **derivative of function** f on the interval I and is denoted f'.

Thus defined function f' belongs to $\mathcal{A}$. By the second principle of expansion it also enters $\mathcal{C}$, where it has the same properties as in $\mathcal{A}$. (That is, properties defined exclusively by the means of one of the $\mathcal{A}$, world $\mathcal{C}$ or, no matter which one.) So for example Lagrange's mean value theorem, proved for a continuous function on an interval $[a, b]$, holds also in the case when the end points belong to $\mathcal{C}$ and not necessarily to $\mathcal{A}$ and so on.

Most propositions in this section are formulated as two assertions. However, we prove only one. The second assertion – in brackets – is dual and its proof is analogous; or the second assertion is an immediate consequence of the first one if we substitute the function $-f$ for the function f.

Proposition 16.25. Let f have a derivative at a point c, possibly improper. If $f'(c) > 0$ (or $f'(c) < 0$) then the function f is increasing (or decreasing) at the point c.

Proof. Since

$$\frac{f(c+\alpha) - f(c)}{\alpha} \doteq f'(c) > 0,$$

it follows that $f(c+\alpha) - f(c)$ has the same sign as α. □

We say that a function f has a **local maximum** (or **minimum**) at a point c if

$$f(c+\alpha) \leq f(c) \qquad (\text{or} \quad f(c+\alpha) \geq f(c)) \qquad \text{for every } \alpha.$$

We say that this **local maximum** (or **minimum**) is **strict**, if these inequalities are strict for every α.

Proposition 16.26. Let f have a derivative at a point c and let it have a local maximum (or minimum) at the point c. Then $f'(c) = 0$.

Proof. If for example $f(c+\alpha) \geq f(c)$ for every α, then

$$f'(c) \doteq \frac{f(c+\alpha) - f(c)}{\alpha} \begin{cases} \geq 0 & \text{for } \alpha > 0, \\ \leq 0 & \text{for } \alpha < 0. \end{cases}$$

Hence $f'(c) = 0$. □

We define the **nth order derivative** of a function f on an interval I by

$$f^{(1)} = f'; \qquad f^{(n+1)} \quad \text{is the derivative of the function } f^{(n)}.$$

Apart from that we set $f^{(0)} = f$.

We say that a function f is **differentiable up to the n-th order** on an interval I if its derivatives

$$f^{(1)}, f^{(2)}, \dots, f^{(n)}$$

exist on this interval.

Proposition 16.27. Let f be differentiable on I up to the 1st order and let $f'(c) = 0$.

(i) If $f'(c+\alpha) > 0$ (or $f'(c+\alpha) < 0$) for every α, then the function f is increasing (or decreasing) at the point c.

(ii) If f' is increasing (or decreasing) at the point c, then the function f has a strict local minimum (or maximum) at the point c.

Proof. Let $\alpha > 0$. Then there exists $0 < \beta < \alpha$ such that

$$f(c+\alpha) - f(c) = \alpha \cdot f'(c+\beta) > 0,$$

since $f'(c+\beta) > 0$, so $f(c+\alpha) > f(c)$. Similarly there exists $-\alpha < \gamma < 0$ such that

$$f(c) - f(c-\alpha) = \alpha \cdot f'(c+\gamma),$$

which is positive in case (i), so $f(c) > f(c-\alpha)$, and negative in case (ii), so $f(c) < f(c-\alpha)$. □

Proposition 16.28. Let f be differentiable on an interval I up to the order $n \geq 1$, $n \in FN$. Let f have nth derivative $f^{(n)}(c) \neq 0$ at a point c, possibly improper. Assume that $f^{(k)}(c) = 0$ for every $k = 1, 2, \ldots, n-1$, $k \in \mathrm{FN}$. Then the following holds:

(i) If n is odd, $f^{(n)}(c) > 0$ (or $f^{(n)}(c) < 0$) then the function f is increasing (or decreasing) at the point c.

(ii) If n is even, $f^{(n)}(c) > 0$ (or $f^{(n)}(c) < 0$) then the function f has a strict local minimum (or maximum) at the point c.

Proof. For $n = 1$ see Proposition 16.25. What follows is a proof by induction. Let us suppose that the proposition holds for $n-1$, where $n \geq 2$. The proposition holds also for function f' since it satisfies all the desired requirements. Let n be odd, then $n-1$ is even, and so by the induction hypothesis the function f' has a strict local minimum at the point c. As $f'(c) = 0$, $f'(c+\alpha) > f'(c) = 0$, and so by Proposition 16.27(i) the function f is increasing in the point c. If n is even, $n-1$ is odd, and so by the induction hypothesis the function f' is increasing at the point c, and because $f'(c) = 0$, the function f has a strict local minimum at the point c by Proposition 16.27(ii). □

Let a function f have a derivative (proper) at a point c. If

$$f(c+\alpha) > f(c) + \alpha \cdot f'(c) \quad (\text{or} \quad f(c+\alpha) < f(c) + \alpha \cdot f'(c)))$$

for every α, we say that the function f is **strictly convex** (or **strictly concave** respectively) at the point c. If

$$\begin{aligned} f(c+\alpha) &> f(c) + \alpha \cdot f'(c) \qquad \text{for every } \alpha > 0 \text{ and} \\ f(c+\alpha) &< f(c) + \alpha \cdot f'(c) \qquad \text{for every } \alpha < 0 \end{aligned}$$

(or vice versa), we say that the function f has a **left inflection** (or **right inflection** respectively) at the point c. Such a point c is called a **point of inflection**.

Proposition 16.29. Let f be differentiable on the interval I up to the order $n-1$ where $n > 1$, $n \in \mathrm{FN}$ and let it have a proper derivative $f^{(n)}(c) \neq 0$ at a point c. Let $f^{(k)}(c) = 0$ for every $1 < k < n$.

(i) If n is even, $f^{(n)}(c) > 0$ (or $f^{(n)}(c) < 0$), then f is strictly convex (or strictly concave) at the point c.

(ii) If n is odd, $f^{(n)}(c) > 0$ (or $f^{(n)}(c) < 0$), then the function f has left inflection (or right inflection) at the point c.

Proof. For $x \in I$ we set $g(x) = f(x) - f(c) - f'(c)(x-c)$. Obviously

$$g'(x) = f'(x) - f'(c), \qquad g^{(k)}(x) = f^{(k)}(x) \quad \text{for } 1 < k < n.$$

Hence

$$g(c) = g'(c) = \cdots = g^{(n-1)}(c) = 0, \qquad g^{(n)}(c) > 0.$$

Therefore the function g satisfies the requirements of the previous proposition. For n even, g has a strict local minimum at the point c and because $g(c) = 0$, $g(c+\alpha) > 0$ for every α, so also

$$f(c+\alpha) > f(c) + \alpha \cdot f'(c)\,.$$

For n odd, g is increasing at the point c and because $g(c) = 0$,

$$\begin{aligned} g(c+\alpha) &> 0 \qquad \text{for } \alpha > 0,\\ g(c+\alpha) &< 0 \qquad \text{for } \alpha < 0, \end{aligned}$$

so

$$\begin{aligned} f(c+\alpha) &> f(c) + \alpha \cdot f'(c) \qquad \text{for } \alpha > 0,\\ f(c+\alpha) &< f(c) + \alpha \cdot f'(c) \qquad \text{for } \alpha < 0. \end{aligned}$$

□

16.8 Limit of a Function at a Point

In this section we assume that functions f, g are defined for all $x \doteq D$, but not necessarily at the point D.

We say that a number A is the **limit of a function** f at a point D, denoted

$$A = \lim_{x \to D} f(x),$$

if for every $x \doteq D$, $x \neq D$ we have

$$A \doteq f(x) \quad (\text{that is, } A = \mathrm{Proj}(f(x)).$$

If the number D is proper then we say that A is the **right-sided** (or **left-sided**) **limit** of the function f at the point D, denoted

$$A = \lim_{x \to D^+} f(x) \qquad (\text{or } A = \lim_{x \to D^-} f(x))$$

if

$$A \doteq f(D + \alpha) \qquad \text{for every} \quad \alpha > 0 \quad (\text{or } \alpha < 0).$$

In that case it is sufficient if f is defined for all $D + \alpha$, where $\alpha > 0$ (or $\alpha < 0$).

Obviously the following holds:

Proposition 16.30. A function f defined on an interval I with an interior point D is continuous at the point D if and only if

$$f(D) = \lim_{x \to D} f(x).$$

Similar propositions hold for right-continuity and left-continuity.

Proposition 16.31. Let $\lim_{x \to D} f(x) = A$, $\lim_{x \to D} g(x) = B$.

(i) Assume that $A + B$ is not an indeterminate expression. Then

$$\lim_{x \to D} (f(x) + g(x)) = A + B.$$

(ii) Assume that $A - B$ is not an indeterminate expression. Then

$$\lim_{x \to D} (f(x) - g(x)) = A - B.$$

(iii) Assume that $A \cdot B$ is not an indeterminate expression. Then

$$\lim_{x \to D} (f(x) \cdot g(x)) = A \cdot B.$$

(iv) Assume that $\frac{A}{B}$ is not an indeterminate expression. Let $g(x) \neq 0$ for every $x \doteq D$, $x \neq D$. Then

$$\lim_{x \to D} \frac{f(x)}{g(x)} = \frac{A}{B}.$$

In this case if $A \neq 0$, $B = 0$, then the following holds anyway:

(a) If $\frac{f(x)}{g(x)} > 0$ (or $\frac{f(x)}{g(x)} < 0$) for every $x \doteq D$, $x \neq D$, then

$$\lim_{x \to D} \frac{f(x)}{g(x)} = \infty \qquad \left(\text{or } \lim_{x \to D} \frac{f(x)}{g(x)} = -\infty \text{ respectively}\right).$$

(b) If $\frac{f(x)}{g(x)} > 0$ for some $x \doteq D$, $x \neq D$ and $\frac{f(x)}{g(x)} < 0$ for other, then $\lim_{x \to D} \frac{f(x)}{g(x)}$ does not exist.

Obviously analogous rules hold also for right-sided and left-sided limits.

In Proposition 16.31(iv) the cases $\frac{0}{0}$, $\frac{\infty}{\infty}$ (or $\frac{-\infty}{\infty}$, $\frac{\infty}{-\infty}$, $\frac{-\infty}{-\infty}$) remained open. When determining the limits in these cases it is often possible to use the so-called l'Hôpital Rule. To make it easier to formulate this rule, let us assume in the remaining part of this section that functions f, g are defined on an open interval (D, C), where $D < C$. The case of an interval (C, D), where $C < D$ is so similar that we will not even state it.

Proposition 16.32. (L'Hôpital rule for the case $\frac{0}{0}$) Let the number D be proper. Let

$$\lim_{x \to D+} f(x) = \lim_{x \to D+} g(x) = 0.$$

Let functions f, g have a derivative at every point of the interval (D, C). Let $g(x) \neq 0$, $g'(x) \neq 0$ for every $x \in (D, C)$. Let there exist $\lim\limits_{x \to D+} \frac{f'(x)}{g'(x)}$. Then

$$\lim_{x \to D+} \frac{f(x)}{g(x)} = \lim_{x \to D+} \frac{f'(x)}{g'(x)}.$$

Proof. Without loss of generality we can set $f(D) = g(D) = 0$. So functions f, g are right-continuous at the point D and in consequence they satisfy the requirements of Cauchy's mean value theorem on every closed interval $[D, x]$, where $x < C$. This means that for every $\alpha > 0$ there exists $0 < \beta < \alpha$ such that

$$\frac{f(D+\alpha)}{g(D+\alpha)} = \frac{f(D+\alpha) - f(D)}{g(D+\alpha) - g(D)} = \frac{f'(D+\beta)}{g'(D+\beta)} \doteq \lim_{x \to D+} \frac{f'(x)}{g'(x)}.$$

□

Before stating the second l'Hôpital Rule, let us prove two auxiliary assertions.

Proposition 16.33. Let the number D be proper. Let $\lim\limits_{x \to D+} f(x) = \infty$. Then for every $v \doteq \infty$ there exists $x \doteq D$, $D < x$, such that $f(x) < v$.

Proof. Assume that there exists $v \doteq \infty$ such that for all $x \doteq D$ with $D < x$, $f(x) \geq v$. Let $m \in \mathrm{IN}$ and let $\delta = \frac{C-D}{m}$, where C is proper number. Clearly $\delta \doteq 0$. Let $n \in \mathrm{N}$ be the largest number such that for each x satisfying $D < x < D + n\delta = y$ we have $f(x) \geq v$. Then $D < \mathrm{Proj}(y)$. Let $D < b < \mathrm{Proj}(y)$, where b belongs to $\mathcal{A}$. The number $f(b)$ is a proper and at the same time $f(b) > v$, which is a contradiction. □

Proposition 16.34. Let the number D be proper. Let

$$\lim_{x \to D+} f(x) = \lim_{x \to D+} g(x) = \infty\,.$$

Then for every $\alpha > 0$ there exists $\beta > 0$ such that

$$\frac{f(D+\beta)}{f(D+\alpha)} \doteq \frac{g(D+\beta)}{g(D+\alpha)} \doteq 0.$$

Proof. Let w be the smaller of numbers $f(D+\alpha)$, $g(D+\alpha)$. Obviously $w \doteq \infty$. So there exists a $v \doteq \infty$ such that $v^2 < w$. According to the previous Proposition 16.33 there exists $\beta > 0$ such that $f(D+\beta) + g(D+\beta) < v$, so $f(D+\beta) < v$, $g(D+\beta) < v$. Hence

$$\frac{f(D+\beta)}{f(D+\alpha)} < \frac{v}{w} < \frac{v}{v^2} = \frac{1}{v} \doteq 0.$$

Analogously for function g. □

Proposition 16.35. (L'Hôpital Rule for the case $\frac{\infty}{\infty}$) Let the number D be proper. Let $\lim_{x \to D+} f(x) = \lim_{x \to D+} g(x) = \infty$. Let functions f, g have derivatives at every point x of the interval (D, C), $g'(x) \neq 0$. Let there exist $\lim_{x \to D+} \frac{f'(x)}{g'(x)}$. Then

$$\lim_{x \to D+} \frac{f(x)}{g(x)} = \lim_{x \to D+} \frac{f'(x)}{g'(x)}.$$

Proof. We can clearly suppose without loss of generality that $g(x) \neq 0$ for every $x \in (D, C)$. Let us choose $\alpha > 0$. By the previous proposition there exists $\beta > 0$ such that

$$\frac{f(D+\beta)}{f(D+\alpha)} \doteq \frac{g(D+\beta)}{g(D+\alpha)} \doteq 0.$$

By Cauchy's mean value theorem there exists $\gamma > 0$ such that

$$\frac{f(D+\alpha) - f(D+\beta)}{g(D+\alpha) - g(D+\beta)} = \frac{f'(D+\gamma)}{g'(D+\gamma)}.$$

Furthermore

$$\frac{f(D+\alpha) - f(D+\beta)}{g(D+\alpha) - g(D+\beta)} = \frac{f(D+\alpha)}{g(D+\alpha)} \cdot \frac{1 - \frac{f(D+\beta)}{f(D+\alpha)}}{1 - \frac{g(D+\beta)}{g(D+\alpha)}} \doteq \frac{f(D+\alpha)}{g(D+\alpha)}.$$

Hence

$$\frac{f'(D+\gamma)}{g'(D+\gamma)} \doteq \frac{f(D+\alpha)}{g(D+\alpha)} \quad \text{so} \quad \lim_{x \to D+} \frac{f'(x)}{g'(x)} = \lim_{x \to D+} \frac{f(x)}{g(x)}.$$

□

Proposition 16.36. Both l'Hôpital Rules hold even when the number D is improper.

Proof. Let for example $D = \infty$. We can clearly suppose without loss of generality that functions f, g are defined on the interval (c, ∞), where $0 < c < \infty$. Then the functions $F(y) = f\left(\frac{1}{y}\right)$, $G(y) = g\left(\frac{1}{y}\right)$ are defined on the interval $(0, \frac{1}{c})$. If we set $x = \frac{1}{y}$, then obviously

$$\lim_{y \to 0+} F(y) = \lim_{y \to 0+} f\left(\frac{1}{y}\right) = \lim_{x \to \infty} f(x).$$

Analogously $\lim\limits_{y\to0+} G(y) = \lim\limits_{x\to\infty} g(x)$. The functions F, G have derivatives at every point y of the interval $\left(0, \frac{1}{c}\right)$ and we can find this derivative using the Chain Rule. So

$$F'(y) = -f'\left(\frac{1}{y}\right)\cdot\frac{1}{y^2}, \quad G'(y) = -g'\left(\frac{1}{y}\right)\cdot\frac{1}{y^2}.$$

Hence

$$\frac{F'(y)}{G'(y)} = \frac{f'(x)}{g'(x)} \quad \text{and so} \quad \lim_{y\to0+}\frac{F'(y)}{G'(y)} = \lim_{x\to\infty}\frac{f'(x)}{g'(x)}.$$

The functions F, G consequently satisfy the conditions for application of the l'Hôpital Rule and therefore the following holds:

$$\lim_{y\to0+}\frac{F(y)}{G(y)} = \lim_{y\to0+}\frac{F'(y)}{G'(y)} = \lim_{x\to\infty}\frac{f'(x)}{g'(x)}.$$

Furthermore

$$\lim_{y\to0+}\frac{F(y)}{G(y)} = \lim_{y\to0+}\frac{f\left(\frac{1}{y}\right)}{g\left(\frac{1}{y}\right)} = \lim_{x\to\infty}\frac{f(x)}{g(x)}.$$

□

Note. The cases $\infty-\infty$ (or $-\infty+\infty$) that remained open in Proposition 16.31(i) and (ii) can be transformed to $\frac{0}{0}$ by setting $h(x) = \frac{1}{f(x)}$, $l(x) = \frac{1}{g(x)}$ on a suitable interval (D, C). Obviously

$$\lim_{x\to D+} h(x) = \lim_{x\to D+} l(x) = 0, \qquad f(x) - g(x) = \frac{l(x) - h(x)}{h(x)\cdot l(x)}.$$

Note. The open case $0\cdot\infty$ from Proposition 16.31(iii) can be transformed to $\frac{0}{0}$ by setting $h(x) = \frac{1}{g(x)}$ on a suitable interval (D, C). Obviously

$$f(x)\cdot g(x) = \frac{f(x)}{h(x)},$$

which is the case $\frac{0}{0}$ since $\lim\limits_{x\to D+} h(x) = 0$.

The case $0\cdot\infty$ can often be transformed to $\frac{\infty}{\infty}$, namely in the case when $f(x) \neq 0$ on a suitable interval (D, C). In that case we set $h(x) = \frac{1}{f(x)}$. Clearly $\lim\limits_{x\to D+} h(x) = \infty$ and

$$f(x)\cdot g(x) = \frac{g(x)}{h(x)},$$

which is the case $\frac{\infty}{\infty}$.

16.9 Taylor's Expansion

In this section I denotes an open interval, $c \in I$, and f denotes a function defined on I which is differentiable on I up to the order $n+1$ (where $n \geq 0$, $n \in \mathrm{FN}$).

For $x \in I$ let

$$Z_{n+1}(x) = f(x) - P_n(x)$$

where

$$P_n(x) = f(c) + f'(c) \cdot \frac{x-c}{1!} + f^{(2)}(c) \cdot \frac{(x-c)^2}{2!} + \cdots + f^{(n)}(c) \cdot \frac{(x-c)^n}{n!}.$$

The polynomial function P_n (where the degree n is in FN) is clearly defined for all x. At the point c, P_n has the first n derivatives equal to those of the function f. Furthermore $P_n(c) = f(c)$.

The function P_n is called **Taylor's polynomial.**

Proposition 16.37. (Taylor's theorem) Let $g(t)$ be a function continuous on the closed interval $\overline{I}$ with end points $c, x \in I$, $x \neq c$. Let g have a non-zero derivative at each inner point of $\overline{I}$. Then there exists an inner point y of the interval $\overline{I}$ such that

$$Z_{n+1}(x) = \frac{(x-y)^n}{n!} \cdot \frac{g(x)-g(c)}{g'(y)} \cdot f^{(n+1)}(y).$$

Proof. Let $x \in I$ be fixed and set

$$F(t) = f(x) - f(t) - f'(t) \cdot \frac{x-t}{1!} - f^{(2)}(t) \cdot \frac{(x-t)^2}{2!} - \cdots - f^{(n)}(t) \cdot \frac{(x-t)^n}{n!}.$$

The function F clearly has a derivative with respect to t at each point of the interval I. Also

$$\begin{aligned} F'(t) = &- f'(t) - \left(-f'(t) + f^{(2)}(t) \cdot \frac{x-t}{1!}\right) - \\ &- \left(-f^{(2)}(t) \cdot \frac{x-t}{1!} + f^{(3)}(t) \cdot \frac{(x-t)^2}{2!}\right) - \cdots - \\ &- \left(-f^{(n)}(t) \cdot \frac{(x-t)^{n-1}}{(n-1)!} + f^{(n+1)}(t) \cdot \frac{(x-t)^n}{n!}\right). \end{aligned}$$

The first entry in every bracket cancels out with the entry in front of it and hence

$$F'(t) = -f^{(n+1)}(t) \cdot \frac{(x-t)^n}{n!}.$$

Clearly $F(x) = 0$, $F(c) = Z_{n+1}(x)$. By Cauchy's mean value theorem there exists a point y in $\overline{I}$ such that

$$\frac{F(x)-F(c)}{g(x)-g(c)} = \frac{F'(y)}{g'(y)}.$$

Hence

$$-\frac{Z_{n+1}(x)}{g(x)-g(c)} = \frac{F'(y)}{g'(y)}, \qquad \text{and so} \quad Z_{n+1}(x) = -\frac{g(x)-g(c)}{g'(y)} \cdot F'(y).$$

Substituting for $F'(y)$ yields the required result. □

Upon taking $g(t) = (x-t)^{n+1}$ we get **Lagrange's form of the remainder**

$$Z_{n+1}(x) = \frac{(x-c)^{n+1}}{(n+1)!} \cdot f^{(n+1)}(y).$$

and upon taking $g(t) = t$ we get **Cauchy's form of the remainder**

$$Z_{n+1}(x) = \frac{(x-y)^n \cdot (x-c)}{n!} \cdot f^{(n+1)}(y).$$

Note. Taylor theorem for $c = 0$ is also called **Maclaurin's theorem.** In that case we have

$$f(x) = f(0) + f'(0) \cdot \frac{x}{1!} + f^{(2)}(0) \cdot \frac{x^2}{2!} + \cdots + f^{(n)}(0) \cdot \frac{x^n}{n!} + Z_{n+1}(x).$$

Here Lagrange's form of the remainder is

$$Z_{n+1}(x) = \frac{x^{n+1}}{(n+1)!} \cdot f^{(n+1)}(sx), \qquad \text{where} \quad 0 < s < 1.$$

and Cauchy's form of the remainder is

$$Z_{n+1}(x) = \frac{x^{n+1}}{n!} \cdot (1-s)^n \cdot f^{(n+1)}(sx), \qquad \text{where} \quad 0 < s < 1.$$

Chapter 17

Elementary Functions and Their Derivatives

17.1 Power Functions

In this section x, y denote positive real numbers. The following assertion about **integer powers** is assumed.

Proposition 17.1. Let p, q be integers, $|p|, |q| \in \mathrm{FN}$.

(i) $1^p = 1,\ \ 0 < x^p$.

(ii) $x^p \cdot y^p = (xy)^p$; $\left(\frac{1}{y}\right)^p = \frac{1}{y^p}$. Hence $\left(\frac{x}{y}\right)^p = \frac{x^p}{y^p}$.

(iii) $x^p \cdot x^q = x^{p+q}$; $x^{-p} = \frac{1}{x^p}$. Hence $\frac{x^p}{x^q} = x^{p-q}$.

(iv) If $0 < p$, $x < y$ then $x^p < y^p$. Hence if $p < 0$, $x < y$ then $y^p < x^p$.

(v) If $1 < x$, $p < q$ then $x^p < x^q$. Hence if $0 < x < 1$, $p < q$ then $x^q < x^p$.

(vi) $(x^p)^q = x^{p \cdot q}$.

Root functions as inverse functions to power functions

Let $n \in \mathrm{FN}$. Let g_n be the function defined for $x \in (0, \infty)$ by $g_n(x) = x^n$.

The function g_n has a derivative at each point $x \in (0, \infty)$ which equals nx^{n-1} (see Proposition 16.8), and hence it is continuous at each such point.

Since $nx^{n-1} > 0$, by Propositions 16.25 and 16.17 the function g_n is increasing on the interval $(0, \infty)$, and hence it is bijective. Since for each $x > 1$ we have $x < g_n(x)$,

$$\lim_{x \to \infty} g_n(x) = \infty$$

and since for each $0 < x < 1$ we have $g_n(x) < x$,

$$\lim_{x \to 0^+} g_n(x) = 0.$$

Consequently by Proposition 16.18 the function g_n maps the interval $(0, \infty)$ bijectively again onto $(0, \infty)$.

It follows that the function g_n^{-1}, *inverse* to the function g_n, exists.

By Proposition 16.22 the function g_n^{-1} also maps the interval $(0, \infty)$ bijectively onto $(0, \infty)$, and it is increasing on this interval. By Proposition 16.23 the function g_n^{-1} is continuous on $(0, \infty)$.

We shall call the function g_n^{-1} a **root function**. For $x > 0$ we write

$$\sqrt[n]{x} = g_n^{-1}(x).$$

Proposition 17.2. Let $n, m \in \mathrm{FN}$ and let x, y be positive real numbers.

(ii') $\sqrt[n]{x} \cdot \sqrt[n]{y} = \sqrt[n]{x \cdot y}; \qquad \sqrt[n]{\frac{1}{y}} = \frac{1}{\sqrt[n]{y}}.$

(vi') $\sqrt[n]{\sqrt[m]{x}} = \sqrt[n \cdot m]{x}$

Proof. Taking powers of both sides we obtain identity by Proposition 17.1(ii) and 17.1(vi). □

Let p be an integer, $|p| \in \mathrm{FN}$, $m \in \mathrm{FN}$. The function defined by

$$x^{\frac{p}{m}} = \sqrt[m]{x^p}.$$

is called a **power function with rational exponent**.

In order to show that this really defines a power function for each rational number (that is, that we do not obtain different results for different representations of the same rational number) we need to show that for each natural number k the numbers

$$u = \sqrt[m]{x^p}, \qquad v = \sqrt[km]{x^{pk}}$$

are equal. Since $u^m = x^p$ and $v^{mk} = x^{pk}$, we have $v^m = x^p$ and thus $u^m = v^m$. Consequently $u = v$.

Considering Proposition 17.1 (i) – (vi) with $\frac{p}{n}$, $\frac{q}{n}$ substituted for p, q and using Proposition 17.2(ii'), (vi') and the above properties of the function g_n, we can easily check that:

Proposition 17.3. Properties (i) – (vi) from Proposition 17.1 remain valid also under the assumption that p, q are rational numbers.

Proposition 17.4. Let x be a fixed real number, $x > 0$.

(i) The sequence $\{\sqrt[n]{x}\}$ has a limit equal to 1.

(ii) If r is a rational number, $r \doteq 0$, then $x^r \doteq 1$.

(iii) If p is a finitely large rational number, then x^p is a finitely large real number.

(iv) If p, q are finitely large rational numbers, $p \doteq q$ then $x^p \doteq x^q$.

Proof. (i) If $1 < x$ then $1 < \sqrt[n]{x} < x$ and if $0 < x < 1$ then $x < \sqrt[n]{x} < 1$. Hence if we choose $m > x$, $m \in \mathrm{FN}$ (or $m > \frac{1}{x}$) then starting from the mth member, the sequence $\{\sqrt[n]{x}\}$ satisfies the conditions of Proposition 15.17.

(ii) If $x^r \doteq 1$ then also $\frac{1}{x^r} \doteq 1$. Thus it suffices to assume that $r > 0$.

Let $m \in \mathbf{IN}$ be such that $r < \frac{1}{m}$. Then

$$\begin{aligned} &1 < x^r < \sqrt[m]{x} \doteq 1 && \text{for } 1 < x, \\ &1 \doteq \sqrt[m]{x} < x^r < 1 && \text{for } 0 < x < 1\,. \end{aligned}$$

(iii) Let $k \in \mathbf{FN}$, $-k < p < k$. Then

$$\begin{aligned} &0 < x^p < x^k && \text{for } 1 < x, \\ &x^p < x^{-k} && \text{for } 0 < x < 1\,. \end{aligned}$$

The numbers x^k a x^{-k} are finitely large.

(iv) We have

$$x^p - x^q = x^p(1 - x^{q-p}).$$

By (iii), x^p is finitely large and by (ii), $1 - x^{q-p} \doteq 0$. □

Let p and $x > 0$ be real numbers belonging to $\mathcal{A}$. Let r be a rational number belonging to $\mathcal{C}$ such that $r \doteq p$. We define the **power function for an arbitrary real exponent** by

$$x^p = \mathrm{Proj}(x^r); \quad \sqrt[p]{x} = x^{\frac{1}{p}} \quad (\text{for} \quad p \neq 0).$$

By Proposition 17.4(iv), x^p does not depend on the choice of a particular $r \doteq p$. By Proposition 17.4(iii), x^p is a proper real number belonging to $\mathcal{A}$.

Using Proposition 17.3 we can easily check that the following proposition holds.

Proposition 17.5. The properties (i)–(vi) from Proposition 17.1 remain valid under the assumption that p, q are real numbers.

Hence the next assertion follows almost directly:

Proposition 17.6. (i) The function x^p is defined for every $x \in (0, \infty)$.

(ii) If $p > 0$ then x^p is increasing on the interval $(0, \infty)$,

$$\lim_{x \to \infty} x^p = \infty, \qquad \lim_{x \to 0+} x^p = 0.$$

(iii) If $p < 0$ then x^p is decreasing on $(0, \infty)$,

$$\lim_{x \to \infty} x^p = 0, \qquad \lim_{x \to 0+} x^p = \infty.$$

(iv) If $p \neq 0$ then $x^{\frac{1}{p}} = \sqrt[p]{x}$ is the inverse function to x^p.

Proposition 17.7. The function x^p, where p is fixed, maps the interval $(0, \infty)$ bijectively onto $(0, \infty)$ and it is continuous on this interval.

Proof. See Sections 16.4, 16.5, 16.6 and and the previous Proposition 17.6. □

Proposition 17.8. The function x^p enters $\mathcal{C}$ and it keeps here all its properties that can be formulated merely by the means of the world $\mathcal{A}$.

Proof. See the second principle of expansion. □

In accordance with the introduction of improper numbers ∞, $-\infty$ in the world $\mathcal{A}$ we define also powers of improper numbers.

$$\begin{aligned} A^\infty &= \infty && \text{for } 1 < A. \\ A^\infty &= 0 && \text{for } 0 \leq A < 1. \\ \infty^A &= \infty && \text{for } 0 < A. \end{aligned}$$

The expressions 1^∞, ∞^0, 0^0 are indeterminate.

17.2 Exponential Function

In this section a denotes a positive real number, $a \neq 1$.

The exponential function with base a is defined for each real number $x \in (-\infty, \infty)$ by

$$f(x) = a^x.$$

Clearly $a^x > 0$ for each x.

Proposition 17.9. (i) If $1 < a$ then the function a^x is increasing on the interval $(-\infty, \infty)$.

(ii) If $a < 1$ then the function a^x is decreasing on the interval $(-\infty, \infty)$.

Proof. This follows from the properties of the power function (Proposition 17.1(v)). □

Proposition 17.10. The function a^x is continuous at each point x.

Proof. We have
$$a^{x+\alpha} - a^x = a^x(a^\alpha - 1).$$
Since a^x is a finitely large number it suffices to show that
$$a^\alpha - 1 \doteq 0.$$
Let $0 < \alpha$. Let $m \in \mathrm{IN}$ be such that $\alpha < \frac{1}{m}$. Then
$$\begin{aligned} 1 < a^\alpha < \sqrt[m]{a} \qquad & \text{for} \quad 1 < a, \\ \sqrt[m]{a} < a^\alpha < 1 \qquad & \text{for} \quad 0 < a < 1. \end{aligned}$$
By Proposition 17.4(i), $\sqrt[m]{a} \doteq 1$ and hence $a^\alpha \doteq 1$.

Since $a^{-\alpha} = \frac{1}{a^\alpha}$, the case of $\alpha < 0$ follows. □

Proposition 17.11. (i) If $1 < a$ then $\lim_{x\to\infty} a^x = \infty$, $\lim_{x\to-\infty} a^x = 0$.

(ii) If $0 < a < 1$ then $\lim_{x\to\infty} a^x = 0$, $\lim_{x\to-\infty} a^x = \infty$.

Proof. (i) By Proposition 15.14 we have $\lim_{n\to\infty} a^n = \infty$. Since a^x is increasing, also $\lim_{x\to\infty} a^x = \infty$. Furthermore
$$\lim_{x\to-\infty} a^x = \lim_{x\to\infty} a^{-x} = \lim_{x\to\infty} \frac{1}{a^x} \doteq \frac{1}{\infty} \doteq 0.$$

(ii) By Proposition 15.14 we have $\lim_{n\to\infty} a^n = 0$. Since a^x is decreasing, also $\lim_{x\to\infty} a^x = 0$. Furthermore
$$\lim_{x\to-\infty} a^x = \lim_{x\to\infty} \frac{1}{a^x} \doteq \frac{1}{0} \doteq \pm\infty.$$
Since $\frac{1}{a^x} > 0$, this limit is $+\infty$. □

Proposition 17.12. The function a^x maps the interval $(-\infty, \infty)$ bijectively onto the interval $(0, \infty)$.

Proof. Let $1 < a$, $0 < y$. By Proposition 17.11 there exist real numbers $u < v$ such that
$$a^u < y < a^v.$$
By Bolzano's intermediate value theorem there exists $x \in (u, v)$ such that $a^x = y$. By Proposition 17.9(i) such x is unique.

The case $0 < a < 1$ is similar. □

Proposition 17.13.
$$\frac{e^\alpha - 1}{\alpha} \doteq 1.$$

Proof. Let $0 < \alpha$. Since e^x is a continuous function,

$$e^\alpha \doteq 1.$$

Since $e > 1$, $e^\alpha - 1 > 0$. Thus there exists $n \in \mathrm{IN}$ such that

$$\frac{1}{n+1} < e^\alpha - 1 < \frac{1}{n}.$$

Hence

$$n < \frac{1}{e^\alpha - 1} < n+1 \qquad \text{and} \qquad 1 + \frac{1}{n+1} < e^\alpha < 1 + \frac{1}{n}.$$

It follows that

$$\left(1 + \frac{1}{n+1}\right)^n < (e^\alpha)^{\frac{1}{e^\alpha - 1}} < \left(1 + \frac{1}{n}\right)^{n+1}.$$

Since the backward projection of both is e, also

$$(e^\alpha)^{\frac{1}{e^\alpha - 1}} = e^{\frac{\alpha}{e^\alpha - 1}} \doteq e,$$

and hence

$$\frac{\alpha}{e^\alpha - 1} \doteq 1.$$

The case of $\alpha < 0$ is similar. □

17.3 Logarithmic Function

In this section a again denotes a positive real number, $a \neq 1$.

The function inverse to a^x on the interval $(-\infty, \infty)$ is called the **logarithmic function to base** a; we denote it $\log_a$.

Hence we have:

Proposition 17.14.

$$a^{\log_a(x)} = x, \qquad \log_a(a^x) = x \qquad (x > 0,\, a > 0,\, a \neq 1).$$

In other words, the logarithm to base a of the number x is the exponent to which the number a has to be raised in order to obtain the number x. We stress that only positive numbers have logarithms.

The following proposition follows from the more general propositions in Section 16.6 and from propositions concerning the exponential function (Section 17.2).

Proposition 17.15.

(i) The function $\log_a$ maps the interval $(0, \infty)$ bijectively onto the interval $(-\infty, \infty)$.

(ii) If $1 < a$ then the function $\log_a$ is increasing on the interval $(0, \infty)$.

(iii) If $0 < a < 1$ then the function $\log_a$ is on the interval interval $(0, \infty)$ decreasing.

(iv) The function $\log_a$ is continuous on the interval $(0, \infty)$.

The following proposition is obvious:

Proposition 17.16. Let $x > 0$.

(i) If $1 < a$ then $\lim\limits_{x \to 0+} \log_a(x) = -\infty$, $\lim\limits_{x \to \infty} \log_a(x) = \infty$.

(ii) If $0 < a < 1$ then $\lim\limits_{x \to 0+} \log_a(x) = \infty$, $\lim\limits_{x \to \infty} \log_a(x) = -\infty$.

Proposition 17.17. Let $x > 0$, $y > 0$.

(i) $\log_a(a) = 1$, $\log_a(1) = 0$;

(ii) $\log_a(x \cdot y) = \log_a(x) + \log_a(y)$,
$\log_a\left(\dfrac{1}{x}\right) = -\log_a(x)$,
$\log_a\left(\dfrac{x}{y}\right) = log_a(x) - \log_a(y)$.

(iii) For every p, $\log_a(x^p) = p \cdot \log_a(x)$.

(iv) For $b > 0$, $b \neq 1$, $\log_a(x) = \log_b(x) \cdot \log_a(b)$.

(v) $\log_{\frac{1}{a}}(x) = -\log_a(x)$.

Proof. The assertion (i) is trivial. For the other ones we can use the fact that $\log_a$ is the function inverse to the exponential function a^x with base a.

(ii) From the definition of logarithmic function and comparing,

$$a^{\log_a(x \cdot y)} = x \cdot y,$$
$$a^{\log_a(x) + \log_a(y)} = a^{\log_a(x)} \cdot a^{\log_a(y)} = x \cdot y.$$

Similarly in the remaining cases and in part (iii).

(iv) Let $u = \log_b(x)$. Then $b^u = x$ and thus

$$\log_a(x) = \log_a(b^u) = u \cdot \log_a(b) = \log_b(x) \cdot \log_a(b).$$

(v) By (iv), since $\log_a\left(\frac{1}{a}\right) = -1$. □

From Propositions 17.15(ii), (iii) and 17.17(i) we have the following

Proposition 17.18.

(i) Let $a > 1$.

If $x > 1$ then $\log_a(x) > 0$.

If $0 < x < 1$ then $\log_a(x) < 0$.

(ii) Let $0 < a < 1$.

If $x > 1$ then $\log_a(x) < 0$.

If $0 < x < 1$ then $\log_a(x) > 0$.

Logarithm to base e is called **natural logarithm** and it is denoted log. Hence we have

Proposition 17.19. Let $x > 0$, $a > 0$, $a \neq 1$.

(i) $a^x = e^{x \cdot \log(a)}$

(ii) $\log_a(x) = \frac{\log(x)}{\log(a)}$, $\quad \log(x) = \frac{\log_a(x)}{\log_a(e)}$.

Proof. See Proposition 17.17(iv). □

17.4 Derivatives of Power, Exponential and Logarithmic Functions

For the derivatives of exponential functions there are the following rules.

Rule 17.20.

$$(e^x)' = e^x.$$

Proof.

$$(e^x)' \doteq \frac{e^{x+\alpha} - e^x}{\alpha} = e^x \cdot \frac{e^\alpha - 1}{\alpha} \doteq e^x.$$

The last equality holds by Proposition 17.13. □

Rule 17.21. For $a > 0$

$$(a^x)' = \log(a) \cdot a^x.$$

Proof. By the rule about the derivative of composite function $a^x = e^{x \cdot \log(a)}$,

$$(a^x)' = \log(a) \cdot e^{x \cdot \log(a)} = \log(a) \cdot a^x.$$

□

For the derivatives of logarithmic functions we have the following rules.

Rule 17.22. For $x > 0$

$$\log'(x) = \frac{1}{x}.$$

Proof. By the rule about the derivatives of inverse functions

$$\log'(x) = \frac{1}{(e^y)'} = \frac{1}{e^y}, \qquad \text{where} \quad x = e^y.$$

Hence $\log'(x) = \frac{1}{x}$. □

Rule 17.23. For $a > 0$ a $x > 0$

$$\log_a'(x) = \log_a(e) \cdot \frac{1}{x}.$$

Proof. By Proposition 17.19(ii), $\log_a(x) = \log_a(e) \cdot \log(x)$. Hence by the rule 17.22

$$\log_a'(x) = \log_a(e) \cdot \log'(x) = \log_a(e) \cdot \frac{1}{x}.$$

□

For the derivatives of power functions we have the following rules.

Rule 17.24. For any p and $x > 0$

$$(x^p)' = p \cdot x^{p-1}.$$

Proof. By Proposition 17.18 we have $x^p = e^{p \cdot \log(x)}$. By the rule about the derivatives of composite functions

$$(x^p)' = p \cdot \log'(x) \cdot e^{p \cdot \log(x)} = p \cdot \frac{1}{x} \cdot e^{p \cdot \log(x)} = p \cdot \frac{1}{x} \cdot x^p = p \cdot x^{p-1}.$$

□

Rule 17.25. For $x > 0$

$$(x^x)' = x^x \cdot \log(e \cdot x).$$

Proof. By Proposition 17.18, $x^x = e^{x \cdot \log(x)}$. By the rule about the derivatives of composite functions

$$(x^x)' = e^{x \cdot \log(x)} \cdot (x \cdot \log(x))' = x^x \cdot \left(\log(x) + x \cdot \frac{1}{x}\right) =$$
$$= x^x \cdot (\log(x) + 1) = x^x \cdot (\log(x) + \log(e)) = x^x \cdot \log(e \cdot x).$$

□

Proposition 17.26. Let I be an open interval, f, g functions defined on I, $f(y) > 0$ for every $y \in I$. Let f, g have derivatives at the point $x \in I$. Then

$$\left(f(x)^{g(x)}\right)' = \left(f(x)^{g(x)}\right) \cdot \left(g'(x) \cdot \log(f(x)) + g(x) \cdot \frac{f'(x)}{f(x)}\right).$$

Proof. For any $x \in I$

$$f(x)^{g(x)} = e^{g(x)\cdot \log(f(x))}.$$

Using the rule about the derivatives of composite functions we obtain

$$\left(e^{g(x)\cdot \log(f(x))}\right)' = e^{g(x)\cdot \log(f(x))}\,(g(x)\cdot \log(f(x)))' =$$
$$= f(x)^{g(x)}\cdot \left(g'(x)\cdot \log(f(x)) + g(x)\cdot \frac{f'(x)}{f(x)}\right).$$

□

17.5 Trigonometric Functions $\sin x$, $\cos x$ and Their Derivatives

The following properties of functions $\sin x$, $\cos x$ are assumed:

(a) The functions $\sin x$ a $\cos x$ are defined for each real number $x \in (-\infty, \infty)$.

(b) For $x, y \in (-\infty, \infty)$

(i) $\sin(x \pm y) = \sin x \cdot \cos y \pm \cos x \cdot \sin y$,

(ii) $\cos(x \pm y) = \cos x \cdot \cos y \mp \sin x \cdot \sin y$,

(iii) $\cos(-x) = \cos x$,

(iv) $\sin(-x) = -\sin x$.

(c) In the world $\mathcal{A}$ there exists a positive number π such that the function $\sin x$ is increasing in the interval $\left[0, \frac{\pi}{2}\right]$,

(i) $\sin 0 = 0$,

(ii) $\sin\frac{\pi}{2} = 1$.

(d) For all α

$\sin\alpha \doteq 0.$

Proposition 17.27.

$$\cos 0 = 1.$$

Proof. From the above properties (b)(i), (c)(i)–(ii) it follows that

$$1 = \sin\left(\frac{\pi}{2} + 0\right) = \sin\frac{\pi}{2}\cdot\cos 0 + \sin 0\cdot\cos\frac{\pi}{2} = \sin\frac{\pi}{2}\cdot\cos 0 = \cos 0.$$

□

Proposition 17.28.

$$\sin^2 x + \cos^2 x = 1.$$

Proof. By properties (b)(ii)–(iv) we obtain for any $x \in (-\infty, \infty)$ that

$$1 = \cos(x - x) = \cos(x) \cdot \cos(-x) - \sin(x) \cdot \sin(-x) =$$
$$= \cos x \cdot \cos x - \sin x \cdot (-\sin x) = \cos^2 x + \sin^2 x.$$

□

Proposition 17.29.

$$\cos \frac{\pi}{2} = 1.$$

Proof. By Proposition 17.28 and property (c)(ii),

$$\cos^2 \frac{\pi}{2} = 1 - \sin^2 \frac{\pi}{2} = 1 - \left(\sin \frac{\pi}{2}\right)^2 = 1 - 1 = 0.$$

□

Obviously $\cos^2 \alpha \doteq 1$ and hence $\cos \alpha \doteq 1$.

The following Propositions 17.30–17.36 are valid for any real number $x \in (-\infty, \infty)$.

Proposition 17.30.

(i) $\sin\left(\frac{\pi}{2} + x\right) = \cos x,$

(ii) $\cos\left(\frac{\pi}{2} + x\right) = -\sin x.$

Proof. By (b)(i), (c)(ii) and Proposition 17.29,

$$\sin\left(\frac{\pi}{2} + x\right) = \sin \frac{\pi}{2} \cdot \cos x + \cos \frac{\pi}{2} \cdot \sin x = \cos x + 0 = \cos x,$$

Similarly in the case (ii). □

Proposition 17.31.

(i) $\sin(\pi + x) = \cos\left(\frac{\pi}{2} + x\right) = -\sin x,$

(ii) $\cos(\pi + x) = -\sin\left(\frac{\pi}{2} + x\right) = -\cos x.$

Proof. Repeatedly using 17.30 we obtain

$$\sin(\pi + x) = \sin\left(\frac{\pi}{2} + \left(\frac{\pi}{2} + x\right)\right) = \cos\left(\frac{\pi}{2} + x\right) = -\sin x.$$

Similarly in the other case. □

Proposition 17.32.

(i) $\sin(2\pi + x) = \sin x,$

(ii) $\cos(2\pi + x) = \cos x.$

Proof. This follows by a repeated use of 17.31. □

Proposition 17.33. Let k be an integer. For any real number $x \in (-\infty, \infty)$ we have

(i) $\sin(2k\pi + x) = \sin x,$

(ii) $\cos(2k\pi + x) = \cos x.$

In other words, the functions $\sin x$ and $\cos x$ are **periodic** with **period** 2π.

Proof. This follows by a repeated use of 17.32. □

Proposition 17.34.

(i) $\sin x + \sin y = 2\sin\left(\frac{x+y}{2}\right) \cdot \cos\left(\frac{x-y}{2}\right),$

(ii) $\sin x - \sin y = 2\cos\left(\frac{x+y}{2}\right) \cdot \sin\left(\frac{x-y}{2}\right),$

(iii) $\cos x + \cos y = 2\cos\left(\frac{x+y}{2}\right) \cdot \cos\left(\frac{x-y}{2}\right),$

(iv) $\cos x - \cos y = -2\sin\left(\frac{x+y}{2}\right) \cdot \sin\left(\frac{x-y}{2}\right).$

Proof. Letting $\overline{x} = \frac{x+y}{2}$, $\overline{y} = \frac{x-y}{2}$ and using property (b) we can easily see that

$$\begin{aligned}
\sin(\overline{x} + \overline{y}) + \sin(\overline{x} - \overline{y}) &= 2\sin\overline{x} \cdot \cos\overline{y}, \\
\sin(\overline{x} + \overline{y}) - \sin(\overline{x} - \overline{y}) &= 2\cos\overline{x} \cdot \sin\overline{y}, \\
\cos(\overline{x} + \overline{y}) + \cos(\overline{x} - \overline{y}) &= 2\cos\overline{x} \cdot \cos\overline{y}, \\
\cos(\overline{x} + \overline{y}) - \cos(\overline{x} - \overline{y}) &= -2\sin\overline{x} \cdot \sin\overline{y}.
\end{aligned}$$

□

The following proposition expresses a property for cosine analogous to (c).

Proposition 17.35.

(i) $\cos\left(\frac{\pi}{2} - x\right) = \sin x.$

(ii) The function $\cos x$ is decreasing on the interval $\left[0, \frac{\pi}{2}\right]$,

$$\cos 0 = 1,$$
$$\cos\tfrac{\pi}{2} = 0.$$

Proof. The proof follows directly from Proposition 17.30 and property (b)(iv). □

Proposition 17.36. For $0 < x < \frac{\pi}{8}$

$$2\sin x < \sin 4x.$$

Proof. By Propositions 17.35 and 17.28,

$$\sin\frac{\pi}{4} = \cos\frac{\pi}{4} = \frac{\sqrt{2}}{2}.$$

By (b)(i),

$$\sin 2x = 2\cos x \cdot \sin x.$$

By 17.35(ii) for $0 < x < \frac{\pi}{4}$,

$$\frac{\sqrt{2}}{2} < \cos x < 1, \qquad \text{hence} \quad \sqrt{2} \cdot \sin x < \sin 2x.$$

It follows that for $0 < x < \frac{\pi}{8}$ we have

$$2\sin x < \sin 4x.$$

□

Proposition 17.37. The function $\sin x$, $\cos x$ are continuous at each point.

Proof. By Proposition 17.30 it suffices to prove merely the continuity of the function $\sin x$. From the assumed property (b)(i),

$$\sin(x + \alpha) = \sin x \cdot \cos\alpha + \cos x \cdot \sin\alpha \doteq \sin x + 0 = \sin x.$$

□

Proposition 17.38. Let k be an integer, $|k| \in \mathrm{FN}$.

(i) $\sin x = 0$ if and only if $x = k\pi$.

(ii) $\cos x = 0$ if and only if $x = \frac{\pi}{2} + k\pi$.

Proof. By Proposition 17.30, $\cos x = \sin\left(\frac{\pi}{2} + x\right)$. Hence it suffices to prove part (i).

Since the function $\sin x$ is periodic with period 2π by Proposition 17.33, it suffices to consider those x for which $0 \leq x \leq 2\pi$. We have

$$\sin\pi = \sin\left(\frac{\pi}{2} + \frac{\pi}{2}\right) = 2\sin\frac{\pi}{2} \cdot \cos\frac{\pi}{2} = 0.$$

By property (c) the function $\sin x$ is increasing on the interval $\left[0, \frac{\pi}{2}\right]$, and hence for $0 < x < \frac{\pi}{2}$

$$0 < \sin x < 1.$$

Since $\cos^2 x = 1 - \sin^2 x$ by Proposition 17.28, for $0 < x < \frac{\pi}{2}$ we have $\cos x \neq 0$.

Since $\sin\left(\frac{\pi}{2} + x\right) = \cos x$ by Proposition 17.30, for $\frac{\pi}{2} < x < \pi$ we have $\sin x \neq 0$.

By Proposition 17.31, $\sin(\pi + x) = -\sin x$ and hence for $\pi < x < 2\pi$ we have $\sin x \neq 0$. □

Trigonometric (goniometric) functions – as suggested already in their name – have been adopted by the infinitesimal calculus from geometry, where they had been introduced as measures of angles (initially just acute angles). These measures were determined by lengths of line segments (sides of right-angled triangles) since the length of a line segments, that is, the shortest distance between two points, can be measured more easily than, for example, the length of a general curve.

For acute angles and, later on, also for obtuse angles or even reflex angles, the above assumed properties (a), (b), (c) and (d) were justified by geometric intuition; in the geometric world it is possible to see directly as evident that they are right. (Hence (a), (b), (c), (d) are characteristic properties of the world $\mathcal{A}$.)

Within the infinitesimal calculus however these measures of angles were adjusted so as to suit the objectives of this mathematical discipline. That means so that they were functions on real numbers and not on angles as such, as in trigonometry. That is why we understand them nowadays as functions of the lengths of the corresponding arcs on the unit circle. Furthermore their domain was enlarged in the obvious way to include all real numbers.

From the propositions proved above it is already apparent that for more detailed knowledge of the behaviour of trigonometric functions it suffices to have knowledge of the behaviour of the function $\sin x$ in the interval $\left[0, \frac{\pi}{2}\right]$ (that is, knowledge of the behaviour of this function on acute angles). However, from the properties (a), (b), (c), (d) we learn little about it.

To capture trigonometric functions $\sin x$ and $\cos x$ fully – in particular, to place the number π on the number line of the world $\mathcal{A}$ precisely – it is necessary to accept the following additional assumed property.

(e) for each α,

$$\frac{\sin \alpha}{\alpha} \doteq 1.$$

The inequality

$$\sin \alpha \leq \alpha$$

can be justified by geometric intuition on the basis of the line segment connecting two points making the shortest connection between the two points. However, the assertion (e) is stronger and its verity cannot be seen as evident in the geometric world $\mathcal{A}$. Hence (e) is a characteristic property of the world $\mathcal{C}$. Its acceptance is justifiable only on the basis of the definition of the length of a circle.

We shall see in Section 19.1 that these properties (a), (b), (c), (d), (e) do uniquely determine the functions $\sin x$ a $\cos x$.

For the derivatives of these functions we have the following rules

Rule 17.39.

$$(\sin x)' = \cos x.$$

Proof.

$$(\sin x)' = \frac{\sin(x+\alpha) - \sin x}{\alpha} = \frac{2}{\alpha} \cdot \cos\left(x + \frac{\alpha}{2}\right) \cdot \sin\frac{\alpha}{2} =$$
$$= \cos\left(x + \frac{\alpha}{2}\right) \cdot \frac{2}{\alpha} \cdot \sin\frac{\alpha}{2} \doteq \cos x.$$

□

Rule 17.40.

$$(\cos x)' = -\sin x.$$

Proof.

$$(\cos x)' = \frac{\cos(x+\alpha) - \cos x}{\alpha} = -\frac{2}{\alpha} \cdot \sin\left(x + \frac{\alpha}{2}\right) \cdot \sin\frac{\alpha}{2} =$$
$$= -\sin\left(x + \frac{\alpha}{2}\right) \cdot \frac{2}{\alpha} \cdot \sin\frac{\alpha}{2} \doteq -\sin x.$$

□

17.6 Trigonometric Functions $\tan x$, $\cot x$ and Their Derivatives

Let k be an integer. We define functions $\tan x$ and $\cot x$ by

(i) $\tan x = \dfrac{\sin x}{\cos x}$, provided $\cos x \neq 0$, that is, for $x \neq \frac{\pi}{2} + k\pi$,

(ii) $\cot x = \dfrac{\cos x}{\sin x}$, provided $\sin x \neq 0$, that is, for $x \neq k\pi$.

The following properties of the functions $\tan x$, $\cot x$ follow directly from the definitions of these functions and from the above properties of the functions $\sin x$, $\cos x$.

Proposition 17.41. The functions $\tan x$, $\cot x$ are continuous at each point where they are defined, that is,

$$\tan x \quad \text{on intervals} \quad \left(\frac{\pi}{2} + k\pi, \frac{\pi}{2} + (k+1)\pi\right),$$
$$\cot x \quad \text{on intervals} \quad (k\pi, (k+1)\pi).$$

Proposition 17.42.

(i) $\displaystyle\lim_{x \to \left(\frac{\pi}{2} + k\pi\right)^+} \tan x = -\infty,$

(ii) $\displaystyle\lim_{x \to \left(\frac{\pi}{2} + k\pi\right)^-} \tan x = \infty,$

(iii) $\lim_{x\to(k\pi)^+} \cot x = \infty,$

(iv) $\lim_{x\to(k\pi)^-} \cot x = -\infty.$

Proposition 17.43.

(i) $\tan(-x) = -\tan x,$

(ii) $\cot(-x) = -\cot x.$

Proposition 17.44. The functions $\tan x$, $\cot x$ are periodic with period π.

The following propositions are immediate consequences of Bolzano's intermediate value theorem.

Proposition 17.45.

(i) The function $\tan x$ maps the interval $\left(\frac{\pi}{2} + k\pi, \frac{\pi}{2} + (k+1)\pi\right)$ bijectively onto the interval $(-\infty, \infty)$.

(ii) The function $\cot x$ maps the interval $(k\,\pi, (k+1)\,\pi)$ bijectively onto the interval $(-\infty, \infty)$.

Proposition 17.46.

$$\tan\frac{\pi}{4} = \cot\frac{\pi}{4} = 1.$$

Proof.

$$\sin\frac{\pi}{4} = \sin\left(\frac{\pi}{2} - \frac{\pi}{4}\right) = \cos\left(-\frac{\pi}{4}\right) = \cos\frac{\pi}{4}.$$

□

For differentiating the functions tangent a cotangent we have the following rules .

Rule 17.47. If $x \neq \frac{\pi}{2} + k\pi$, where k is an integer, then

$$(\tan x)' = \frac{1}{\cos^2 x}.$$

Proof. From the definition of $\tan x$ and by the Quotient Rule,

$$(\tan x)' = \left(\frac{\sin x}{\cos x}\right)' = \frac{\cos x \cdot \cos x - \sin x \cdot (-\sin x)}{\cos^2 x} =$$

$$= \frac{\sin^2 x + \cos^2 x}{\cos^2 x} = \frac{1}{\cos^2 x}.$$

□

Rule 17.48. If $x \neq k\pi$, where k is an integer, then

$$(\cot x)' = -\frac{1}{\sin^2 x}.$$

Proof. From the definition of $\cot x$ and by the Quotient Rule,

$$(\cot x)' = \left(\frac{\cos x}{\sin x}\right)' = \frac{(-\sin x)\cdot \sin x - \cos x \cdot \cos x}{\sin^2 x} = \\ = -\frac{\sin^2 x + \cos^2 x}{\sin^2 x} = -\frac{1}{\sin^2 x}.$$

□

Proposition 17.49.

(i) The function $\tan x$ is increasing on each interval

$$\left(\frac{\pi}{2} + k\pi, \frac{\pi}{2} + (k+1)\pi\right).$$

(ii) The function $\cot x$ is decreasing on each interval

$$(k\pi, (k+1)\pi).$$

Proof. The derivative of the function $\tan x$ is positive at each point of the given interval. The derivative of the function $\cot x$ is negative. □

17.7 Cyclometric Functions and Their Derivatives

Trigonometric functions are not bijective. Hence they do not have inverse functions. However, if we limit their domains to some suitable interval, we obtain functions that do have inverses.

The following are suitable intervals:

(i) For the function $\sin x$ the interval $\left[\frac{\pi}{2} + k\pi, \frac{\pi}{2} + (k+1)\pi\right]$, on which the function $\sin x$ is increasing if k is odd, and decreasing if k is even.

(ii) For the function $\cos x$ the interval $[k\pi, (k+1)\pi]$, on which the function $\cos x$ is increasing if k is odd, and decreasing if k is even.

(iii) For the function $\tan x$ the interval $\left(\frac{\pi}{2} + k\pi, \frac{\pi}{2} + (k+1)\pi\right)$, on which the function $\tan x$ is increasing.

(iv) For the function $\cot x$ the interval $(k\pi, (k+1)\pi)$, on which the function $\cot x$ is decreasing.

If we fix an integer k then we define cyclometric functions on the appropriate intervals as the inverse functions to the corresponding trigonometric functions:

$$\begin{aligned} y &= \arcsin x && \text{if and only if} && x = \sin y, \\ y &= \arccos x && \text{if and only if} && x = \cos y, \\ y &= \arctan x && \text{if and only if} && x = \tan y, \\ y &= \text{arccot x} && \text{if and only if} && x = \cot y. \end{aligned}$$

The following properties of cyclometric functions are simple consequences of the corresponding properties of the trigonometric functions.

Proposition 17.50. Cyclometric functions are continuous at each point at which they are defined.

Proposition 17.51. The functions $\arcsin x$, $\arccos x$ are defined on the interval $[-1, 1]$. For odd k they are increasing on this interval and for even k decreasing.

Proposition 17.52. The functions $\arctan x$, $\mathrm{arccot\,x}$ are defined on the interval $(-\infty, \infty)$. The function $\arctan x$ is increasing on this interval and the function $\mathrm{arccot\,x}$ decreasing.

Proposition 17.53.

(i) $\lim_{x\to\infty} \arctan x = \frac{\pi}{2} + (k+1)\pi$,

(ii) $\lim_{x\to-\infty} \arctan x = \frac{\pi}{2} + k\pi$,

(iii) $\lim_{x\to\infty} \mathrm{arccot\,x} = \mathrm{k}\pi$,

(iv) $\lim_{x\to-\infty} \mathrm{arccot\,x} = (\mathrm{k}+1)\pi$.

For differentiating cyclometric functions there are the following rules.

Rule 17.54.

$$(\arcsin x)' = \frac{\pm 1}{\sqrt{1-x^2}},$$

where the sign + applies for odd k and the sign − for even k.

Proof. Let $x = \sin y$. By the rules for differentiating inverse and trigonometric functions,

$$(\arcsin x)' = \frac{1}{(\sin y)'} = \frac{1}{\cos y} = \frac{\pm 1}{\sqrt{1-\sin^2 y}} = \frac{\pm 1}{\sqrt{1-x^2}}.$$

□

Rule 17.55.

$$(\arccos x)' = \frac{\mp 1}{\sqrt{1-x^2}},$$

where the sign − applies for odd k and the sign + for even k.

Proof. Let $x = \cos y$. By the rules for differentiating inverse and trigonometric functions,

$$(\arccos x)' = \frac{1}{(\cos y)'} = -\frac{1}{\sin y} = \frac{\mp 1}{\sqrt{1-\cos^2 y}} = \frac{\mp 1}{\sqrt{1-x^2}}.$$

□

Rule 17.56.

$$(\arctan x)' = \frac{1}{1+x^2}.$$

Proof. Let $x = \tan y$. By the rules for differentiating inverse and trigonometric functions,

$$(\arctan x)' = \frac{1}{(\tan y)'} = \cos^2 y = \frac{\cos^2 y}{\sin^2 y + \cos^2 y} = \frac{1}{\frac{\cos^2 y + \sin^2 y}{\cos^2 y}} =$$
$$= \frac{1}{1+\tan^2 y} = \frac{1}{1+x^2},$$

since $\sin^2 y + \cos^2 y = 1$, and thus $1 + \tan^2 x = \frac{1}{\cos^2 x}$. □

Rule 17.57.

$$(\operatorname{arccot} \mathrm{x})' = \frac{-1}{1+\mathrm{x}^2}.$$

Proof. Let $x = \cot y$. By the rules for differentiating inverse and trigonometric functions,

$$(\operatorname{arccot} \mathrm{x})' = \frac{1}{(\cot y)'} = -\sin^2 y = \frac{-\sin^2 y}{\sin^2 y + \cos^2 y} = \frac{-1}{\frac{\sin^2 y + \cos^2 y}{\sin^2 y}} =$$
$$= \frac{-1}{1+\cot^2 y} = \frac{-1}{1+x^2},$$

since again $\sin^2 y + \cos^2 y = 1$, and hence $1 + \cot^2 x = \frac{1}{\sin^2 x}$. □

Chapter 18

Numerical Series

18.1 Convergence and Divergence

If $\{a_n\}$ is a sequence of real numbers belonging to the world $\mathcal{A}$ then

$$\sum a_n$$

denotes the **infinite series**

$$a_1 + a_2 + a_3 + \cdots .$$

We say that a real number A is the **sum of the series** $\sum a_n$, denoted

$$A = \sum_{n=1}^{\infty} a_n,$$

if for each $m \in \mathrm{IN}$,

$$A = \mathrm{Proj}\left(\sum_{n=1}^{m} a_n\right).$$

We say that the series $\sum a_n$ is **convergent**, if it has a sum that is a *proper number*.

Otherwise we say that the series $\sum a_n$ is **divergent**. If its sum is ∞ (or $-\infty$) then we say that the series **diverges** to ∞ (or $-\infty$ respectively). A series that has no sum is called **oscillating**.

If the series $\sum a_n$ has a sum then $\sum\limits_{n=1}^{\infty} a_n$ also denotes the number that is its sum. If the series is oscillating then $\sum\limits_{n=1}^{\infty} a_n$ does not denote any number.

The following five Propositions clearly do not need proving.

Proposition 18.1. If we set

$$s_n = \sum_{m=1}^{n} a_m,$$

then the sequence $\{s_n\}$ has a limit if and only if the series $\sum a_n$ has a sum. In that case

$$\lim_{n\to\infty} s_n = \sum_{n=1}^{\infty} a_n.$$

The sequence $\{s_n\}$ is called the **sequence of partial sums** of the series $\sum a_n$.

Proposition 18.2. (Bolzano's criterion for convergence) The series $\sum a_n$ is convergent if and only if for each $m \in$ IN, $k \in$ N,

$$\sum_{n=0}^{k} a_{m+n} \doteq 0.$$

Proposition 18.3. Let

$$A = \sum_{n=1}^{\infty} a_n = \sum_{n=1}^{\infty} b_n.$$

Let $\{c_n\}$ be a sequence such that for each n, $a_n \le c_n \le b_n$. Then

$$A = \sum_{n=1}^{\infty} c_n.$$

Proposition 18.4. Let series $\sum a_n$, $\sum b_n$ have sums and let a, b be proper real numbers. Assume that the expression

$$a \cdot \sum_{n=1}^{\infty} a_n + b \cdot \sum_{n=1}^{\infty} b_n$$

is not indeterminate. Then the series $\sum (a \cdot a_n + b \cdot b_n)$ has a sum and

$$\sum_{n=1}^{\infty} (a \cdot a_n + b \cdot b_n) = a \cdot \sum_{n=1}^{\infty} a_n + b \cdot \sum_{n=1}^{\infty} b_n.$$

Proposition 18.5. Let $\sum a_n$ be convergent. Then

$$\lim_{n\to\infty} a_n = 0.$$

In other words, for each $n \in$ IN we have $a_n \doteq 0$.

Proof. By Bolzano's criterion, for $n \in \mathrm{IN}$ we have

$$a_n + a_{n+1} + a_{n+2} \doteq 0, \qquad a_{n+1} + a_{n+2} \doteq 0.$$

Hence $a_n \doteq 0$. □

The series $\sum q^n$ where q is a real number is called a **geometric series**. The number q is called the **quotient** of this geometric series.

Proposition 18.6.

(i) If $|q| < 1$ then the series $\sum q^n$ converges and

$$\sum_{n=1}^{\infty} q^n = \frac{q}{1-q}, \qquad \text{so} \qquad \sum_{n=0}^{\infty} q^n = \frac{1}{1-q}.$$

(ii) If $1 \leq q$ then $\sum q^n = \infty$.

(iii) If $q \leq -1$ then the series $\sum q^n$ oscillates.

Proof. With $q = 1$ we have

$$1 + 1 + 1 + \cdots, \qquad \text{hence} \quad \sum_{n=1}^{\infty} q^n = \infty.$$

Let $q \neq 1$. Let $s_n = \sum_{m=0}^{n} q^m$. Since

$$(1-q) \cdot s_n = (1-q) \cdot (1 + q + \cdots + q^n) = 1 - q^{n+1},$$

we have $s_n = \frac{1-q^{n+1}}{1-q}$. By Proposition 15.14 for $|q| < 1$,

$$\lim_{n \to \infty} q^n = 0.$$

Hence

$$\sum_{n=0}^{\infty} q^n = \frac{1}{1-q}, \qquad \text{so} \quad \sum_{n=1}^{\infty} q^n = \frac{q}{1-q}.$$

For $q > 1$

$$\lim_{n \to \infty} q^n = \infty, \qquad \text{so} \quad \sum_{n=0}^{\infty} q^n = \lim_{n \to \infty} s_n = \infty.$$

Finally if $q \leq -1$ then for n even we have $q^n \geq 1$ and for n odd we have $q^n \leq -1$. Consequently the sequence $\{s_n\}$ as well as the series $\sum q^n$ oscillates.
□

Abel's Partial Summation Formula Let $a_1, a_2, \ldots, a_n, b_1, b_2, \ldots, b_n$ be real numbers. Let

$$\begin{aligned} s_1 &= a_1, \\ s_2 &= a_1 + a_2, \\ s_3 &= a_1 + a_2 + a_3, \\ &\vdots \\ s_n &= a_1 + a_2 + \cdots + a_n. \end{aligned}$$

Then

$$\begin{aligned} & a_1 \cdot b_1 + a_2 \cdot b_2 + \cdots + a_n \cdot b_n = \\ & \quad = b_1 \cdot s_1 + b_2 \cdot (s_2 - s_1) + \cdots + b_n \cdot (s_n - s_{n-1}) = \\ & \quad = s_1 \cdot (b_1 - b_2) + s_2 \cdot (b_2 - b_3) + \cdots + s_{n-1} \cdot (b_{n-1} - b_n) + s_n \cdot b_n. \end{aligned}$$

Proposition 18.7. (Abel's lemma) Let $\{a_n\}$ and $\{b_n\}$ be sequences. Let

$$b_1 \geq b_2 \geq b_3 \cdots \geq b_n \geq 0.$$

Let c be a real number such that for each $j = 1, 2, \ldots, n$ we have

$$s_j \leq c \ \ (\text{or } c \leq s_j), \qquad \text{where } s_j = a_1 + a_2 + \cdots + a_j.$$

Then

$$a_1 b_1 + a_2 b_2 + \cdots + a_n b_n \leq c\, b_1 \ \ (\text{or } c\, b_1 \leq a_1 b_1 + a_2 b_2 + \cdots + a_n b_n \text{ respectively}).$$

Proof. For $n = 1$ it holds obviously. Let $1 < n$. Then

$$\begin{aligned} & a_1 \cdot b_1 + a_2 \cdot b_2 + \cdots + a_n \cdot b_n = \\ & \quad = s_1 \cdot (b_1 - b_2) + s_2 \cdot (b_2 - b_3) + \cdots + s_{n-1} \cdot (b_{n-1} - b_n) + s_n \cdot b_n \leq \\ & \qquad \qquad \leq c \cdot (b_1 - b_2) + c \cdot (b_2 - b_3) + \cdots + c \cdot b_n \leq c \cdot b_1, \end{aligned}$$

since

$$\begin{aligned} b_1 &\geq b_1 - b_2 \geq 0, \\ b_1 &\geq b_2 - b_3 \geq 0, \\ &\vdots \\ b_1 &\geq b_{n-1} - b_n \geq 0, \\ b_1 &\geq b_n. \end{aligned}$$

The other case in brackets is similar. □

Proposition 18.8. (First Abel's theorem) Let $\{b_n\}$ be a non-increasing sequence of non-negative numbers, $\lim_{n\to\infty} b_n = 0$. Let $\sum a_n$ be a series such that the sequence $\{s_n\}$, where $s_n = \sum_{m=1}^{n} a_m$, is bounded from below and from above. Then the series

$$\sum_{n=1}^{\infty} a_n \cdot b_n$$

converges.

Proof. Let c be a proper real number such that for each n we have $-c < s_n < c$. Let $m \in \mathrm{IN}$, $k \in \mathrm{N}$. By Abel's lemma,

$$-c \cdot b_m \leq a_m \cdot b_m + \cdots + a_{m+k} \cdot b_{m+k} \leq c \cdot b_m.$$

Since $b_m \doteq 0$, we have $c{\cdot}b_m \doteq 0$ and hence the series $\sum a_n \cdot b_n$ satisfies Bolzano's criterion of convergence. □

Proposition 18.9. Let $\{a_n\}$ be a non-increasing sequence of non-negative numbers such that $\lim_{n\to\infty} a_n = 0$. Then the series

$$\sum(-1)^{n+1}a_n \qquad \text{(that is,} \quad a_1 - a_2 + a_3 - a_4 + a_5 - \cdots)$$

converges and

$$a_1 - a_2 \leq \sum_{n=1}^{\infty}(-1)^{n+1}a_n \leq a_1.$$

Proof. Since the sequence of partial sums of the series $\sum(-1)^{n+1}$ is bounded from below and from above, the series $\sum(-1)^{n+1}a_n$ converges by the First Abel's theorem. Since for each $m > 2$,

$$a_1 - a_2 \leq \sum_{n=1}^{m}(-1)^{n+1}a_n \leq a_1,$$

we also have

$$a_1 - a_2 \leq \sum_{n=1}^{\infty}(-1)^{n+1}a_n \leq a_1.$$

□

Proposition 18.10. Let $0 < r$. Then

(i) The series

$$\sum(-1)^{n+1}\frac{1}{n^r} \qquad \text{(that is,} \quad 1 - \frac{1}{2^r} + \frac{1}{3^r} - \frac{1}{4^r} + \cdots)$$

converges. In particular,the series

$$1 - \frac{1}{2} + \frac{1}{3} - \frac{1}{4} + \cdots$$

converges.

(ii) The series

$$\sum(-1)^{n+1}\frac{1}{(2n-1)^r} \qquad \text{(that is,} \quad 1-\frac{1}{3^r}+\frac{1}{5^r}-\frac{1}{7^r}+\cdots)$$

converges. In particular, the series

$$1-\frac{1}{3}+\frac{1}{5}-\frac{1}{7}+\cdots$$

converges.

Proof. By Proposition 18.9, or by Proposition 17.6 we have

$$\lim_{x\to\infty}\frac{1}{x^r}=0, \qquad \text{hence also} \qquad \lim_{n\to\infty}\frac{1}{n^r}=0.$$

□

Proposition 18.11. (Second Abel's theorem) Let $\{b_n\}$ be a non-increasing sequence of non-negative numbers and let $\sum a_n$ be a convergent series. Then the series

$$\sum a_n\cdot b_n$$

converges.

Proof. Let $m\in$ IN, $k\in$ N. Let α be the largest of the numbers

$$|a_m+a_{m+1}+\cdots+a_{m+n}|, \qquad \text{where} \quad n=1,2,\ldots,k.$$

Since the series $\sum a_n$ satisfies Bolzano's criterion of convergence, $\alpha \doteq 0$. By Abel's lemma

$$-\alpha\cdot b_m\le a_m\cdot b_m+\cdots+a_{m+k}\cdot b_{m+k}\le\alpha\cdot b_m.$$

Since $\alpha\cdot b_m\doteq 0$, the series $\sum a_n\cdot b_n$ satisfies Bolzano's criterion of convergence.
□

Proposition 18.12. Let $\{b_n\}$ be a non-increasing sequence of non-negative numbers. Let $0<r$. Then the series

$$\sum b_n\cdot(-1)^{n+1}\frac{1}{n^r} \qquad \text{and} \qquad \sum b_n\cdot(-1)^{n+1}\frac{1}{(2n-1)^r}$$

converge.

Proof. By the Second Abel's theorem and Proposition 18.10. □

Proposition 18.13. Let the series $\sum a_n$ be convergent. Then also the series

$$\sum a_n\cdot\frac{n+1}{n} \qquad \text{and} \qquad \sum a_n\cdot\left(1+\frac{1}{n}\right)^{n+1}$$

converge.

Proof. By the Second Abel's theorem, since the sequence $\{1+\frac{1}{n}\}$ is non-increasing and by Proposition 15.18 also the sequence $\{(1+\frac{1}{n})^{n+1}\}$ is non-increasing. □

18.2 Series with Non-negative Terms

In this section $\sum a_n$ and $\sum b_n$ denote series with non-negative terms. That is,

$$a_n \geq 0, \qquad b_n \geq 0 \qquad \text{for each } n.$$

Since the sequence of partial sums of a series with non-negative terms is non-decreasing, the following propositions are immediate consequences of similar assertions about non-decreasing sequences contained in Section 15.2.

Proposition 18.14. Every series with non-negative terms has a sum (and this sum is a non-negative number, possibly ∞).

Proposition 18.15. Let $a_n \leq b_n$ for each n. Then

$$\sum_{n=1}^{\infty} a_n \leq \sum_{n=1}^{\infty} b_n.$$

Hence if the series $\sum b_n$ converges then also $\sum a_n$ converges. If the series $\sum a_n$ diverges then also the series $\sum b_n$ diverges.

If moreover $\sum a_n$ converges and there exists m such that $a_m < b_m$ then

$$\sum_{n=1}^{\infty} a_n < \sum_{n=1}^{\infty} b_n,$$

since $(b_m - a_m) + \sum\limits_{n=1}^{\infty} a_n \leq \sum\limits_{n=1}^{\infty} b_n$.

Proposition 18.16. Let $\{h(n)\}_n$ be a sequence of natural numbers belonging to the world $\mathcal{A}$ and such that for each n, $h(n) < h(n+1)$. Let $h(0) = 1$. Let

$$b_n = \sum_{m=h(n-1)}^{h(n)-1} a_m.$$

Then

$$\sum_{n=1}^{\infty} a_n = \sum_{n=1}^{\infty} b_n.$$

Proof. Let $n \in \mathrm{IN}$. Then also $h(n) \in \mathrm{IN}$ and

$$\sum_{k=1}^{n} b_k = \sum_{k=1}^{h(n)-1} a_k.$$

□

Proposition 18.17. Let $0 \leq r$.

(i) If $1 < r$ then the series $\sum \frac{1}{n^r}$ converges and

$$\sum_{n=1}^{\infty} \frac{1}{n^r} \leq \frac{2^r}{2^r - 2}.$$

(ii) If $0 \leq r \leq 1$ then

$$\sum_{n=1}^{\infty} \frac{1}{n^r} = \infty.$$

Proof. (i) Let

$$h(n) = 2^n, \qquad b_n = \sum_{m=h(n-1)}^{h(n)-1} \frac{1}{m^r}.$$

If $h(n-1) \leq m < h(n)$ then $(h(n-1))^r \leq m^r$ and hence

$$b_n \leq \left(\frac{2}{2^r}\right)^{n-1}.$$

Let $q = \frac{2}{2^r}$. Then $0 < q < 1$ and hence by 18.4 and by the previous Proposition 18.16,

$$\sum_{n=1}^{\infty} \frac{1}{n^r} = \sum_{n=1}^{\infty} b_n \leq \sum_{n=0}^{\infty} q^n = \frac{2^r}{2^r - 2}.$$

(ii) Since for $0 \leq r \leq 1$ we have $n^r \leq n$,

$$\sum_{n=1}^{\infty} \frac{1}{n} \leq \sum_{n=1}^{\infty} \frac{1}{n^r}.$$

Thus it suffices to show that $\sum \frac{1}{n} = \infty$. Let

$$h(n) = 2^n + 1, \qquad c_n = \sum_{m=h(n-1)}^{h(n)-1} \frac{1}{m}.$$

By the previous Proposition 18.16 clearly

$$\sum_{n=1}^{\infty} \frac{1}{n} = 1 + \sum_{n=1}^{\infty} c_n.$$

If $h(n-1) \leq m < h(n)$ then $\frac{1}{2^n} \leq \frac{1}{m}$. Hence

$$c_n \geq \frac{2^{n-1}}{2^n} = \frac{1}{2},$$

which means that

$$\sum_{n=1}^{\infty} c_n \geq \frac{1}{2} + \frac{1}{2} + \frac{1}{2} + \cdots, \qquad \text{hence} \qquad \sum_{n=1}^{\infty} \frac{1}{n} = \infty.$$

□

How quickly does the **harmonic series**, that is, the series

$$\sum \frac{1}{n} = 1 + \frac{1}{2} + \frac{1}{3} + \cdots ,$$

diverge to infinity is illustrated by the following assertion.

Proposition 18.18. Let $s_n = 1 + \frac{1}{2} + \cdots + \frac{1}{n}$. Then for each $n > 1$

$$s_n - \log(n) < 1.$$

Proof. We shall show that for each n

$$s_n - \log(n) > s_{n+1} - \log(n+1), \qquad \text{so} \qquad -\log(n) > \frac{1}{n+1} - \log(n+1).$$

Equivalently,

$$\log\left(1 + \frac{1}{n}\right) > \frac{1}{n+1} \qquad \text{that is,} \qquad \log\left(1 + \frac{1}{n}\right)^{n+1} > 1\,.$$

This is true since by Proposition 15.18 the sequence $\left\{\left(1 + \frac{1}{n}\right)^{n+1}\right\}$ is non-increasing and its limit is e. Hence for each n

$$\left(1 + \frac{1}{n}\right)^{n+1} > e\,,$$

that is,

$$\log\left(1 + \frac{1}{n}\right)^{n+1} > \log(e) = 1.$$

For $n = 1$ we have $s_1 - \log(1) = 1$ and so for $n > 1$,

$$s_n - \log(n) < 1.$$

□

Proposition 18.19. (Cauchy's criterion of convergence)

(i) Assume that for each $n \in \mathrm{IN}$,

$$\sqrt[n]{a_n} < q, \qquad \text{where} \quad 0 < q < 1.$$

Then

$$\sum_{n=1}^{\infty} a_n < \infty.$$

(i') Assume that

$$\lim_{n\to\infty} \sqrt[n]{a_n}$$

exists and it is less than 1. Then

$$\sum_{n=1}^{\infty} a_n < \infty.$$

(ii) Let $m \in \mathrm{IN}$ be such that

$$\sqrt[m]{a_m} > 1.$$

Then

$$\sum_{n=1}^{\infty} a_n = \infty.$$

(ii') Assume that

$$\lim_{n\to\infty} \sqrt[n]{a_n}$$

exists and it is greater than 1. Then

$$\sum_{n=1}^{\infty} a_n = \infty.$$

Proof. (i) Let k be the smallest natural number such that for each $n > k$ we have $\sqrt[n]{a_n} < q$. Clearly $k \in \mathrm{FN}$.

If $\frac{1}{m} < 1 - q$ then $\sqrt[n]{a_n} < 1 - \frac{1}{m}$ for each $n > k$.

Let j be the smallest natural number such that for each $n > k$ we have $\sqrt[n]{a_n} < 1 - \frac{1}{j}$. By the rule of defined objects $j \in \mathrm{FN}$. Thus we can assume $q = 1 - \frac{1}{j}$ so q belongs to the world $\mathcal{A}$. Hence

$$\sum_{n=1}^{\infty} a_n = \sum_{n=1}^{k} a_n + \sum_{n=k+1}^{\infty} a_n \leq \sum_{n=1}^{k} a_n + \sum_{n=k+1}^{\infty} q^n < \infty.$$

(i') If $\lim\limits_{n\to\infty} \sqrt[n]{a_n} < 1$ then there exists q such that

$$\lim_{n\to\infty} \sqrt[n]{a_n} < q < 1.$$

Hence for each $n \in \mathrm{IN}$, $\sqrt[n]{a_n} < q$.

(ii) If $\sqrt[m]{a_m} \geq 1$ then $a_m \geq 1$ and hence it is not the case that $\lim\limits_{n\to\infty} a_n = 0$. By Proposition 18.5, $\sum a_n$ does not converge.

(ii') This case can be reduced to (ii) similarly as (i') to (i) above. □

Proposition 18.20. Let H be a bijective mapping of the class FN onto FN. (Hence H belongs to the world $\mathcal{A}$.) Let $b_n = a_{H(n)}$. Then

$$\sum_{n=1}^{\infty} b_n = \sum_{n=1}^{\infty} a_n.$$

Proof. For each m

$$\sum_{n=1}^{m} a_{H(n)} \le \sum_{n=1}^{k} a_n,$$

where k is the largest of the numbers $H(n)$ for $n \le m$. Hence

$$\sum_{n=1}^{\infty} b_n \le \sum_{n=1}^{\infty} a_n.$$

The other inequality obtains upon replacing the mapping H by H^{-1}, the inverse to H. □

18.3 Convergence Criteria for Series with Positive Terms

In this section $\sum a_n$, $\sum b_n$ denote series with positive terms.

Proposition 18.21. Assume that for each $n \in \mathrm{IN}$ we have

$$\frac{a_{n+1}}{a_n} \le \frac{b_{n+1}}{b_n}.$$

If the series $\sum b_n$ converges then also the series $\sum a_n$ converges (and hence if $\sum\limits_{n=1}^{\infty} a_n = \infty$ then $\sum\limits_{n=1}^{\infty} b_n = \infty$).

Proof. Let k be the smallest natural number such that for each $n \ge k$,

$$\frac{a_{n+1}}{a_n} \le \frac{b_{n+1}}{b_n}.$$

Clearly $k \in \mathrm{FN}$. For $n > k$ we have

$$\frac{a_n}{a_k} = \frac{a_n}{a_{n-1}} \cdot \frac{a_{n-1}}{a_{n-2}} \cdot \dots \cdot \frac{a_{k-1}}{a_k} \le \frac{b_n}{b_{n-1}} \cdot \frac{b_{n-1}}{b_{n-2}} \cdot \dots \cdot \frac{b_{k-1}}{b_k} = \frac{b_n}{b_k},$$

and hence $a_n \le \frac{a_k}{b_k} \cdot b_n$. Consequently

$$\sum_{n=1}^{\infty} a_n = \sum_{n=1}^{k-1} a_n + \sum_{n=k}^{\infty} a_n \le \sum_{n=1}^{k-1} a_n + \frac{a_k}{b_k} \cdot \sum_{n=k}^{\infty} b_n < \infty.$$

□

Proposition 18.22. (i) Assume that for each $n \in \mathrm{IN}$ we have

$$\frac{a_{n+1}}{a_n} \ge 1.$$

Then $\sum\limits_{n=1}^{\infty} a_n = \infty$.

(ii) Assume that $\lim\limits_{n\to\infty} \frac{a_{n+1}}{a_n}$ exist and that it is greater than 1. Then $\sum\limits_{n=1}^{\infty} a_n = \infty$.

Proof. (i) Since $\frac{1}{1} \leq \frac{a_{n+1}}{a_n}$, we have

$$a_1(1+1+1+\cdots) \leq \sum_{n=1}^{\infty} a_n$$

by Proposition 18.21.

(ii) If $\lim\limits_{n\to\infty} \frac{a_{n+1}}{a_n} = v > 1$ then for each $n \in \mathrm{IN}$,

$$\frac{a_{n+1}}{a_n} > \frac{1+v}{2},$$

which reduces this case to (i). □

Proposition 18.23. (Jensen's criterion of convergence)

(i) Let $w > 0$ and assume that for each $n \in \mathrm{IN}$ we have

$$b_n \cdot \frac{a_n}{a_{n+1}} - b_{n+1} > w.$$

Then $\sum\limits_{n=1}^{\infty} a_n < \infty$.

(ii) Assume that

$$\lim_{n\to\infty} \left(b_n \cdot \frac{a_n}{a_{n+1}} - b_{n+1} \right)$$

exists and that it is a positive number. Then $\sum\limits_{n=1}^{\infty} a_n < \infty$.

Proof. (i) Let k be the smallest natural number such that for each $n \geq k$,

$$b_n \cdot \frac{a_n}{a_{n+1}} - b_{n+1} > w.$$

Clearly $k \in \mathrm{FN}$. Let m be the smallest natural number such that for each $n \geq k$,

$$b_n \cdot \frac{a_n}{a_{n+1}} - b_{n+1} > \frac{1}{m}.$$

By the rule of defined objects $m \in \mathrm{FN}$. Let $w = \frac{1}{m}$. For $n \geq k$ we have

$$a_{n+1} < \frac{1}{w} \cdot (b_n \cdot a_n - b_{n+1} \cdot a_{n+1}).$$

Hence

$$a_{k+1} < \frac{1}{w} \cdot (b_k \cdot a_k - b_{k+1} \cdot a_{k+1}),$$
$$a_{k+2} < \frac{1}{w} \cdot (b_{k+1} \cdot a_{k+1} - b_{k+2} \cdot a_{k+2}),$$
$$\vdots$$
$$a_{k+n} < \frac{1}{w} \cdot (b_{k+n-1} \cdot a_{k+n-1} - b_{k+n} \cdot a_{k+n}).$$

It follows that

$$a_{k+1} + a_{k+2} + \cdots + a_{k+n} < \frac{1}{w} \cdot (b_k \cdot a_k - b_{k+n} \cdot a_{k+n}) < \frac{1}{w} \cdot b_k \cdot a_k.$$

Hence

$$\sum_{n=k+1}^{\infty} a_n \leq \frac{1}{w} \cdot (b_k \cdot a_k), \quad \text{so} \quad \sum_{n=1}^{\infty} a_n = \sum_{n=1}^{k} a_n + \sum_{n=k+1}^{\infty} a_n < \infty.$$

(ii) If

$$\lim_{n\to\infty} \left(b_n \cdot \frac{a_n}{a_{n+1}} - b_{n+1} \right) = v > 0,$$

then for each $n \in \mathrm{IN}$,

$$b_n \cdot \frac{a_n}{a_{n+1}} - b_{n+1} > \frac{v}{2},$$

which reduces this case to (i). □

Proposition 18.24. (D'Alembert's criterion of convergence)

(i) Let there exist $0 < q < 1$ such that for each $n \in \mathrm{IN}$,

$$\frac{a_{n+1}}{a_n} < q.$$

Then $\sum\limits_{n=1}^{\infty} a_n < \infty$.

(ii) Let there exist

$$\lim_{n\to\infty} \frac{a_{n+1}}{a_n} < 1.$$

Then $\sum\limits_{n=1}^{\infty} a_n < \infty$.

Proof. Letting $b_n = 1$ for each $n \in \mathrm{IN}$ in Jensen's Criterion and $w = \frac{1}{q} - 1$, we obtain d'Alembert's Criterion. □

Proposition 18.25. (Raabe's Criterion of Convergence) Let $r > 1$ be such that for each $n \in \mathrm{IN}$,

$$n \cdot \left(\frac{a_n}{a_{n+1}} - 1 \right) > r.$$

Then $\sum\limits_{n=1}^{\infty} a_n < \infty$.

Proof. Letting $b_n = n$ for each $n \in \mathrm{IN}$ in Jensen's Criterion and $r = 1 + w$ we obtain Raabe's Criterion. □

18.4 Absolutely and Non-absolutely Convergent Series

We say that the series $\sum a_n$ is **absolutely convergent** if the series $\sum |a_n|$ is convergent.

We say that the series $\sum a_n$ is **non-absolutely convergent** if it is convergent and it is not absolutely convergent.

Proposition 18.26. If the series $\sum a_n$ is absolutely convergent then it is convergent.

Proof. If $m \in \mathrm{IN}$ then for each $k \in \mathrm{N}$,

$$\sum_{n=0}^{k} |a_{m+n}| \doteq 0,$$

but

$$\left| \sum_{n=0}^{k} a_{m+n} \right| \leq \sum_{n=0}^{k} |a_{m+n}| \doteq 0.$$

Hence $\sum a_n$ converges. □

The **series of non-negative** (or **non-positive**) **terms** of the series $\sum a_n$ is the series

$$\sum a_n^+ \qquad (\text{or} \qquad \sum a_n^- \ ),$$

which we obtain from the series $\sum a_n$ by replacing every negative term (or positive term respectively) by zero.

Proposition 18.27. The series $\sum a_n^+$, $\sum a_n^-$ have sums, which may be improper numbers. Furthermore

$$\sum_{n=1}^{\infty} a_n^- \leq 0 \leq \sum_{n=1}^{\infty} a_n^+.$$

Proof. The series $\sum a_n^+$ and $\sum(-a_n^-)$ are series of non-negative terms and hence, by Proposition 18.14, they have sums. Furthermore

$$\sum_{n=1}^{\infty} a_n^- = -\sum_{n=1}^{\infty}(-a_n^-).$$

Consequently, $\sum\limits_{n=1}^{\infty} a_n^- \leq 0$. □

Proposition 18.28. Assume that the expression $\sum\limits_{n=1}^{\infty} a_n^+ + \sum\limits_{n=1}^{\infty} a_n^-$ is not indeterminate (that is, of the form $\infty - \infty$). Then

$$\sum_{n=1}^{\infty} a_n = \sum_{n=1}^{\infty} a_n^+ + \sum_{n=1}^{\infty} a_n^- .$$

Proof. Clearly for each m, also for $m \in$ IN,

$$\sum_{n=1}^{m} a_n = \sum_{n=1}^{m} a_n^+ + \sum_{n=1}^{m} a_n^- .$$

□

Proposition 18.29. The series $\sum a_n$ is absolutely convergent if and only if both the series $\sum a_n^+$, $\sum a_n^-$ are convergent.

Proof. Clearly for each m we have

$$\sum_{n=1}^{m} |a_n| = \sum_{n=1}^{m} a_n^+ - \sum_{n=1}^{m} a_n^- .$$

Hence if both the above series are convergent then also the series $\sum |a_n|$ is convergent. Now assume that $\sum |a_n|$ is convergent. Then by Proposition 18.15,

$$\sum_{n=1}^{\infty} a_n^+ \leq \sum_{n=1}^{\infty} |a_n| < \infty,$$

$$\sum_{n=1}^{\infty} -a_n^- \leq \sum_{n=1}^{\infty} |a_n| < \infty.$$

□

The following proposition is an immediate consequence of the two previous propositions.

Proposition 18.30. The series $\sum a_n$ is non-absolutely convergent or oscillating if and only if

$$\sum_{n=1}^{\infty} a_n^+ = \infty, \qquad \sum_{n=1}^{\infty} a_n^- = -\infty.$$

Proposition 18.31. Let $\sum a_n$ be an absolutely convergent series. Let H be a bijective mapping of the class FN onto itself. (Thus H belongs to the world $\mathcal{A}$.) Then the series $\sum a_{H(n)}$ converges and

$$\sum_{n=1}^{\infty} a_{H(n)} = \sum_{n=1}^{\infty} a_n.$$

Proof. By Proposition 18.20 we clearly have

$$\sum_{n=1}^{\infty} a_{H(n)} = \sum_{n=1}^{\infty} a^+_{H(n)} + \sum_{n=1}^{\infty} a^-_{H(n)} = \sum_{n=1}^{\infty} a^+_n + \sum_{n=1}^{\infty} a^-_n = \sum_{n=1}^{\infty} a_n.$$

□

Concerning rearranging terms of a non-absolutely convergent series, we have the well known Riemann theorem, which follows from the subsequent lemma.

Lemma 18.32. Let $\sum b_n$ be a series of non-negative numbers, $\sum c_n$ a series of non-positive numbers,

$$\sum_{n=1}^{\infty} b_n = \infty, \quad \sum_{n=1}^{\infty} c_n = -\infty, \quad \lim_{n\to\infty} b_n = 0, \quad \lim_{n\to\infty} c_n = 0.$$

Then the following hold:

(i) Let A be a real number (possibly improper). Then the terms of the series $\sum b_n$, $\sum c_n$ may be arranged into one series $\sum a_n$ such that

$$A = \sum_{n=1}^{\infty} a_n.$$

(ii) The terms of the series $\sum b_n$, $\sum c_n$ may be arranged into a series that oscillates.

Proof. (i) Assume first that $A = \infty$. Let $k(n)$ be the smallest natural number such that

$$b_1 + b_2 + \cdots + b_{k(n)} > n - c_1 - c_2 - \cdots - c_n.$$

For each n there exists such a number $k(n)$ since $\sum\limits_{n=1}^{\infty} b_n = \infty$. The series

$$b_1 + b_2 + \cdots + b_{k(1)} + c_1 + b_{k(1)+1} + \cdots + b_{k(2)} + c_2 + \cdots$$

clearly exhausts all the terms of the series $\sum b_n$, $\sum c_n$ and its sum is ∞. The case of $-\infty$ is similar.

Now let A be a proper number. Then we compose the series $\sum a_n$ as follows. First we take subsequent terms of the series $\sum b_n$ until their sum exceeds the number A. Then we add terms of the series $\sum c_n$ until the total sum drops below A, then add terms of the series $\sum b_n$ until the total sum exceeds A etc. In this manner we gradually exhaust all the terms of the series $\sum b_n$ and $\sum c_n$.

When we reach IN with this process, then since there we have

$$b_n \doteq 0, \qquad c_n \doteq 0,$$

we always exceed A only by an infinitesimal quantity and drop below A also only by an infinitesimal quantity. This means that the sum of the resulting series is equal to A.

(ii) If we wish to arrange the terms of the series $\sum b_n$ and $\sum c_n$ so that we obtain an oscillating series then we take terms of the series $\sum b_n$ until their sum exceeds the number 1, then add terms of the series $\sum c_n$ until the total sum drops below -1 and we continue in this manner. □

If we leave out of the series $\sum a_n^+$ all those terms that are equal to 0, for which $a_n < 0$, and from the series $\sum a_n^-$ all those terms that are equal to 0, for which $a_n > 0$, then using Lemma 18.32 we can easily see that the following proposition holds.

Proposition 18.33. (Riemann's theorem) Let $\sum a_n$ be a non-absolutely convergent series.

(i) Let A be a real number (possibly improper). Then there exists a bijective mapping H of the class of natural numbers FN onto itself such that

$$\sum_{n=1}^{\infty} a_{H(n)} = A.$$

(ii) There exists a bijective mapping H of the class of natural numbers FN onto itself such that the series $\sum a_{H(n)}$ is oscillating.

Chapter 19

Series of Functions

19.1 Taylor and Maclaurin Series

In this section f denotes a function defined on an open interval I that has derivatives of all orders on I; $c, x \in I$, $c \neq x$.

Similarly as in Section 16.9 we define

$$P_n(x) = \sum_{m=0}^{n} f^{(m)}(c) \cdot \frac{(x-c)^m}{m!}, \qquad \text{where} \quad f^{(0)}(c) = f(c).$$

Then

$$f(x) = P_n(x) + Z_{n+1}(x).$$

Proposition 19.1. If $\lim_{n \to \infty} Z_n(x) = 0$, then

$$f(x) = \lim_{n \to \infty} P_n(x) = \sum_{n=0}^{\infty} f^{(n)}(c) \cdot \frac{(x-c)^n}{n!}.$$

In this case we say that

$$\sum f^{(n)}(c) \cdot \frac{(x-c)^n}{n!}$$

is the **Taylor series** (or **Taylor expansion**) of the function f.

Obviously, for such x the Taylor series converges.

In the special case of $c = 0$ the Taylor series has the form

$$\sum f^{(n)}(0) \cdot \frac{x^n}{n!}$$

and it is called the **Maclaurin series**.

If we choose to express $Z_n(x)$ in its Lagrange's form, that is,

$$Z_n(x) = \frac{(x-c)^n}{n!} \cdot f^{(n)}(y),$$

where y is an element of the open interval with end points x, c, then by Proposition 15.16,

$$\lim_{n\to\infty} \frac{(x-c)^n}{n!} = 0,$$

and hence whether or not $\lim\limits_{n\to\infty} Z_n(x) = 0$, is decided merely by the factor $f^{(n)}(y)$ for $n \in \mathrm{IN}$.

Proposition 19.2. Assume that for each $n \in \mathrm{IN}$ and for each y from the open interval with endpoints x and c, $f^{(n)}(y)$ is a finitely large real number. Then

$$f(x) = \sum_{n=0}^{\infty} f^{(n)}(c) \cdot \frac{(x-c)^n}{n!}.$$

Proof. In this case for each $n \in \mathrm{IN}$,

$$Z_n(x) = f^{(n)'}(y) \cdot \frac{(x-c)^n}{n!} \doteq 0.$$

□

We remark that for $c = 0$ under the above assumptions we have

$$f(x) = \sum_{n=0}^{\infty} f^{(n)}(0) \cdot \frac{x^n}{n!}.$$

19.2 Maclaurin Series of the Exponential Function

Since the exponential function e^x has the derivative e^x at each point again, it has all derivatives at the point x equal to e^x. Furthermore, for each x, e^x is a finitely large real number. By Proposition 19.2 the function e^x thus has a Taylor series for each $c \neq x$.

If we set $c = 0$ then

$$e^0 = 1,$$

and hence the Maclaurin series of the function e^x has the following form.

Proposition 19.3. For each x,

$$e^x = 1 + \frac{x}{1!} + \frac{x^2}{2!} + \cdots = \sum_{n=0}^{\infty} \frac{x^n}{n!}.$$

Hence in particular

$$e = 1 + \frac{1}{1!} + \frac{1}{2!} + \frac{1}{3!} + \cdots = \sum_{n=0}^{\infty} \frac{1}{n!},$$

$$e^{-1} = 1 - \frac{1}{1!} + \frac{1}{2!} - \frac{1}{3!} + \cdots = \sum_{n=0}^{\infty} \frac{(-1)^n}{n!}.$$

Proposition 19.4. The number e is irrational.

Proof. We shall prove this by contradiction. Assume that

$$e = \frac{n}{k},$$

where k, n are natural numbers. Then

$$e^{-1} = \frac{k}{n}.$$

By Maclaurin theorem,

$$Z_n(-1) = (-1)^n \cdot \frac{1}{n!} \cdot e^y,$$

where $-1 < y < 0$ and hence $0 < e^y < 1$. Consequently,

$$0 \neq |Z_n(-1)| < \frac{1}{n!}.$$

However, we also have

$$Z_n(-1) = \frac{k}{n} - 1 + \frac{1}{1!} - \frac{1}{2!} + \cdots + \frac{(-1)^n}{(n-1)!}.$$

If we express the right hand side as a fraction with the denominator $n!$, we obtain a number $\frac{p}{n!}$, where p is an integer, $p \neq 0$, and hence a number with an absolute value greater or equal to $\frac{1}{n!}$, contradiction. □

19.3 Maclaurin Series of Functions $\sin x$, $\cos x$

Higher derivatives of function $\sin x$ are, in order,

$$\cos x, \qquad -\sin x, \qquad -\cos x, \qquad \sin x, \qquad \cos x, \qquad \ldots$$

Higher derivatives of function $\cos x$ are, in order,

$$-\sin x, \qquad -\cos x, \qquad \sin x, \qquad \cos x, \qquad -\sin x, \qquad \ldots$$

Hence we have for each x and for each n

$$|f^{(n)}(x)| \leq 1,$$

where $f(x)$ is $\sin x$ or $\cos x$.

Since $\sin 0 = 0$, $\cos 0 = 1$, the corresponding Maclaurin series have the following forms: for each x,

Proposition 19.5.

(i) $$\sin x = \frac{x}{1!} - \frac{x^3}{3!} + \frac{x^5}{5!} - \cdots = \sum_{n=0}^{\infty} (-1)^n \frac{x^{2n+1}}{(2n+1)!},$$

(ii) $$\cos x = 1 - \frac{x^2}{2!} + \frac{x^4}{4!} - \frac{x^6}{6!} + \cdots = \sum_{n=0}^{\infty} (-1)^n \frac{x^{2n}}{(2n)!}.$$

Proposition 19.6. The above Maclaurin series for functions $\sin x$, $\cos x$ are absolutely convergent for each x.

Proof. For $0 \le x$ we have $0 \le x^n$, and hence the Maclaurin series for the function e^x (see Proposition 19.3) is absolutely convergent. For $x < 0$, both the series of negative and positive terms of the Maclaurin series for e^x are convergent since these series are selected from the series for e^x, where $x > 0$. Hence also this Maclaurin series for e^x is absolutely convergent. Since the series of positive and negative terms of Maclaurin series for $\sin x$, $\cos x$ are selected from Maclaurin series for e^x, these series are also absolutely convergent.

□

19.4 Powers of Complex Numbers

The addition, subtraction, multiplication and division of complex numbers can be carried out using familiar rules. However, there was no obvious way of introducing powers of complex numbers.

Powers of complex numbers should have the basic properties for calculation (i)-(iii), with (i) replaced by

(i′) $$1^p = 1,$$

not $x^p > 0$ as before, see page 267, since that would have no meaning in the complex domain.

In particular, it should be the case that

$$(c + \mathrm{i}d)^{a+\mathrm{i}b} = (c + \mathrm{i}d)^a \cdot (c + \mathrm{i}d)^{\mathrm{i}b},$$

where a, b, c, d are real numbers and i is the **imaginary unit**, $\mathrm{i}^2 = -1$.

This expression offers the possibility to split the problem of working out complex powers, and work out powers of complex numbers to a real exponent on the one hand, and powers of complex numbers to a purely imaginary exponent on the other hand separately

If we substitute the complex number $\mathrm{i}x$ for the real number x into the Maclaurin series for e^x, we obtain

$$e^{\mathrm{i}x} = \cos x + \mathrm{i} \cdot \sin x.$$

Since the complex number $c + id$ can be written in the form

$$r(\cos x + \mathrm{i} \cdot \sin x) = r \cdot e^{\mathrm{i}x}, \qquad \text{where} \quad r = \sqrt{c^2 + d^2}$$

(and $r = 0$ if and only if $c + \mathrm{i} \cdot d = 0$), we can define the required powers of complex numbers in the following way.

Let $0 < r$. We set

$$\left(r \cdot e^{\mathrm{i}x}\right)^a = r^a \cdot e^{\mathrm{i}xa} = r^a \cdot (\cos xa + \mathrm{i} \sin xa),$$
$$\left(r \cdot e^{\mathrm{i}x}\right)^{ib} = r^{ib} \cdot e^{\mathrm{i}x \cdot ib} = r^{ib} \cdot e^{-xb}.$$

Thus it remains to define r^{ib}, that is, the power of a positive real number to a purely imaginary exponent. Since $r = e^{\log r}$, we set

$$r^{ib} = e^{ib \cdot \log r} = \cos(b \cdot \log r) + \mathrm{i} \sin(b \cdot \log r).$$

We can easily verify that powers of complex numbers defined in this way have properties (i'), (ii), (iii).

Substituting 2π for x into the expression

$$e^{\mathrm{i}x} = \cos x + \mathrm{i} \sin x$$

yields the famous **Euler's Identity**

$$e^{2\pi \mathrm{i}} = 1.$$

We remark that the expression of a complex number $c + id$ as $r \cdot e^{\mathrm{i}x}$ is not unique, and hence powers in the complex domain are not uniquely determined. Clearly, for each integer k,

$$e^{\mathrm{i}x} = e^{\mathrm{i}(x + 2k\pi)}.$$

In the Section 17.1 we defined x^p only for $0 < x$. Taking powers in complex domain as defined above makes it possible to define as a special case also $(-r)^p$, where r, p are real numbers, $r > 0$. We have

$$(-r)^p = r^p \cdot e^{\mathrm{i}\pi p} = r^p \cdot (\cos(\pi p) + \mathrm{i} \cdot \sin(\pi p)).$$

Also, $\sin(\pi p) = 0$ if and only if p is an integer. In other words, $(-r)^p$ is a real number just when p is an integer.

19.5 Maclaurin Series of the Function $\log(1+x)$ for $-1 < x \le 1$

Proposition 19.7. For $-1 < x$ let $f(x) = \log(1+x)$. Then

$$
\begin{aligned}
f(0) &= \log(1) = 0, \\
f'(x) &= \frac{1}{1+x}, & f'(0) &= 1, \\
f^{(2)}(x) &= \frac{-1}{(1+x)^2}, & f^{(2)}(0) &= -1!, \\
f^{(3)}(x) &= \frac{2}{(1+x)^3}, & f^{(3)}(0) &= 2!, \\
&\vdots \\
f^{(n)}(x) &= (-1)^{n-1} \cdot \frac{(n-1)!}{(1+x)^n}, & f^{(n)}(0) &= (-1)^{n-1} \cdot (n-1)!,
\end{aligned}
$$

and by Taylor's theorem for $-1 < x$, Maclaurin series of the function f has the form

$$
\begin{aligned}
\log(1+x) &= \frac{x}{1} - \frac{x^2}{2} + \frac{x^3}{3} - \cdots + (-1)^{n-1}\frac{x^n}{n} + Z_{n+1}(x) = \\
&= \sum_{n=1}^{\infty} (-1)^{n+1} \cdot \frac{x^n}{n!}.
\end{aligned}
$$

Proof. For $0 < x \le 1$ we shall use the *Lagrange's form* of the remainder. Since

$$
f^{(n+1)}(y) = (-1)^n \cdot \frac{n!}{(1+y)^{n+1}}, \qquad \text{where} \quad 0 \le y < x \le 1,
$$

we have

$$
\begin{aligned}
Z_{n+1}(x) &= \frac{x^{n+1}}{(n+1)!} \cdot (-1)^n \cdot \frac{n!}{(1+y)^{n+1}} = \\
&= (-1)^n \cdot \frac{x^{n+1}}{n+1} \cdot \frac{1}{(1+y)^{n+1}}.
\end{aligned}
$$

Hence

$$
|Z_{n+1}(x)| \le \frac{1}{n+1}, \qquad \text{since} \quad \frac{1}{(1+y)^{n+1}} < 1
$$

for $|x| \le 1$.

For $-1 < x < 0$ we shall use the *Cauchy's form* of the remainder. In this case there again exists y such that $-1 < x < y < 0$ and

$$Z_{n+1}(x) = \frac{(x-y)^n}{n!} \cdot x \cdot (-1)^n \cdot \frac{n!}{(1+y)^{n+1}} = \\ = (-1)^n \cdot \frac{(x-y)^n \cdot x}{(1+y)^{n+1}} = \\ = (-1)^n \cdot x^{n+1} \cdot \frac{(1-\frac{y}{x})^n}{(1+y)^n} \cdot \frac{1}{1+y}.$$

Since $0 > y > x > -1$, we have

$$0 < \left(\frac{1-\frac{y}{x}}{1+y}\right)^n < 1, \qquad \text{and hence} \quad |Z_{n+1}(x)| < \frac{|x^{n+1}|}{1+y}.$$

Hence $Z_{n+1}(x) \doteq 0$ for each $n \in \mathrm{IN}$.

This shows that for $-1 < x \leq 1$ we have

$$\log(1+x) = \frac{x}{1} - \frac{x^2}{2} + \frac{x^3}{3} - \frac{x^4}{4} + \cdots = \sum_{n=1}^{\infty} (-1)^{n+1} \cdot \frac{x^n}{n}.$$

When $|x| < 1$ then the above series is absolutely convergent since if we set

$$a_n = \frac{|x^n|}{n} \qquad \text{then} \quad \frac{a_{n+1}}{a_n} = |x| \cdot \left(1 - \frac{1}{n+1}\right),$$

and so $\lim\limits_{n\to\infty} \frac{a_{n+1}}{a_n} = 0$. Hence d'Alembert's criterion of convergence can be used.

If $1 < x$ then the above series is divergent since the absolute values of its terms increase and hence it is not the case that $\lim\limits_{n\to\infty} \frac{x^n}{n} = 0$.

If $x = 1$ then the above series is, in accord with Proposition 18.17 and the First Abel's theorem, non-absolutely convergent. □

If $-1 < x < 1$ then

$$\log(1-x) = -\frac{x}{1} - \frac{x^2}{2} - \frac{x^3}{3} - \cdots$$

and this series is also absolutely convergent. If we subtract this series from the series for $\log(1+x)$, we obtain

$$\log\left(\frac{1+x}{1-x}\right) = 2 \cdot \left(\frac{x}{1} + \frac{x^3}{3} + \frac{x^5}{5} + \cdots\right).$$

Hence if $y > 0$ and if we set $x = \frac{y-1}{y+1}$, we have $|x| < 1$, $y = \frac{1+x}{1-x}$. It follows that

$$\log y = 2 \cdot \left(\frac{x}{1} + \frac{x^3}{3} + \frac{x^5}{5} + \cdots\right) = 2 \sum_{n=0}^{\infty} \frac{x^{2n+1}}{2n+1}.$$

19.6 Maclaurin Series of the Function $(1+x)^r$ for $|x|<1$

Proposition 19.8. For $-1<x<1$ we set $f(x)=(1+x)^r$. Then

$$\begin{aligned}
f(0)&=1,\\
f'(x)&=r\cdot(1+x)^{r-1}, & f'(0)&=r,\\
f^{(2)}(x)&=r\cdot(r-1)\cdot(1+x)^{r-2}, & f^{(2)}(0)&=r\cdot(r-1),\\
f^{(3)}(x)&=r\cdot(r-1)\cdot(r-2)\cdot(1+x)^{r-3}, & f^{(3)}(0)&=r\cdot(r-1)\cdot(r-2),\\
&\vdots\\
f^{(n)}(x)&=r\cdot(r-1)\cdot(r-2)\cdot\dots\cdot(r-n+1)\cdot(1+x)^{r-n},\\
& & f^{(n)}(0)&=r\cdot(r-1)\cdot(r-2)\cdot\dots\cdot(r-n+1),
\end{aligned}$$

and by Taylor's theorem, Maclaurin series of f has the following form

$$(1+x)^r=\binom{r}{0}+\binom{r}{1}\cdot x+\binom{r}{2}\cdot x^2+\dots+\binom{r}{n}\cdot x^n+Z_{n+1}(x).$$

Proof. If $r=0,1,2,\dots$ then

$$(1+x)^r=\binom{r}{0}+\binom{r}{1}\cdot x+\binom{r}{2}\cdot x^2+\dots+\binom{r}{r}\cdot x^r,$$

and hence by the Binomial theorem $Z_n(x)=0$ for $n>r$.

Hence in what follows we assume that $r\neq 0,1,2,\dots$ and $x\neq 0$. In that case the Cauchy's form of remainder is

$$\begin{aligned}
Z_{n+1}(x)&=\frac{(x-y)^n}{n!}\cdot x\cdot(1+y)^{r-n-1}\cdot r\cdot(r-1)\cdot\dots\cdot(r-n)=\\
&=x^{n+1}\cdot\left(1-\frac{y}{x}\right)^n\cdot(1+y)^{r-n-1}\cdot\binom{r}{n}\cdot(r-n).
\end{aligned}$$

The number $(1+y)^{r-1}$ lies in the interval with end points 1 and $(1+x)^{r-1}$, hence

$$(1+y)^{r-1}\le\max\{1,(1+x)^{r-1}\}.$$

It is easy to see that

$$\left(1-\frac{y}{x}\right)^n(1+y)^{-n}<1.$$

Hence

$$|Z_{n+1}(x)|\le|x|^{n+1}\cdot\max\{1,(1+x)^{r-1}\}\cdot\left|\binom{r}{n}\right|\cdot|r-n|=Q_{n+1}(x).$$

Thus it suffices to prove that $\lim\limits_{n\to\infty}Q_n(x)=0$, and for this end by Proposition 15.15(ii) it suffices to show that

$$\lim_{n\to\infty}\frac{Q_{n+1}(x)}{Q_n(x)}<1.$$

Since

$$\frac{Q_{n+1}(x)}{Q_n(x)} = |x| \cdot \left| \frac{r-n}{n} \right| = |x| \cdot \left| 1 - \frac{r}{n} \right|,$$

we have

$$\lim_{n \to \infty} \frac{Q_{n+1}(x)}{Q_n(x)} = |x| < 1.$$

□

In accordance with the Binomial theorem, this shows the following proposition.

Proposition 19.9. For every real number r and every real number $|x| < 1$,

$$(1+x)^r = \sum_{n=0}^{\infty} \binom{r}{n} \cdot x^n.$$

Proposition 19.10. Let $r \neq 0, 1, 2, \ldots$ Then the series

$$\sum \binom{r}{n} \cdot x^n$$

is absolutely convergent for $|x| < 1$ and divergent for $1 < |x|$.

Proof. Let

$$a_n = \left| \binom{r}{n} \cdot x^n \right|.$$

Then

$$\frac{a_{n+1}}{a_n} = |x| \cdot \left| \frac{r-n}{n+1} \right| = |x| \cdot \left| \frac{r+1}{n+1} - 1 \right|.$$

Hence $\lim\limits_{n \to \infty} \frac{a_{n+1}}{a_n} = |x|$.

Consequently, if $|x| < 1$ then the series $\sum a_n$ converges by d'Alembert's criterion. If $1 < |x|$ then this series diverges and its terms, starting from that a_m for which

$$|x| \cdot \left| \frac{r+1}{m+1} - 1 \right| > 1,$$

increase. Hence it is not the case that $\lim\limits_{n \to \infty} a_n = 0$. □

The series

$$\sum \binom{r}{n} \cdot x^n$$

is called the **Binomial Series**. The following proposition generalizes the Binomial theorem.

Proposition 19.11. Let $|x| < |y|$. Then

$$(y + x)^r = \sum_{n=0}^{\infty} \binom{r}{n} \cdot x^n \cdot y^{r-n}.$$

Proof. Since $\left|\frac{x}{y}\right| < 1$, we have

$$\begin{aligned}(y + x)^r &= y^r \cdot \left(1 + \frac{x}{y}\right)^r = \\ &= y^r \cdot \sum_{n=0}^{\infty} \binom{r}{n} \cdot \left(\frac{x}{y}\right)^n = \sum_{n=0}^{\infty} \binom{r}{n} \cdot x^n \cdot y^{r-n}.\end{aligned}$$

□

19.7 Binomial Series $\sum \binom{r}{n} x^n$ for $x = \pm 1$

In this section we assume that $r \neq 0, 1, 2, \ldots$

Proposition 19.12.

(i) If $r < -1$ then

$$\lim_{n\to\infty} \left|\binom{r}{n}\right| = \infty.$$

(ii) If $r = -1$ then

$$\binom{r}{n} = (-1)^n.$$

(iii) If $-1 < r$ then

$$\lim_{n\to\infty} \binom{r}{n} = 0.$$

Proof. (i) We have

$$\begin{aligned}\left|\binom{r}{n}\right| &= \left|\frac{-r}{1} \cdot \frac{-r+1}{2} \cdot \ldots \cdot \frac{-r+n-1}{n}\right| = \\ &= \left|\left(1 + \frac{-r-1}{1}\right) \cdot \ldots \cdot \left(1 + \frac{-r-1}{n}\right)\right| \geq \\ &\geq 1 + \frac{-r-1}{1} + \frac{-r-1}{2} + \cdots + \frac{-r-1}{n} = \\ &= 1 + (-r-1) \cdot \left(1 + \frac{1}{2} + \frac{1}{3} + \cdots + \frac{1}{n}\right),\end{aligned}$$

since $0 < -r - 1$.

(ii) Clearly

$$\binom{-1}{n} = \frac{(-1)\cdot(-2)\cdot\dots\cdot(-n)}{1\cdot 2\cdot\dots\cdot n} = (-1)^n.$$

(iii) Let

$$a_n = \frac{r+1}{n}.$$

Clearly $a_n > 0$. Let k be the smallest natural number such that $k > r+1$. Then for $n \geq k$

$$1 - a_n = \frac{n-r-1}{n} > 0, \qquad \frac{a_n}{1-a_n} = \frac{r+1}{n-r-1} > 0.$$

Let $b_n = \binom{r}{n}$. Then

$$\left|\frac{1}{b_n}\right| = \left|\frac{1}{-r}\cdot\dots\cdot\frac{k-1}{-r+k-2}\right|\cdot\left|\frac{k}{-r+k-1}\cdot\dots\cdot\frac{n}{-r+n-1}\right|.$$

The first factor is positive and the second equals

$$\left|\left(1+\frac{a_k}{1-a_k}\right)\cdot\dots\cdot\left(1+\frac{a_n}{1-a_n}\right)\right| \geq 1 + \frac{a_k}{1-a_k} + \dots + \frac{a_n}{1-a_n} =$$
$$= 1 + (r+1)\cdot\left(\frac{1}{k-r-1} + \dots + \frac{1}{n-r-1}\right) \geq$$
$$\geq (r+1)\cdot\left(\frac{1}{k^+} + \dots + \frac{1}{n^+}\right),$$

where m^+ denotes the smallest natural number greater than $m-r-1$. This shows

$$\lim_{n\to\infty}\left|\frac{1}{b_n}\right| = \infty \qquad \text{and hence} \qquad \lim_{n\to\infty} b_n = 0.$$

□

The following proposition captures the behaviour of binomial series for $x = -1$.

Proposition 19.13. (i) For $r < 0$ the series $\sum (-1)^n \cdot \binom{r}{n}$ diverges.

(ii) For $r > 0$,

$$(1-1)^r = \sum_{n=0}^{\infty} (-1)^n \cdot \binom{r}{n} = 0.$$

Proof. By Proposition 15.5,

$$\sum_{n=0}^{m}(-1)^n \cdot \binom{r}{n} = (-1)^m \cdot \binom{r-1}{m}.$$

For $r < 0$ we have $r - 1 < -1$ so by the previous proposition 19.12(i) it is not the case that

$$\lim_{m\to\infty} \binom{r-1}{m} = 0,$$

and hence the series diverges.

For $0 < r$ we have $-1 < r - 1$ so by the previous proposition 19.12(iii),

$$\lim_{m\to\infty} \binom{r-1}{m} = 0, \qquad \text{and hence also} \quad \sum_{n=0}^{\infty}(-1)^n \cdot \binom{r}{n} = 0.$$

□

The following proposition captures the behaviour of binomial series for $x = 1$.

Proposition 19.14. For $-1 < r$,

$$2^r = (1+1)^r = \sum_{n=0}^{\infty} \binom{r}{n}.$$

Proof. For Lagrange's remainder of the series $(1+x)^r$ we have

$$Z_{n+1}(x) = \frac{x^{n+1}}{(n+1)!} \cdot f^{(n+1)}(y),$$

where $0 < y < x = 1$, and hence

$$Z_{n+1}(1) = \binom{r}{n+1} \cdot (1+y)^{r-n-1}.$$

Since

$$\lim_{n\to\infty} (1+y)^{r-n} = 0, \qquad \text{it follows that} \quad \lim_{n\to\infty} Z_{n+1}(x) = 0.$$

□

The following proposition captures the behaviour of the binomial series $\sum \binom{r}{n}$ (for $x = 1$) in more detail.

Proposition 19.15. (i) For $0 < r$ the series $\sum \binom{r}{n}$ is absolutely convergent.

(ii) For $-1 < r < 0$ the series $\sum \binom{r}{n}$ is non-absolutely convergent.

(iii) For $r \le -1$ the series $\sum \binom{r}{n}$ is divergent.

Proof. (i) Let $a_n = \left|\binom{r}{n}\right|$. Then

$$\frac{a_n}{a_{n+1}} = \frac{n+1}{|r-n|},$$

and hence for $n > r$ we have

$$\frac{a_n}{a_{n+1}} = \frac{n+1}{n-r} = 1 + \frac{r+1}{n-r}, \qquad \text{so} \quad \frac{a_n}{a_{n+1}} - 1 = \frac{r+1}{n-r}.$$

Hence

$$n \cdot \left(\frac{a_n}{a_{n+1}} - 1\right) = (r+1) \cdot \frac{n}{n-r} > r+1.$$

By Raabe's criterion it follows that for $m > r$, $\sum\limits_{n=m}^{\infty} a_n < \infty$, and hence also $\sum\limits_{n=1}^{\infty} a_n < \infty$.

(ii) For even n we have $\binom{r}{n} > 0$, for odd n we have $\binom{r}{n} < 0$. Hence

$$\sum_{n=0}^{\infty} \left|\binom{r}{n}\right| = \sum_{n=0}^{\infty} (-1)^n \cdot \binom{r}{n}.$$

Since $r < 0$, the series $\sum \binom{r}{n}$ diverges by Proposition 19.13(i).

(iii) By Proposition 19.12(i), (ii) for $r \le -1$ it is not the case that

$$\lim_{n \to \infty} \binom{r}{n} = 0,$$

and hence our series $\sum \binom{r}{n}$ diverges. □

19.8 Series Expansion of the Function $\arctan x$ for $|x| \le 1$

Proposition 19.16. The series

$$\sum (-1)^{n+1} \cdot \frac{x^{2n-1}}{2n-1} \qquad \left(\text{that is,} \quad \frac{x}{1} - \frac{x^3}{3} + \frac{x^5}{5} - \cdots\right)$$

is

(i) absolutely convergent for $|x| < 1$,

(ii) non-absolutely convergent for $|x| = 1$,

(iii) divergent for $|x| > 1$.

Proof. (ii) For $x = \pm 1$ this series converges by the first Abel's theorem. However, if the series

$$1 + \frac{1}{3} + \frac{1}{5} + \cdots$$

converged then by Proposition 18.15 also the series

$$\frac{1}{2} + \frac{1}{4} + \frac{1}{6} + \cdots,$$

would converge and hence so would the series

$$1 + \frac{1}{2} + \frac{1}{3} + \frac{1}{4} + \cdots,$$

contradiction.

(i), (iii) Let $x \neq \pm 1$. Let

$$a_n = \frac{x^{2n-1}}{2n-1}.$$

Then

$$\frac{a_{n+1}}{a_n} = x^2 \cdot \frac{2n-1}{2n+1} < x^2.$$

It follows by d'Alembert's criterion that for $|x| < 1$ the given series is absolutely convergent. If $|x| > 1$ then

$$\frac{a_{n+1}}{a_n} > 1 \qquad \text{pro} \quad n > \frac{x^2+1}{x^2-1},$$

and hence it is not the case that $\lim_{n\to\infty} a_n = 0$. Consequently the given series diverges. □

Proposition 19.17. For each $|x| \leq 1$

$$\arctan x = \frac{x}{1} - \frac{x^3}{3} + \frac{x^5}{5} - \cdots = \sum_{n=1}^{\infty} (-1)^{n+1} \cdot \frac{x^{2n-1}}{2n-1}.$$

Proof. Let

$$f_n(x) = \sum_{k=1}^{n} (-1)^{k+1} \cdot \frac{x^{2k-1}}{2k-1}.$$

Furthermore, let

$$g_n(x) = \arctan x - f_n(x).$$

It suffices to prove that for each $|x| < 1$ and for each $n \in \mathrm{IN}$

$$g_n(x) \doteq 0.$$

The function $g_n(x)$ has a derivative at each point, namely

$$g_n'(x) = (\arctan x)' - f_n'(x) = \frac{1}{1+x^2} - \sum_{k=1}^{n} (-1)^{k+1} \cdot x^{2k-2}.$$

Since
$$\frac{1}{1+x^2} = 1 - x^2 + x^4 - x^6 + x^8 - \cdots,$$
we have
$$g_n'(x) = \sum_{k=n}^{\infty}(-1)^k \cdot x^{2k} = (-1)^n \cdot \frac{x^{2n}}{1+x^2},$$
as the series is geometric, with the initial term $(-1)^{n+1} \cdot x^{2n}$ and quotient $-x^2$.

Hence for each $|x| < 1$,
$$|g_n'(x)| < x^{2n} < 1.$$
Since $\arctan 0 = 0$ and $f_n(0) = 0$, we have also $g_n(0) = 0$. By the Lagrange Mean Value theorem,
$$|g_n(x)| = |g_n(x) - g_n(0)| = |x| \cdot |g_n'(y_n)|, \qquad \text{where} \quad |y_n| < |x|.$$
Hence if $|x| < 1$ then
$$|g_n(x)| \leq |x| \cdot |y_n|^{2n} < |x| \cdot |x|^{2n}.$$
If moreover $n \in \mathrm{IN}$ then $g_n(x) \doteq 0$.

Thus it remains to consider the case $x = 1$ (the case $x = -1$ is similar). Let u be a real number such that $0 < 1 - u < 1$. Then
$$g_n(1) = g_n(1) - g_n(1-u) + g_n(1-u) - g_n(0) = u \cdot g_n'(y_n) + (1-u) \cdot g_n'(z_n),$$
where
$$0 < z_n < 1 - u < y_n < 1.$$
For $n \in \mathrm{IN}$ we have by the previous case $g_n'(z_n) \doteq 0$, so
$$(1-u) \cdot g_n'(z_n) = \alpha.$$
Hence
$$g_n(1) = u \cdot g_n'(y_n) + \alpha.$$
Since $g_n'(y_n) < 1$ (because $|g_n'(x)| < 1$ for each $|x| < 1$),
$$g_n(1) < u + \alpha.$$
This holds for each u such that $0 < u < 1$. Hence $g_n(1) \doteq 0$. □

Proposition 19.18. (Leibniz Series for $\frac{\pi}{4}$)
$$\frac{\pi}{4} = 1 - \frac{1}{3} + \frac{1}{5} - \frac{1}{7} + \frac{1}{9} - \cdots = \sum_{n=1}^{\infty}(-1)^{n+1} \cdot \frac{1}{2n-1}.$$

Proof. By Proposition 17.46,
$$\tan\left(\frac{\pi}{4}\right) = 1 \qquad \text{and hence} \quad \arctan(1) = \frac{\pi}{4}.$$
The rest follows by Proposition 19.17. □

19.9 Uniform Convergence

In this section $\{f_n\}$ denotes a sequence of real functions defined on an interval I and c denotes an inner point of the interval I. The expression

$$\sum f_n$$

denotes the **series of functions**

$$f_1 + f_2 + f_3 + \cdots .$$

The expression

$$\sum f_n(x) \qquad \text{for} \quad x \in I$$

denotes the series of numbers

$$f_1(x) + f_2(x) + f_3(x) + \cdots .$$

We say that the **sequence** $\{f_n\}$ is **uniformly convergent** at a point c if for any α and any $m, n \in \mathrm{IN}$

$$f_m(c+\alpha) \doteq f_n(c+\alpha).$$

We say that the **series** $\sum f_n$ is **uniformly convergent** at the point c, if for any α and any $m, n \in \mathrm{IN}$, $m < n$,

$$\sum_{k=m}^{n} f_k(c+\alpha) \doteq 0.$$

Proposition 19.19. Let

$$s_n = \sum_{k=1}^{n} f_k .$$

Then the series $\sum f_n$ is uniformly convergent at a point c if and only if the sequence of functions $\{s_n\}$ is uniformly convergent at the point c.

Hence propositions about the uniform convergence of sequences directly imply similar propositions about the uniform convergence of series, which thus need no further proofs.

Proposition 19.20. Let $\{f_n\}$ be a sequence of functions continuous at a point c which is uniformly convergent at the point c. Then also for any $n \in \mathrm{IN}$ and any α

$$f_n(c) \doteq f_n(c+\alpha).$$

Proof. Let α be fixed. Then for any $n \in \mathrm{FN}$,

$$|f_n(c) - f_n(c+\alpha)| < \frac{1}{n}.$$

There exists $m \in \mathrm{IN}$ such that

$$|f_m(c) - f_m(c+\alpha)| < \frac{1}{m}, \qquad \text{hence} \quad f_m(c) \doteq f_m(c+\alpha).$$

However, for any $n \in \mathrm{IN}$ we also have

$$f_n(c) \doteq f_m(c) \doteq f_m(c+\alpha) \doteq f_n(c+\alpha).$$

□

Proposition 19.21. Let $\{f_n\}$ be a sequence of functions continuous at a point c.

(i) Assume that the sequence $\{f_n\}$ is uniformly convergent at the point c. Then the sequence $\{f_n(c)\}$ is convergent and

$$\lim_{n\to\infty} f_n(c) = \operatorname{Proj}\left(f_m(c+\alpha)\right)$$

for any α and for any $m \in \mathrm{IN}$.

(ii) Assume that the series $\sum f_n$ is uniformly convergent at the point c. Then the series $\sum f_n(c)$ is convergent and

$$\sum_{n=1}^{\infty} f_n(c) = \operatorname{Proj}\left(\sum_{n=1}^{m} f_n(c+\alpha)\right)$$

for any α and for any $m \in \mathrm{IN}$.

Proof. Let β be fixed and let

$$d = \operatorname{Proj}\left(f_n(c+\beta)\right) \qquad \text{for any} \quad n \in \mathrm{IN}.$$

By Proposition 19.20 for any $n \in \mathrm{IN}$

$$f_n(c+\beta) \doteq f_n(c), \qquad \text{and so} \quad d = \lim_{n\to\infty} f_n(c),$$

since $d \doteq f_n(c)$ for $n \in \mathrm{IN}$. For any α and any $n \in \mathrm{IN}$, $f_n(c+\alpha) \doteq f_n(c)$, hence $f_n(c+\alpha) \doteq d$ and consequently

$$d = \operatorname{Proj}\left(f_n(c+\alpha)\right).$$

□

Proposition 19.22. Let $\{f_n\}$ be a sequence of functions continuous at a point c.

(i) Assume that the sequence $\{f_n\}$ is uniformly convergent at the point c. Assume that for any $x \in I$, the sequence $\{f_n(x)\}$ is convergent,

$$f(x) = \lim_{n\to\infty} f_n(x).$$

Then the function f is continuous at the point c.

(ii) Assume that the series $\sum f_n$ is uniformly convergent at the point c. Assume that for any $x \in I$ the series $\sum f_n(x)$ is convergent,

$$f(x) = \sum_{n=1}^{\infty} f_n(x).$$

Then the function f is continuous at the point c.

Proof. Assume that β is such that it is not the case that $f(c) \doteq f(c+\beta)$, and let for example

$$f(c) < f(c+\beta).$$

Then there exist real numbers p, r such that

$$f(c) < p < r < f(c+\beta).$$

By Proposition 19.21 for any $n \in \mathrm{IN}$ and any α,

$$f(c) \doteq f_n(c+\alpha) \qquad \text{and hence} \quad f_n(c+\alpha) < r.$$

Let k be the smallest natural number such that for any $n > k$, $|x| < \frac{1}{k}$ we have $c + x \in I$, $f_n(c+x) < r$. Clearly $k \in \mathrm{FN}$ since if $m \in \mathrm{IN}$, $|x| < \frac{1}{m}$ then $f_m(c+x) < r$. That means that for any $|x| < \frac{1}{k}$, x belonging to the world $\mathcal{A}$ we have

$$f(c+x) \leq r.$$

By the first principle of expansion this property holds also in the world $\mathcal{C}$ and hence

$$f(c+\beta) \leq r,$$

contradiction. □

Proposition 19.23. (Interchanging limits, interchanging limit and sum) Let a sequence $\{f_n\}$ (or a series $\sum f_n$ respectively) be uniformly convergent at a point c. Assume that for any $x \in I$, the sequence $\{f_n(x)\}$ (or the series $\sum f_n(x)$ respectively) converges. Assume that for any $n \in \mathrm{FN}$ there exists $\lim_{x \to c} f_n(x)$. Then

$$\lim_{n\to\infty} \lim_{x\to c} f_n(x) = \lim_{x\to c} \lim_{n\to\infty} f_n(x)$$

$$\left(\text{or } \sum_{n=1}^{\infty} \lim_{x\to c} f_n(x) = \lim_{x\to c} \sum_{n=1}^{\infty} f_n(x) \quad \text{respectively}\right).$$

Proof. We modify the functions f_n so that at the point c we set

$$f_n(c) = \lim_{x\to c} f_n(x),$$

and we apply the previous Proposition 19.22. □

Proposition 19.24. (Interchanging limit and differentiation, interchanging sum and differentiation) Let $\{f_n\}$ be a sequence such that for any $x \in I$ and any n, the derivatives $f_n'(x)$ are defined. Assume that for any such x the sequences $\{f_n(x)\}$, $\{f_n'(x)\}$ (or the series $\sum f_n(x)$, $\sum f_n'(x)$ respectively) converge,

$$f(x) = \lim_{n\to\infty} f_n(x) \qquad (\text{or} \quad f(x) = \sum_{n=1}^{\infty} f_n(x) \ ),$$

$$g(x) = \lim_{n\to\infty} f_n'(x).$$

Assume that the sequence $\{f_n'\}$ (or the series $\sum f_n'$ respectively) is uniformly convergent at the point c. Then

$$f'(c) = g(c) \qquad (\text{or} \quad f'(c) = \sum_{n=1}^{\infty} f_n'(c) \ ).$$

Moreover the sequence $\{f_n\}$ (or the series $\sum f_n$ respectively) is uniformly convergent at the point c.

Proof. Let $\overline{I}$ be the interval of all real numbers x such that $x+c \in I$. For $x \in \overline{I}$ we set

$$h_n(x) = \frac{f_n(c+x) - f_n(c)}{x}, \qquad h(x) = \frac{f(c+x) - f(c)}{x}.$$

Clearly

$$\lim_{n\to\infty} h_n(x) = h(x), \qquad \lim_{x\to 0} h_n(x) = f_n'(c), \qquad \lim_{x\to 0} h(x) = f'(c).$$

We shall show that the sequence $\{h_n(x)\}$ is uniformly convergent at the point 0. We have

$$h_n(x) - h_m(x) = \frac{1}{x} \cdot \left((f_n(c+x) - f_m(c+x)) - (f_n(c) - f_m(c))\right).$$

By Lagrange's mean value theorem there exists y such that $0 < y < x$ (or $x < y < 0$) and

$$h_n(x) - h_m(x) = \pm(f_n'(c+y) - f_m'(c+y)) \doteq 0.$$

It follows that for $m, n \in \mathrm{IN}$ and $x \doteq 0$ we have $h_n(x) \doteq h_m(x)$. By the previous proposition 19.23,

$$\lim_{n\to\infty} \lim_{x\to 0} h_n(x) = \lim_{x\to 0} \lim_{n\to\infty} h_n(x),$$

and hence

$$\lim_{n\to\infty} f_n'(c) = \lim_{x\to 0} h(x) = f'(c).$$

Since for each $m, n \in \mathrm{IN}$

$$f_n(c+\alpha) - f_m(c+\alpha) = \alpha \cdot (h_n(\alpha) - h_m(\alpha)) + f_n(c) - f_m(c) \doteq 0,$$

the sequence $\{f_n\}$ is uniformly convergent at the point c. $\square$

Let $\{g_n\}$ be another sequence of functions defined on I and let $\{a_n\}$ be a sequence of real numbers. ($\{g_n\}$ and $\{a_n\}$ belong to the world $\mathcal{A}$.) We say that the series $\sum g_n$ (or $\sum a_n$ respectively) **majorises** $\sum f_n$ on I if for any $x \in I$ and any n,

$$|f_n(x)| \leq g_n(x) \qquad (\text{or} \quad |f_n(x)| \leq a_n \text{ respectively }).$$

Proposition 19.25. Assume that the series $\sum g_n$ majorises the series $\sum f_n$ on I and that the series $\sum g_n$ is uniformly convergent at the point c. Then also the series $\sum f_n$ is uniformly convergent at the point c.

Proof. Let $m < n$. Then

$$\left|\sum_{k=m}^{n} f_k(c+\alpha)\right| \leq \sum_{k=m}^{n} |f_k(c+\alpha)| \leq \sum_{k=m}^{n} g_k(c+\alpha).$$

For $m, n \in \mathrm{IN}$ the last member of this inequality is an infinitely small number, and consequently so is the first. It follows that the series $\sum f_n$ is convergent. $\square$

Proposition 19.26. Assume that the series $\sum a_n$ converges and majorises the series $\sum f_n$ on I. Then the series $\sum f_n$ is uniformly convergent at each inner point of the interval I belonging to the world $\mathcal{A}$.

Proof. Let $m < n$. Then

$$\left|\sum_{k=m}^{n} f_k(c+\alpha)\right| \leq \sum_{k=m}^{n} |f_k(c+\alpha)| \leq \sum_{k=m}^{n} a_k.$$

If c is an inner point of interval I, $m, n \in \mathrm{IN}$, $m, n \in \mathrm{IN}$ then again the last member of this inequality is an infinitely small number, and consequently so is the first. It follows that the series $\sum f_n$ is uniformly convergent at the point c. $\square$

Note. (i) All the above assertions can be stated and proved in the obvious way also for the limit, continuity and derivative from the right or left. In these cases c can also be the left or right end point of the interval I respectively.

(ii) the uniform convergence from left or right at a point c can be defined in the obvious way and the corresponding propositions stated and proved accordingly.

(iii) The uniform convergence of a sequence $\{f_n\}$ (or of a series $\sum f_n$ respectively) is independent of changes to finitely many terms of the sequence $\{f_n\}$ (or of the series $\sum f_n$ respectively).

Appendix to Part III – Translation Rules

The subject of our study in this part was the geometric world $\mathcal{A}$. We were interested in sequences and series of real numbers and functions belonging to the world $\mathcal{A}$, their relations and properties.

However, we studied them not only within the world $\mathcal{A}$ alone but also – and in fact above all – using the world $\mathcal{C}$, into which the world $\mathcal{A}$ expanded. For we have formulated various relations between objects of the world $\mathcal{A}$, properties of these objects and assertions about them using notions that capture phenomena showing only in this expansion. We kept stepping out of the world $\mathcal{A}$ to the world $\mathcal{C}$ also during proofs.

Still, many such notions and assertions can be formulated merely within the world $\mathcal{A}$. In other words, they can be translated into the $\epsilon\delta$-analysis. Assertions thus translated hold not only in the world $\mathcal{A}$, but by the first principle of expansion also in the world $\mathcal{C}$. For these notions and assertions are expressed so simply that their translation to $\epsilon\delta$-analysis can be effected using one of the two translation rules given below. Although these translations can be carried out mechanically, it does not mean that the proofs of these assertions can be translated just as mechanically. If we wished to remain within the geometric world $\mathcal{A}$ even during these proofs, we would need to proceed in the way usually adopted by the textbooks of $\epsilon\delta$-analysis.

In order to conform to the terminology of the contemporary $\epsilon\delta$-analysis, we shall use letters ϵ and δ in this appendix as variables for positive real numbers belonging to the world $\mathcal{A}$. On the other hand, in accordance with the terminology that we have adopted hitherto α and β denote variables for infinitely small non-zero real numbers belonging to the world $\mathcal{C}$.

We shall state two basic rules for eliminating variables α, β using variables ϵ, δ.

Let $\phi(x, K)$, or $\phi(x, y, K)$ respectively, be formulae describing some properties or relations within the world $\mathcal{A}$ alone and hence also within the world $\mathcal{C}$ alone. The letters x, y are variables for real numbers belonging to the world $\mathcal{A}$ and hence also to the world $\mathcal{C}$. To keep our formulae simple, the variables x, y only run through non-zero numbers.The letter K stands for some finite collection of constants denoting objects belonging to the world $\mathcal{A}$ and consequently also to the world $\mathcal{C}$.

In the following two rules, the formula on the right concerns only the world $\mathcal{A}$ (or, if preferred, the world $\mathcal{C}$), whilst the formula on the left is a formula of the infinitesimal calculus, operating simultaneously in both the geometric worlds $\mathcal{A}$ and $\mathcal{C}$ since it involves quantifying infinitely small numbers.

First translation rule

(a) $(\exists\beta)\phi(\beta,K) \equiv (\forall\epsilon)(\exists|y|<\epsilon)\,\phi(y,K).$

(b) $(\forall\alpha)\phi(\alpha,K) \equiv (\exists\delta)(\forall|x|<\delta)\,\phi(x,K).$

Proof. It suffices to prove the implication from right to left in both (a) and (b) since each is equivalent to the implication from left to right in the other. Assume in both cases that the right hand side of the equivalence holds.

(a) Then the right hand side of the equivalence holds also in the world $\mathcal{C}$, and hence even for an infinitely small ϵ there exists $|y|<\epsilon$ such that

$$\phi(y,K).$$

This y is also infinitely small. Hence

$$(\exists\beta)\phi(\beta,K).$$

(b) Let m be the smallest natural number such that for any $|x|<\frac{1}{m}$ we have $\phi(x,K)$. Such a number exists since the right hand side of the equivalence (b) holds. By the rule of defined objects, $m \in \mathrm{FN}$ (that is, m belongs to the world $\mathcal{A}$ even after the expansion of $\mathcal{A}$ to $\mathcal{C}$). The property

$$(\forall|x|<\frac{1}{m})\,\phi(x,K)$$

of objects K thus holds even in the world $\mathcal{C}$, and since for any $|x|<\frac{1}{m}$ we have $\phi(x,K)$, also

$$(\forall\alpha)\,\phi(\alpha,K).$$

□

Second translation rule

(a) $(\exists\beta)(\forall\alpha)\phi(\alpha,\beta,K) \equiv (\exists\delta)(\forall\epsilon)(\exists|y|<\epsilon)(\forall|x|<\delta)\,\phi(x,y,K).$

(b) $(\forall\alpha)(\exists\beta)\phi(\alpha,\beta,K) \equiv (\forall\epsilon)(\exists\delta)(\forall|x|<\delta)(\exists|y|<\epsilon)\,\phi(x,y,K).$

Proof. Since (a) and (b) are equivalent, it suffices to prove (a).

First assume that the right hand side of the equivalence (a) holds. Let n be the smallest natural number such that

$$(\forall\epsilon)(\exists|y|<\epsilon)\left(\forall|x|<\frac{1}{n}\right)\phi(x,y,K).$$

Such n exists and by the rule of defined objects we have $n \in \mathrm{FN}$; hence n remains fixed when $\mathcal{A}$ expands to $\mathcal{C}$. By the first translation rule (case (a)),

$$(\exists\beta)(\forall|x| < \frac{1}{n})\phi(x, \beta, K).$$

Since for each α we have $|\alpha| < \frac{1}{n}$,

$$(\exists\beta)(\forall\alpha)\,\phi(\alpha, \beta, K).$$

Now assume that the left hand side of the equivalence (a) holds. Let $\overline{\beta}$ be such that

$$(\forall\alpha)\phi(\alpha, \overline{\beta}, K).$$

Let

$$m = \min\{n; (\forall|x| < \frac{1}{n})\,\phi(x, \overline{\beta}, K)\}.$$

Since for each $n \in \mathrm{IN}$, n belongs to the set with minimum m, we have $m \in \mathrm{FN}$ and hence m remains fixed when $\mathcal{A}$ expands to $\mathcal{C}$.

Let $\overline{\delta} = \frac{1}{m}$. Then $\overline{\delta}$ belongs to the world $\mathcal{A}$ and

$$(\forall|x| < \overline{\delta})\,\phi(x, \overline{\beta}, K).$$

Hence

$$(\exists\beta)(\forall|x| < \overline{\delta})\,\phi(x, \beta, K).$$

By the first translation rule (case (a)) this is equivalent to the formula

$$(\forall\epsilon)(\exists|y| < \epsilon)(\forall|x| < \overline{\delta})\,\phi(x, y, K).$$

This property is expressed merely by the means of the world $\mathcal{A}$, and hence in this world

$$(\exists\delta)(\forall\epsilon)(\exists|y| < \epsilon)(\forall|x| < \delta)\,\phi(x, y, K).$$

□

Most of the fundamental concepts of the $\epsilon\delta$-analysis arose when eliminating variables for infinitely small real numbers according to the case (b) of the second translation rule. One such important concept is the concept of a *limit*, either of a sequence or of a function, at a given point.

We shall illustrate this on the example of

$$\lim_{x \to c} f(x) = b,$$

where f is a function defined on some interval I, c is an inner point of I and f, c, b belong to the world $\mathcal{A}$.

The definition of $\lim\limits_{x \to c} f(x) = b$ has the following form in infinitesimal calculus:

$$(\forall\alpha)(\exists\beta)(|f(c + \alpha) - b| \leq \beta).$$

By the second translation rule this is equivalent to the formula

$$(\forall\epsilon)(\exists\delta)(\forall|x| < \delta)(\exists|y| < \epsilon)(|f(c+x) - b| \leq y),$$

that is,

$$(\forall\epsilon)(\exists\delta)(\forall|x| < \delta)(|f(c+x) - b| < \epsilon).$$

If $\phi(x_1, \ldots, x_n, y_1, \ldots, y_n, K)$ is a formula describing some relation in the world $\mathcal{A}$ of a similar nature as $\phi(x, y, K)$ above, then the formulae

$$(\forall\alpha_1, \ldots, \alpha_n)(\exists\beta_1, \ldots, \beta_n)\phi(\alpha_1, \ldots, \alpha_n, \beta_1, \ldots, \beta_n, K),$$
$$(\exists\alpha_1, \ldots, \alpha_n)(\forall\beta_1, \ldots, \beta_n)\,\phi(\alpha_1, \ldots, \alpha_n, \beta_1, \ldots, \beta_n, K)$$

can also be translated to the $\epsilon\delta$-analysis in a manner similar to the second translation rule.

Since infinitely large numbers are inverse values of infinitely small numbers, it is clear that similar translation rules can be found also for elimination of variables running through infinitely large numbers.

The material for this first part on infinitesimal calculus was chosen so that all was translatable to the $\epsilon\delta$-analysis using the above rules. However, we would be mistaken if we thought that formulae with three alternating quantifiers (that is, $\exists\forall\exists$ or $\forall\exists\forall$) over infinitely small numbers can also be translated to $\epsilon\delta$-analysis in the way suggested by both the rules above. These concepts – when they concern merely the world $\mathcal{A}$ albeit utilizing infinitely small numbers – are translatable to $\epsilon\delta$-analysis but only by the rather contorted, unintuitive and artificial means of set theory.

Part IV

Making Real Numbers Discrete

Introduction

A geometric world contains not only geometric objects but also all that they show us. Along with line segments it contains also their lengths, which makes it possible to include in this world also real numbers and hence also various relations between them; in particular, real functions, sequences of numbers or functions, and so on. In fact, mathematical analysis has been considered by some mathematicians in the past to be a sort of an extension of analytic geometry, and such conception of it suits also our aims. Briefly put, we will include in the geometric world all that can be modelled in it in some plausible way; all about which we can gain some intuition in the geometric world. We are doing this not just because we can but because calculations in infinitesimal calculus take place mainly in the domain of numbers and the choice and justification of some of the rules for the calculations are often guided by geometric intuition. For this reason it is advantageous to include both these facets of infinitesimal calculus in one subject of study. Besides, so to speak, the more we manage to squeeze in the geometric world, the more comprehensive range of applications of infinitesimal calculus we will gain.

A suitable and enormously powerful tool for investigating the ancient (Euclidean) geometric world are the real numbers. The class Real of all these numbers does lie as far as the geometric horizon but classes Real and Real^n (where $n \in \text{FN}$) of all ordered n-tuples of real numbers perfectly capture the structure of the ancient geometric world nevertheless.

We have two Euclidean geometric worlds, the ancient ($\mathcal{A}$) and the classical ($\mathcal{C}$). As long as we investigate just one of them, it does not signify which one it is. Results obtained in one of them apply also in the other. But we can also study them simultaneously, namely in the relation in which they are placed. That is, in the relation of expansion of $\mathcal{A}$ to $\mathcal{C}$, since the classical geometric world arose by expanding the intuitive ancient geometric world to classical infinity.

If X is a class belonging to the world $\mathcal{A}$ then $\text{Ex}(X)$ denotes the class that is the expansion of the class X. By the second principle of expansion[1], $X \subseteq \text{Ex}(X)$. By the first principle of expansion, Ex(Real) (or $\text{Ex}(\text{Real}^n)$), where $n \in \text{FN}$) is the class of all real numbers (or all ordered n-tuples of real numbers respectively) of the world $\mathcal{C}$.

In the previous part, Infinitesimal Calculus Reaffirmed, the main subject of study was the world $\mathcal{A}$, while the world $\mathcal{C}$ played an important but supporting role. Now our main subject of study will be the duality of the worlds $\mathcal{A}$, $\mathcal{C}$.

[1]See the start of Part III.

Chapter 20

Expansion of the Class Real of Real Numbers

20.1 Subsets of the Class Real

If w is a set, $\emptyset \neq w \subseteq \text{Real}$, then $\min w$ (or $\max w$ respectively) denotes the smallest (or the largest) number belonging to the set w.

Furthermore, $l(w)$ (or $r(w)$ respectively) denotes the set of all $x \in w$ such that

$$\min w \leq x \leq s(w) \qquad (\text{or} \quad s(w) \leq x \leq \max w \ ),$$

where

$$s(w) = \frac{1}{2}(\min w + \max w).$$

Proposition 20.1. Let a set w be a subclass of the class Real. Then the set w is finite.

Proof. Assume that the set w is infinite. We form a sequence of sets $\{w_n\}_{n \in \text{FN}}$ as follows. We set $w_0 = w$. If $l(w_n)$ is infinite (or finite respectively) then we set

$$w_{n+1} = l(w_n) \qquad (\text{or } w_{n+1} = r(w_n) \ ).$$

Clearly for each $n \in \text{FN}$ the set w_n is infinite, $w_{n+1} \subseteq w_n$ and from a certain n up we have:

$$\max(w_n) - \min(w_n) \leq \frac{1}{n}.$$

There exists a number y such that for each n, $y \in w_n$.[1] Let

$$w'_n = w_n - \{y\}.$$

For similar reasons as above, there exists z such that for each n

$$z \in w'_n \subseteq w_n \subseteq w.$$

[1] Axiom of Prolongation, Part II, sections 6.2, 6.4.

Clearly $y \neq z$, $y \doteq z$, which is a contradiction since by Proposition 14.14 no two distinct numbers belonging to Real are infinitely close.. □

The above propositions highlights the continuity of line segments (and consequently also real numbers) in the ancient Greek geometric world. For in there, a line segment can be divided into at most finitely many line segments. In Book VI of Physics, Aristotle[2] wrote

> ... nothing that is continuous can be composed of indivisibles: e.g. a line cannot be composed of points, the line being continuous and the point indivisible.

20.2 Third Principle of Expansion

In the classical geometric world alone the semiset FN does not exist. We can only bring it in from the ancient Greek geometric world, namely when we investigate both the worlds $\mathcal{A}$, $\mathcal{C}$ in their duality . Upon doing this, the semiset FN immediately extends to the class N. These considerations motivate the following principle of expansion:

Third principle of expansion. Let H be a class in the world $\mathcal{A}$. Then the class Ex(H) is revealed. [3]

In particular, the classes Ex(Real), Ex(Real^n) are revealed.

By an **$\mathcal{A}$-sequence** we understand a stable sequence $\{a_n\}_{n \in \mathrm{FN}}$ of objects belonging to the class Ex(H), where H is some class from the world $\mathcal{A}$.

Proposition 20.2. For every $\mathcal{A}$-sequence $\{a_n\}_{n \in \mathrm{FN}}$ there exists in the world $\mathcal{C}$ an extension of it, $\{a_n\}_{n \leq m}$, which is a set.

We remark that if $\{a_n\}_{n \leq m}$, $\{a'_n\}_{n \leq m'}$ are extensions of an $\mathcal{A}$-sequence $\{a_n\}_{n \in \mathrm{FN}}$ then the smallest natural number p such that $p \leq m, m'$ and for each $n \leq p$, $a_n = a_{n'}$ is infinitely large, since for each $n \in \mathrm{FN}$, $a'_n = a_n$.

In the remaining part of this section, all $\mathcal{A}$-sequences will mean $\mathcal{A}$-sequences of real numbers belonging to $\mathcal{C}$.

The following proposition is obvious.

Proposition 20.3. Let $\{A_n\}_{n \leq m}$ be an extension of an $\mathcal{A}$-sequence $\{a_n\}_{n \in \mathrm{FN}}$. Let p be the largest number $p \leq m$ such that for each $n \leq p$, A_n is a number. Then $p \in \mathrm{IN}$ and for each $n \in \mathrm{FN}$, $A_n = a_n$.

On the basis of this argument, we can assume that any $\mathcal{A}$-sequence $\{a_n\}_{n \in \mathrm{FN}}$ has an extension $\{a_n\}_{n \leq m}$, such that all its members are numbers.

If all members of an $\mathcal{A}$-sequence $\{a_n\}_{n \in \mathrm{FN}}$ are natural numbers then for similar reasons we may assume about its extension $\{a_n\}_{n \leq m}$ that all a_n with $n \leq m$ are natural numbers.

[2] See Aristotle, *Phys.* 231a24–26, trans. R. P. Hardie and R. K. Gaye.

[3] See Part II – New Theory of Sets and Semisets, section 6.4.

Proposition 20.4 (First basic assertion about $\mathcal{A}$-sequences of numbers)**.** Let $\{a_n\}_{n\in\mathrm{FN}}$ be an $\mathcal{A}$-sequence such that for each $n \in \mathrm{FN}$, $a_n \doteq 0$. Then there exists an extension $\{a_n\}_{n\leq p}$ where p is a natural number, of this sequence, such that for each $n \leq p$, $a_n \doteq 0$ (clearly $p \in \mathrm{IN}$).

Proof. Let $\{a_n\}_{n\leq m}$ be an extension of the $\mathcal{A}$-sequence $\{a_n\}_{n\in\mathrm{FN}}$. If $|a_n| \leq \frac{1}{n}$ for each $n \leq m$ then set $p = m$. If it is not the case then let p be the smallest natural number such that $\frac{1}{p+1} < |a_{p+1}|$. Clearly $p \in \mathrm{IN}$ since for each $n \in \mathrm{FN}$, $|a_n| \leq \frac{1}{n}$. □

Proposition 20.5 (Second basic assertion about $\mathcal{A}$-sequences)**.** Let $\{a_n\}_{n\in\mathrm{FN}}$ be an $\mathcal{A}$-sequence such that all its members are infinitely large positive numbers. Then there exists an infinitely large natural number q such that for all $n \in \mathrm{FN}$, $q < a_n$.

Proof. For $n \in \mathrm{FN}$ let $b_n = \frac{1}{a_n}$. The $\mathcal{A}$-sequence $\{b_n\}_{n\in\mathrm{FN}}$ clearly satisfies the assumptions of Proposition 20.4. Hence there exists an extension of it such that for each $n \leq p$, $b_n \doteq 0$. Let b be the largest of the numbers $|b_n|$, where $b_n \neq 0$ and $n \leq p$. For $n \in \mathrm{FN}$ we have $\frac{1}{a_n} \leq b$ and hence $\frac{1}{b} \leq a_n$, and $\frac{1}{b}$ is an infinitely large number. Hence it suffices to choose q to be the largest natural number such that $q < \frac{1}{b}$. □

By a fluent extension of an $\mathcal{A}$-sequence $\{a_n\}_{n\in\mathrm{FN}}$ we mean such an extension $\{a_n\}_{n\leq p}$ of it where for each $n \leq p$, $n \in \mathrm{IN}$,

$$a_n \doteq a_p.$$

If an $\mathcal{A}$-sequence $\{a_n\}_{n\in\mathrm{FN}}$ has a fluent extension $\{a_n\}_{n\leq p}$ then we say that numbers a_n, where $n \in \mathrm{IN}$, $n \leq p$, are **limit numbers** of the $\mathcal{A}$-sequence $\{a_n\}_{n\in\mathrm{FN}}$.

Assume that an $\mathcal{A}$-sequence $\{a_n\}_{n\in\mathrm{FN}}$ has a fluent extension. We say that a number A is a **limit** of this sequence, denoted $A = \lim\limits_{n\to\infty} a_n$, if

$$A = \mathrm{Proj}(a_p),$$

where a_p is some (and hence any) limit number of this sequence.

Proposition 20.6. Every $\mathcal{A}$-sequence has at most one limit.

Proof. Let $A = \lim\limits_{n\to\infty} a_n$, $B = \lim\limits_{n\to\infty} a_n$. Any two fluent extensions of $\{a_n\}_{n\in\mathrm{FN}}$ must agree up to some $p \in IN$ so $A = B = Proj(a_p)$. □

We say that a sequence $\{a_n\}_{n\in\mathrm{FN}}$ is **convergent** (or that the sequence **converges**) if it has a limit and this limit is a proper number. Otherwise we say that the sequence $\{a_n\}_{n\in\mathrm{FN}}$ is **divergent**. If its limit is ∞ (or $-\infty$) then we say that the sequence **diverges** to ∞ (or to $-\infty$ respectively). A sequence that has no limit (proper or improper) is called **oscillatory**.

Chapter 21

Infinitesimal Arithmetics

In this entire chapter x, y, z denote real numbers belonging to the world $\mathcal{C}$.

21.1 Orders of Real Numbers

We say that a **number** x is **of a lower order** than a number y, denoted $x \prec y$, if

$$y \neq 0 \qquad \text{a} \qquad \frac{x}{y} \doteq 0.$$

It is easy to see that the following hold.

Proposition 21.1.

(a) If $x \prec y$ and $y \prec z$, then $x \prec z$.

(b) The following are equivalent:

$$x \prec y, \quad x \prec |y|, \quad |x| \prec y, \quad |x| \prec |y|.$$

(c) If $y \neq 0$, $x \doteq 0$ then $x \cdot y \prec y$.

(d) If $0 \leq x \leq y \prec z$ then $x \prec z$.

Proposition 21.2. Let S be a nonempty finite set in the world $\mathcal{C}$. Let f, g be functions defined on the set S, which take real numbers as their values. Assume that for each $s \in S$, $0 < g(s)$ and $f(s) \prec g(s)$. Then

$$\sum_{s \in S} f(s) \prec \sum_{s \in S} g(s).$$

Proof. Let z denote the largest of the numbers $\frac{|f(s)|}{g(s)}$, where $s \in S$. Clearly $z \doteq 0$ and for each $s \in S$

$$|f(s)| \leq z g(s).$$

Moreover

$$\left|\sum_{s\in S} f(s)\right| \le \sum_{s\in S} |f(s)| \le z \sum_{s\in S} g(s) \prec \sum_{s\in S} g(s).$$

By Proposition 21.1(ii)

$$\sum_{s\in S} f(s) \prec \sum_{s\in S} g(s).$$

□

We say that **numbers** x, y are **of the same order**, denoted $x \sim y$, if

$$\text{neither} \quad x \prec y \qquad \text{nor} \quad y \prec x.$$

Furthermore we define $0 \sim 0$.

21.2 Near-Equality

We say that **numbers** x, y are **near-equal**, denoted $x \ddot{=} y$, if

$$\text{either} \quad x = y = 0, \qquad \text{or} \quad x, y \neq 0 \quad \text{and} \quad \frac{x}{y} \doteq 1$$

(and hence also $\frac{y}{x} \doteq 1$).

The following is easily checked.

Proposition 21.3.

(a) If $x \ddot{=} y$ and $y \ddot{=} z$ then $x \ddot{=} z$.

(b) $x \ddot{=} y$ if and only if $-x \ddot{=} -y$.

(c) If $x, y \neq 0$, then $x \ddot{=} y$ if and only if $\frac{1}{x} \ddot{=} \frac{1}{y}$.

(d) If $x \ddot{=} y$ and $u \ddot{=} v$ then

$$x \cdot u \ddot{=} y \cdot v.$$

(e) $x \ddot{=} y$ if and only if there exists $\alpha \doteq 0$ such that

$$y = x + \alpha \cdot x.$$

(f) If $0 \neq x \in \text{Real}$ then

$$x \ddot{=} y \qquad \text{if and only if} \qquad x \doteq y.$$

Proposition 21.4. Let $x, y \neq 0$, $x \ddot{=} u$, $y \ddot{=} v$ and assume that $\frac{x+y}{y}$ is not an infinitely small number (we do not have $x + y \prec y$, so $x + y \neq 0$). Then

$$u + v \ddot{=} x + y.$$

Proof. We have

$$\begin{aligned} u &= x + \alpha \cdot x, \\ v &= y + \beta \cdot y = y + \alpha \cdot y + \gamma \cdot y, \end{aligned}$$

where $\gamma = \beta - \alpha \doteq 0$. Since $\frac{y}{x+y}$ is not infinitely large,

$$\gamma \cdot \frac{y}{x+y} \doteq 0$$

and the result follows by Proposition 21.3 (v). □

Proposition 21.5. Let S be a nonempty finite set in the world $\mathcal{C}$. Let f, g be functions defined on the set S, which assume real numbers as their values. Assume that for each $s \in S$, $0 \leq f(s), g(s)$, $f(s) \ddot{=} g(s)$. Then

$$\sum_{s \in S} f(s) \ddot{=} \sum_{s \in S} g(s).$$

Proof. For each $s \in S$

$$g(s) = f(s) + \alpha(s) \cdot f(s), \qquad \text{where} \quad \alpha(s) \doteq 0.$$

Let α be the largest of the numbers $|\alpha(s)|$, where $s \in S$. Clearly $\alpha \doteq 0$ and

$$\sum_{s \in S} g(s) = \sum_{s \in S} f(s) + \sum_{s \in S} \alpha(s) \cdot f(s) \leq \sum_{s \in S} f(s) + \alpha \cdot \sum_{s \in S} f(s).$$

Similarly

$$\sum_{s \in S} g(s) \geq \sum_{s \in S} f(s) - \alpha \cdot \sum_{s \in S} f(s)$$

and consequently there exists a number $\beta \doteq 0$, such that

$$\sum_{s \in S} g(s) = \sum_{s \in S} f(s) + \beta \cdot \sum_{s \in S} f(s).$$

□

Chapter 22

Discretisation of the Ancient Geometric World

22.1 Grid

A set $M \subset \mathrm{Ex}(\mathrm{Real})$ belonging to the world $\mathcal{C}$ is called a **grid** if the following holds:

(i) The set M is finite from the point of view of $\mathcal{C}$.[1]

(ii) For every proper real number r belonging to $\mathcal{A}$ there exists an $x \in M$ in the world $\mathcal{C}$ such that
$$r = \mathrm{Proj}(x).$$

If M is a grid and $M \subseteq K$ where K is a finite set of real numbers in the world $\mathcal{C}$ then K is clearly also a grid.

Let M be a grid, $x, y \in M$, $x < y$ such that there exists no $z \in M$ for which $x < z < y$ holds. Then we say that points x, y are the **neighbouring points of the grid** M. The point x is the **lower neighbour** of the point y and the point y is the **upper neighbour** of the point x.

Proposition 22.1. Let M be a grid. Then the following holds:

(i) If $x < y$ are neighbouring points of the grid M and x, y are finitely large real numbers then $x \doteq y$.

(ii) If u is the least and v the greatest element of the grid M then
$$\mathrm{Proj}(u) = -\infty, \qquad \mathrm{Proj}(v) = \infty.$$

(iii) The number of elements of the grid M is an infinite natural number.

[1] "Finite from the point of view of $\mathcal{C}$" means that there is a bijective function (a set) mapping the M onto some $[\gamma]$ where γ is a natural number in $\mathcal{C}$. [Ed]

Proof. (i) If we had $\mathrm{Proj}(x) < \mathrm{Proj}(y)$ then there would exist a real number r belonging to $\mathcal{A}$ such that $\mathrm{Proj}(x) < r < \mathrm{Proj}(y)$. Since M is a grid, there would exists $z \in M$ such that $\mathrm{Proj}(z) = r$. Using Proposition 14.18 we can see that $x < z < y$, which is a contradiction.

(ii) Let $\mathrm{Proj}(v) < \infty$. Then there exists an $x \in M$ such that $\mathrm{Proj}(x) = \mathrm{Proj}(v) + 1$. Using Proposition 14.18, $v < x$, which is a contradiction. The case of $-\infty$ is analogous.

(iii) Let k be the number of elements of the grid M. If $k \in \mathrm{FN}$, then there would have to exist one number among numbers $1, 2, 3, \ldots, k, k+1$ that is not a projection of any element of the set M, which is a contradiction. □

In what follows, the letter M (possibly with an index) denotes a grid. Futhermore, $D(M)$ denotes the set of all distances between two neighbouring points of the grid M.

Let $\varepsilon(M)$ denote the smallest and $\delta(M)$ the largest distance between two neighbouring points of the grid M. The number $\frac{\delta(M)}{\varepsilon(M)}$ is a **measure of uniformity of the grid** M.

If $M \subseteq M_1$ then we say that **the grid M_1 is denser than the grid** M.

The following proposition is trivial.

Proposition 22.2. For all grids M, M_1 the following holds:

(i) $\frac{\delta(M)}{\varepsilon(M)} \geq 1$.

(ii) If K is a finite set of real numbers, then $M \cup K$ is a grid.

(iii) If $M \subseteq M_1$, then

$$\delta(M_1) \leq \delta(M), \qquad \varepsilon(M_1) \leq \varepsilon(M).$$

Proposition 22.3. Let M be a grid and $1 < c$. Then it is possible to generate a grid M_c such that

$$M \subseteq M_c \qquad \text{and} \qquad \frac{\delta(M_c)}{\varepsilon(M_c)} < c.$$

Proof. We shall write ε for $\varepsilon(M)$ and δ for $\delta(M)$. Let $0 < \eta$ be small, in particular

$$\eta < \frac{\varepsilon^2(c-1)}{4\delta + \varepsilon} \qquad \text{and} \qquad \eta < \frac{\varepsilon}{2}.$$

Let p_x (for $x \in D(M)$) and q be natural numbers such that for all $x \in D(M)$,

$$\left| \frac{p_x}{q} - x \right| \leq \eta\,.$$

Note that, consequently, for any $x, y \in D(M)$,

$$\frac{p_y}{p_x} = \frac{\frac{p_y}{q}}{\frac{p_x}{q}} \leq \frac{\delta + \eta}{\varepsilon - \eta} \leq 4\frac{\delta}{\varepsilon}$$

and

$$\left|\frac{1}{q}-\frac{x}{p_x}\right| \leq \frac{\eta}{p_x}.$$

A grid M_c is generated by inserting $p_x - 1$ equally distant points between every two neighbouring points of the grid M, the distance of which is x. Obviously $M \subseteq M_c$. This means that we have replaced the line segment whose end points are neighbouring in the grid M, and the length of which is x, by line segments whose lengths are $\frac{x}{p_x}$. Since for any $x, y \in D(M)$

$$\left|\frac{\frac{x}{p_x}}{\frac{y}{p_y}}-1\right| = \left|\frac{\frac{x}{p_x}-\frac{y}{p_y}}{\frac{y}{p_y}}\right| \leq \frac{\frac{\eta}{p_x}+\frac{\eta}{p_y}}{\frac{y}{p_y}} = \frac{\eta\frac{p_y}{p_x}+\eta}{y} \leq \frac{\eta\left(4\frac{\delta}{\varepsilon}+1\right)}{\varepsilon} < c-1,$$

we must have in particular

$$\frac{\delta(M_c)}{\varepsilon(M_c)} - 1 \; < \; c-1$$

so the result follows. $\square$

We say that a **grid** M is **almost uniform**, if $\frac{\delta(M)}{\varepsilon(M)} \doteq 1$.

Proposition 22.4. Every grid M can be made denser to an almost uniform grid.[2]

Proof. Let $c = 1 + \gamma$, where $0 < \gamma \doteq 0$. By Proposition 22.3 the grid M_c is an almost uniform enlargement of the grid M. $\square$

22.2 Fourth Principle of Expansion

We say that a **grid** M is **full** if

$$\mathit{Real} \subseteq M.$$

If M is a full grid and M' a grid such that $M \subseteq M'$ then M' is also a full grid.

Fourth Principle of Expansion. There exists a full grid.

The principle of expansion stated above is different in nature from the previous three principles of expansion. The intuition that has led us to accept them fails in this fourth case. We might even worry that this principle not only fails for our duality of worlds, but that accepting it could lead to a contradiction, which would force us to accept its negation. Our reluctance whether to accept this principle or not is justifiable. To wit, it can be neither proved nor disproved. Our decision as to which of the two options to choose therefore has to be based – so to speak – on external circumstances, and these point to the acceptance of the principle.

[2] In other words, for every grid M there is an almost uniform grid M' such that $M \subseteq M'$.

Scholion

Those who continue to base the consistency of a mathematical theory exclusively on the existence of its model in classical set theory can now follow a brief outline of how this method can help us see the *consistency of the Fourth Principle of Expansion.*

Let R be the set of all real numbers and A the set of all finite subsets of the set R. For $x \in A$ let

$$\overline{x} = \{u; u \in A \ \&\ x \subseteq u\}.$$

Obviously, for $x, y \in A$

$$\overline{x \cup y} \subseteq \overline{x} \cap \overline{y}.$$

Let P denote the set of all $q \subseteq A$ such that there exists $x \in A$ satisfying $\overline{x} \subseteq q$. It is straightforward to check that P is a non-trivial filter on the set A. Let Q be an ultrafilter on the set A such that $P \subseteq Q$. The ultraproduct of the set R over the ultrafilter Q on the set A is a model of the world $\mathcal{C}$, while the set of all constants k_a, where $a \in R$, is a model of the world $\mathcal{A}$.

The function $d(x) = x$ for every $x \in A$ is a finite set in the sense of the thus-created ultraproduct. At the same time, the set

$$\{z; z \in A \ \&\ k_a(z) \in d(z)\} = \overline{\{a\}} \in Q.$$

The function d is therefore a model of the set M whose existence is stated in the Fourth Principle of Expansion.

The Fourth Principle of Expansion (unlike its negation and in contrast to passive indifference) brings a positive incentive into our duality of worlds. Its acceptance clarifies the distribution of numbers from the world $\mathcal{A}$ amongst the numbers in the world $\mathcal{C}$ and makes the distribution more transparent with the help of full grids; thus it enables easier orientation in the relations between the two worlds. To wit, by accepting it we achieve *discretisation* not only of the geometric line but of any continuum in the ancient Greek geometric world since continua in it can be captured by pairs, triples (or n-tuples) of real numbers.

This principle also facilitates infinitesimal considerations concerning backward projections. For that matter, mathematicians who were working on infinitesimal calculus in the eighteenth and nineteenth centuries sometimes implicitly used this principle.

We can accept the Fourth Principle of Expansion because we placed the horizon of the classical geometric world entirely within the known land of the geometric horizon (where it in fact belongs) and not as far as somewhere in absolute infinity. Thanks to this, we can now manipulate this world and position it in relation to the geometric world as we need it.

By Proposition 22.4 any full grid can be enlarged to an almost uniform full grid. The fourth principle of expansion makes Real into a semiset.

22.3 Radius of Monads of a Full Almost-Uniform Grid

The letter M denotes some fixed full almost-uniform grid.

Furthermore, $\min M$ (or $\max M$) denotes the smallest (or the largest) number belonging to the set M. Clearly $\min M, \max M \notin \mathrm{Real}$.

The letters t, s denote elements of the set M and $t + \mathrm{d}t$, where $t < \max M$, denotes the smallest of the numbers belonging to M which are larger than t.

We define

$$\begin{aligned} \mathrm{d}_0 t &= 0, \\ \mathrm{d}_1 t &= \mathrm{d}t, \\ \mathrm{d}_{\alpha+1} t &= \mathrm{d}_\alpha t + \mathrm{d}(t + \mathrm{d}_\alpha t). \end{aligned}$$

If $\max M = t + \mathrm{d}_\alpha t$ then $\mathrm{d}_\beta t$ for $\beta \geq \alpha + 1$ is left undefined.

Proposition 22.5. (i) For each $t \in \mathrm{Real}$,

$$\mathrm{d}_\alpha t \ddot{=} \alpha \cdot \mathrm{d}t.$$

(ii) For each $s, t \in \mathrm{Real}$

$$\mathrm{d}_\alpha s \ddot{=} \alpha \cdot \mathrm{d}s \ddot{=} \alpha \cdot \mathrm{d}t \ddot{=} \mathrm{d}_\alpha t.$$

Proof. (i) If $\beta < \alpha$ then

$$\mathrm{d}_{\beta+1} t - \mathrm{d}_\beta t = \mathrm{d}(t + \mathrm{d}_\beta t) \ddot{=} \mathrm{d}t.$$

By Proposition 21.5

$$\mathrm{d}_\alpha t = \sum_{0 \leq \beta < \alpha} (\mathrm{d}_{\beta+1} t - \mathrm{d}_\beta t) \ddot{=} \alpha \cdot \mathrm{d}t.$$

(ii) For similar reasons, $\mathrm{d}_\alpha s \ddot{=} \alpha \cdot \mathrm{d}s$. Since $\mathrm{d}s \ddot{=} \mathrm{d}t$, also $\alpha \cdot \mathrm{d}s \ddot{=} \alpha \cdot \mathrm{d}t$. Then see Proposition 21.3(i). □

The letter ϑ denotes the natural number corresponding to the number of elements of the set M. Clearly $\vartheta \in \mathrm{IN}$.

By the **radius of the monad of a real number** $t \in \mathrm{Real}$ **on a grid** M, denoted $R_M(t)$ (briefly just $R(t)$) we mean the class of all the natural numbers α for which

$$\mathrm{d}_\alpha(t) \doteq 0.$$

Proposition 22.6. For any real numbers $t, s \in \mathrm{Real}$,

(i) $R(t)$ is a cut on the set of all natural numbers smaller than ϑ.

(ii) $R(t)$ is a π-class.

(iii) $R(t) = R(s)$.

Proof. (i) By Proposition 22.5(i), $R(t)$ is the class of all α such that $\alpha \cdot \mathrm{d}t \doteq 0$.

(ii) Let $b_n(t)$ denote the set of all α such that

$$\alpha \cdot \mathrm{d}t < \frac{1}{n}.$$

Clearly

$$R(t) = \bigcap_{n \in \mathrm{FN}} b_n.$$

(iii) By Proposition 22.5(ii), $\alpha \cdot \mathrm{d}t \doteqdot \alpha \cdot \mathrm{d}s$. □

What we have just proved makes it possible to write just R_M in place of $R_M(t)$, $R_M(s)$, and to talk about the **radius of monads of real numbers on the grid** M.

Editor's Closing Note

In the Czech version of New Infinitary Mathematics, some sections on integration follow the above material. However, Vopěnka did not include them in the English version.

Bibliography

Aquinas, Thomas. *The Summa Theologica of St. Thomas Aquinas Part I.* Literally translated by Fathers of the English Dominican Province. Second and revised edition, Burns Oates & Washbourne, 1920.

Aquinas, Thomas. *Contra Gentiles: On the Truth of the Catholic Faith.* Translated by Anton C. Pegis. New York: Hanover House, 1955–57.

Aristotle. *Physics.* Translated by R. P. Harding and R. K. Gaye. Internet Classics Archive.

Arriaga, Rodrigo. *Cursus Philosophicus.* Antwerp, 1632.

Augustine. *The City of God.* Translated by Marcus Dods. From Nicene and Post-Nicene Fathers, First Series, Vol. 2. Edited by Philip Schaff. Buffalo, NY: Christian Literature Publishing, 1887.

Bendiek, J. "Ein Brief Georg Cantors an P. Ignatius Jeiler O.F.M." *Franziskanische Studien* 47 (1965).

Bernays, Paul. *Axiomatic Set Theory.* Amsterdam: North Holland, 1958.

Bolzano, Bernard. *Theory of Science.* Translated by Paul Rusnock and Rolf George. Oxford: Oxford University Press, 2014.

Bolzano, Bernard. *Paradoxes of the Infinite.* Translated by Donald A. Steele. London: Routledge, 1950.

Bolzano, Bernard. "Rein analytischer Beweis des Lehrsatzes dass zwischen je zwey Werthen, die ein entgegengesetzetes Resultat gewähren, wenigstens eine reelle Wurzel der Gleichung liege." *Abhandlungen der k. Gesellschaft der Wissenschaften.* Prag: Gottlieb Haase, 1817.

Brouček, Vilém Antonín. *Domus Sapientiae Doctoris subtilis Joannis Duns Scoti.* 1663.

Bruno, Giordano. *De l'infinito universo et mondi.* Venezia, 1584.

Bruno, Giordano. *Dialoghi italiani: dialoghi metafisici e dialoghi morali.* Firenze: Sansoni, 1958.

Cantor, Georg. "Über die Ausdehnung eines Satzes aus der Theorie der trigonometrischen Reihen." *Mathematische Annalen* 5 (1872): 123–132.

Cantor, Georg. "Über eine Eigenschaft des Inbegriffes aller reelen algebraischen Zahlen." *Journal für die Reine und Angewandte Mathematik* 77 (1874): 258–262.

Cantor, Georg. "Ueber unendliche, lineare Punktmannichfaltigkeiten." *Mathematische Annalen* 15, no. 1 (1879): 1–7; 17, no. 3 (1880): 355–358; 20, no. 1 (1882): 113–121; 21, no. 1 (1883): 51–58; 23, no. 4 (1884): 553–488.

Cantor, Georg. *Grundlagen einer allgemeinen Mannigfaltigkeitslehre.* Leipzig: Teubner, 1883.

Cantor, Georg. "Über eine elementare Frage der Mannigfaltigkeitslehre." *Jahresbericht der Deutsch. Math. Vereing.* 1 (1890–1891): 75–76.

Cantor, Georg. "Beiträge zur Begründung der transfiniten Mengenlehre II." *Mathematische Annalen* 49 (1897): 207–246.

Cantor, Georg. *Contributions to the Founding of the Theory of Transfinite Numbers.* Translated by P. E. B. Jourdain. New York: Dover Poblications, 1915.

Cantor, Georg. *Briefe.* Edited by Herbert Meschkowski and Winfried Nilson. Berlin, New York: Springer Verlag, 1991.

Cohen, Paul. "The Independence of the Continuum Hypothesis." *Proceedings of the National Academy of Sciences* 50 (1963): 1143–1148.

d'Alembert, Jean le Rond. *Essai sur les éléments de philosophie.* Amsterdam: Z. Chatelain et fils, 1759.

Dauben, Joseph Warren. *Georg Cantor: His Mathematics and Philosophy of the Infinite.* Princeton: Princeton University Press, 1979.

Dedekind, Richard. *Was sind und was sollen die Zahlen?* Braunsweig: Friedrich Vieweg und Sohn, 1888.

Dehn, Max. "Über den Rauminhalt." *Mathematische Annalen* 55, no 3 (1901): 465–478.

Euclid. *Elements.* Translated by T. L. Heath. The Thirteen Books of Euclid's Elements. Dover Publications, 1956.

Euler, Leonhard. *Introduction to Analysis of the Infinite.* Translated by John D. Blanton. New York: Springer Verlag, 1998–1990.

Euler, Leonhard. *Foundations of Differential Calculus.* Translated by John D. Blanton. New York: Springer Verlag, 2000.

Fraenkel, Adolf. "Zu den Grundlagen der Cantor-Zermeloschen Mengenlehre." *Mathematische Annalen* 86 (1922).

Frege, Gottlob. "On the Foundation of Geometry." *The Philosophical Review* 69, no. 1(1960): 3–17. Translated by M. E. Szabo from "Über die Grundlagen der Geometrie." *Jahresbericht der deutschen Mathematiker-Vereinigung* 12 (1903): 319–324, 368–375.

Galilei, Galileo. *Discorsi e dimonstrazioni matematiche, intorno a due nuove scienze.* Leiden: Louis Elsevier, 1638.

Gödel, Kurt. "Die Vollständigkeit der Axiome des logischen Funktionenkalküls," *Monatshefte für Mathematik und Physik* 37 (1930): 349–360.

Gödel, Kurt. "Über formal unentscheidbare Sätze der Principia Mathematica und verwandter Systeme I." *Monatschefte für Mathematik und Physik* 38 (1931): 173–198.

Gödel, Kurt. *Consistency of the Axion of Choice and of the Generalized Continuum Hypothesis with the Axioms of Set Theory.* Princeton University Press, 1940.

Hadamard, Jacques. "Cinq lettres sur la théorie des ensembles." *Bulletin de la Société mathématique de France* 33 (1905): 261–273.

Hausdorff, Felix. *Grundzüge der Mengenlehre.* Leipzig: Veit and Company, 1914.

Heidegger, Martin. *Die Frage nach dem Ding zu Kants Lehre von transzendentalen Grundsätzen.* Tübingen: Max Niemeyer Verlag, 1962.

Heidegger, Martin. *Being and Time.* Oxford, Blackwell, 1967.

Hermite, Charles. *Cours d'Analyse de l'École Polytechnique. Première Partie.* Paris: Gauthier-Villars, 1873.

Heyting, Arend. "Die formalen Regeln der intuitionistischen Logik." In *Sitzungsberichte der Preussischen Akademie der Wissenschaften,* 42–56. Berlin, 1930.

Hilbert, David. *Grundlagen der Geometrie.* Leipzig: B. G. Teubner, 1903.

Hilbert, David. "Über die Grundlagen der Logik und der Arithmetik." In *Verhandlungen des dritten Internationalen Mathematiker-Kongresses in Heidelberg vom 8. bis 13. August 1904.* Edited by A. Krazer, 174–185. Leipzig: Teubner, 1904.

Hilbert, David. "Über das Unendliche." *Mathematische Annalen* 95, no. 1 (1926): 161–190.

Hilbert, David. *Gesammelte Abhandlungen.* Berlin: Julius Springer, 1932.

Hilbert, David. "Mathematische Probleme – Vortrag, gehalten auf dem internationalen Mathematiker-Kongress zu Paris 1900," *Nachrichten von der Königl. Gesellschaft der Wissenschaften zu Göttingen.* Mathematisch-Physikalische Klasse. Heft 3 (1900): 253–297.

Hilbert, David. "Über den Zahlbegriff." *Jahresbericht der Deutschen Mathematiker-Vereinung* 8 (1900): 180–184.

Hobbes, Thomas. *Hobbes's Leviathan: Reprinted from the Edition of 1651.* Oxford: Clarendon, 1909.

l'Hopital, Guillaume. *Analyse des Infiniment Petits pour l'Intelligence des Lignes Courbes.* Paris: Montalant, 1696.

Husserl, Edmund. *Vorlesungen zur Phänomenologie des inneren Zeitbewusstseins.* Tübingen: M. Niemeyer, 1980.

Jung, Carl Gustav. *Psychology and Religion: West and East.* Collected Works of C. G. Jung 11. Edited by G. Adler, M. Fordham and Sir H. Read. Routledge, 1970.

Kanamori, Akihiro. *The Higher Infinite.* Berlin – Heidelberg – New York: Springer Verlag, 1994.

Kuratowski, Casimir. "Sur la notion de l'ordre dans la Théorie des Ensembles." *Fundamenta Mathematicae* 2, no.1 (1921): 161–171.

Lagrange, Joseph Louis. *Mécanique analytique.* Paris: Gauthier-Villars et fils, 1888.

Leibniz, Gottfried Wilhelm. *Opera omnia.* Edited by Louis Dutens. Vol. 2. Part 1. Geneva, 1768.

Lindemann, Ferdinand von. "Über die Zahl π." *Mathematische Annalen* 20 (1882): 213–225.

Locke, John. *An Essay Concerning Human Understanding.* In *The Works of John Locke in Nine Volumes.* London: Rivington, 1824.

Löwenheim, Leopold. "Über Möglichkeiten im Relativkalkül," *Mathematische Annalen* 76 (1915): 447–470.

Meschkowski, Herbert. *Das Problem des Unendlichen: mathematische und philosophische Texte von Bolzano, Gutberlet, Cantor, Dedekind.* München: Deutscher Taschenbuch Verlag, 1974.

Mirimanov, Dimitrij. "Les antinomies de Russell et de Burali-Forti et le probléme fondamental de la théorie des ensembles." *L'Enseigment Mathématique* 19 (1917): 37–52.

Moore, Gregory. *Zermelo's Axiom of Choice: Its Origins, Development and Influence.* Springer-Verlag, 1982.

Neumann, John von. "Eine Axiomatisierung der Mengenlehre." *Journal für die reine und angewandte Mathematik* 154 (1925): 219–240.

Neumann, John von. "Zur Einführung der transfiniten Zahlen." *Acta Scientiarum Mathematicarum* 1, no. 4 (1923).

Newton, Isaac. *The Method of Fluxions and Infinite Series: With Its Applications to the Geometry of Curve-Lines.* London: Henry Woodfall, 1736.

Newton, Isaac. *Philosophiæ naturalis principia mathematica.* London, 1687.

Pascal, Blaise. *Pascal's Pensées; or, Thoughts on Religion.* Translated by Gertrude Burford Rawlings. Mount Vernon, N.Y.: Peter Pauper Press, 1900.

Pascal, Blaise. *Of the Geometrical Spirit.* Translated by Orlando Williams Wight. In *The Harvard Classics: Blaise Pascal.* New York: Colier and Son, 1910.

Plato. *The Republic.* Translated by B. Jowett. Project Gutenberg, August 2008.

Poincaré, Henri. *Science and Method.* Translated by Francis Maitland. London: T. Nelson & sons, 1914.

Poincaré, Henri. "Réflexions sur les deux notes précédentes." *Acta mathematica* 32 (1909): 195–200.

Pospíšil, Bedřich. "Remark on Bicompact Spaces," *Annals of Mathematics* 38 (1937): 845–846.

Robinson,Abraham. *Non-standard Analysis.* Revised ed. Princeton – New York: Princeton University Press, 1996.

Russ, Steve. "Bolzano's Analytical Programme," *Mathematical Intelligencer* 14, no. 3 (1992): 45-53.

Russel, Bertrand. *Introduction to Mathematical Philosophy.* London: George Allen & Unwin, Ltd., 1919.

Russel, Bertrand. "Les Paradoxes de la logique." *Revue de métaphysique et de morale* 14, no. 5 (1906): 627–650.

Senfleben, Johann. *Philosophia Aristotelica.* Typis Universitatis Carolo-Ferdinandeae in Coll. Soc. Jesu ad S. Clementem, 1685.

Skolem, Thoralf. "Einige Bemerkungen zur axiomatischen Begründung der Mengenlehre." *Wissenschaftliche Vorträge auf dem Fünften Kongress der Skandinavischen Mathematiker in Helsingfors*, 1922.

Skolem, Thoralf. "Über die Unmöglichkeit einer vollständingen Charakterisiernug der Zahlenreihe mittels eines endlichen Axiomensystems." *Norks*

matematisk fornings skriften 2, no. 10 (1933): 73–82.
Skolem, Thoralf. "Über die Nicht-charakterisierbarkeit der Zahlenreihe mittels endlich oder abzählbar unendlich vieler Aussagen mit ausschliesslich Zahlenvariablen." *Fundamenta Mathematicae* 23 (1934): 150–161.
Sochor, Antonín. "Petr Vopěnka (born 16. 5. 1935)." *Ann. Pure Appl. Logic* 109 (2001): 1–8.
Thomas, Ivor. *Greek Mathematical Works 1.* Loeb Classical Library: William Heinemann Ltd., 1980.
Spinoza, Benedictus de. *Ethics.* Translated by R. H. M. Elwes. London: G. Bell and Sons, 1887.
Vopěnka, Petr. "A Method of Constructing a Non-standard Model in the Bernays-Gödel Axiomatic Set Theory." *Dokl. Akad. nauk SSSR* 143, no. 1 (1962): 11–12.
Vopěnka, Petr. "The Independence of Continuum Hypothesis." *American Mathematical Society Translations,* Series 2, 57, no. 2 (1964).
Vopěnka, Petr. "The Limits of Sheaves and Applications on Construction of Models." *Bull. Acad. Polon. Sci. Sér. Sci. Math. Astronom. Phys.* 13 (1965): 189–192.
Vopěnka, Petr. "On ∇-model of Set Theory." *Bull. Acad. Polon. Sci. Sér. Sci. Math. Astronom. Phys.* 13 (1965): 267–272.
Vopěnka, Petr. "Properties of ∇-model." *Bull. Acad. Polon. Sci. Sér. Sci. Math. Astronom. Phys.* 13 (1965): 441–444.
Vopěnka, Petr. "∇-models in Which the Generalized Continuum Hypothesis Does Not Hold." *Bull. Acad. Polon. Sci. Sér. Sci. Math. Astronom. Phys.* 14 (1966): 95–99.
Vopěnka, Petr. "The Limits of Sheaves Over Extremally Disconnected Compact Hausdorff Spaces." *Bull. Acad. Polon. Sci. Sér. Sci. Math. Astronom. Phys.* 15 (1967): 1–4.
Vopěnka, Petr, and Petr Hájek. *The Theory of Semisets.* Prague: North Holland and Academia, 1972.
Vopěnka, Petr. *Mathematics in the Alternative Set Theory.* Leipzig: B. G. Teubner, 1979.
Vopěnka, Petr. *Podivuhodný květ českého baroka.* Praha: Karolinum, 1998.
Vopěnka, Petr. *Úhelný kámen evropské vzdělanosti a moci.* Praha: Práh, 2000.
Vopěnka, Petr. *Meditace o základech vědy.* Praha: Práh, 2001.
Vopěnka, Petr. *Vyprávění o kráse novobarokní matematiky.* Praha: Práh, 2004.
Vopěnka, Petr. *Calculus infinitesimalis pars prima.* Kanina: OPS, 2010.
Vopěnka, Petr. *Úvod do klasické teorie množin.* Plzeň – Praha: Západočeská univerzita v Plzni & Fragment, 2011.
Vopěnka, Petr. *Velká iluze matematiky XX. století a nové základy.* Plzeň – Praha: Západočeská univerzita v Plzni & Nakladatelství KONIÁŠ, 2011.
Vopěnka, Petr. *Prolegomena k nové infinitní matematice.* Praha: Karolinum, 2013.
Vopěnka, Petr. *Nová infinitní matematika.* Praha: Karolinum, 2013.
Weber, H. "Leopold Kronecker." *Jahresbericht der Deutschen Mathematiker Vereinigung* 2, Reimer, 1893.

Zermelo, Ernst. "Beweis, dass jede Menge wohlgeordnet werden kann." *Mathematische Annalen* 59 (1904): 514–516.

Zermelo, Ernst. "Untersuchungen über die Grundlagen der Mengenlehre," *Mathematische Annalen* 65 (1908): 261–281.

Zorn, Max. "A Remark on Method in Transfinite Algebra." *Bulletin of the American Mathematical Society* 41, no. 10 (1935): 667–670.